Multivariate t Distributions and Their Applications

Almost all of the results available in the literature on multivariate t distributions published in the last 50 years are now collected together in this comprehensive volume. Because these distributions are becoming more prominent in many applications, this book is a must for any serious researcher or consultant working in multivariate analysis and statistical distributions. Much of this material has never appeared in book form.

The first part of the book emphasizes theoretical results of a probabilistic nature. In the second part of the book, these are supplemented by a variety of statistical aspects. Various generalizations and applications are dealt with in the final chapters. The material on estimation and regression models is of special value for practitioners in statistics and economics. A comprehensive bibliography of more than 350 references is included.

Samuel Kotz is Professor and Senior Research Scholar in the Department of Engineering Management and Systems Engineering at George Washington University. In addition to holding many distinguished visiting positions at prominent universities, he has authored a number of books and more than 100 articles, is Editor-in-Chief and founder of the *Encyclopedia of Statistical Sciences*, and holds three honorary doctorates.

Saralees Nadarajah is Professor of Mathematics at the University of South Florida. He has made important contributions to distribution theory, extreme value theory and its applications in environmental modeling, branching processes, and sampling theory. He has authored more than 80 papers and two books.

Multivariate t Distributions and Their Applications

Samuel Kotz
George Washington University

Saralees Nadarajah
University of South Florida

CAMBRIDGE
UNIVERSITY PRESS

CAMBRIDGE
UNIVERSITY PRESS

Shaftesbury Road, Cambridge CB2 8EA, United Kingdom

One Liberty Plaza, 20th Floor, New York, NY 10006, USA

477 Williamstown Road, Port Melbourne, VIC 3207, Australia

314–321, 3rd Floor, Plot 3, Splendor Forum, Jasola District Centre, New Delhi – 110025, India

103 Penang Road, #05–06/07, Visioncrest Commercial, Singapore 238467

Cambridge University Press is part of Cambridge University Press & Assessment, a department of the University of Cambridge.

We share the University's mission to contribute to society through the pursuit of education, learning and research at the highest international levels of excellence.

www.cambridge.org
Information on this title: www.cambridge.org/9780521826549

First published 2004

A catalogue record for this publication is available from the British Library

Library of Congress Cataloging in Publication data
Kotz, Samuel.
Multivariate t distributions and their applications / Samuel Kotz,
Saralees Nadarajah.
p. cm.
Includes bibliographical references and index.
ISBN 0-521-82654-3
1. Multivariate analysis. 2. Distribution (Probability theory) I. Nadarajah, Saralees. II. Title.
QA278.K635 2004
519.5′35-dc21 2003055353

ISBN 978-0-521-82654-9 Hardback

Contents

List of Illustrations

Preface

Multivariate t distributions have attracted somewhat limited attention of researchers for the last 70 years in spite of their increasing importance in classical as well as in Bayesian statistical modeling. These distributions have been perhaps unjustly overshadowed – during all these years – by the multivariate normal distribution. Both the multivariate t and the multivariate normal are members of the general family of elliptically symmetric distributions. However, we feel that it is desirable to focus on these distributions separately for several reasons:

- Multivariate t distributions are generalizations of the classical univariate Student t distribution, which is of central importance in statistical inference. The possible structures are numerous, and each one possesses special characteristics as far as potential and current applications are concerned.
- Application of multivariate t distributions is a very promising approach in multivariate analysis. Classical multivariate analysis is soundly and rigidly tilted toward the multivariate normal distribution while multivariate t distributions offer a more viable alternative with respect to real-world data, particularly because its tails are more realistic. We have seen recently some unexpected applications in novel areas such as cluster analysis, discriminant analysis, multiple regression, robust projection indices, and missing data imputation.
- Multivariate t distributions for the past 20 to 30 years have played a crucial role in Bayesian analysis of multivariate data. They serve by now as the most popular prior distribution (because elicitation of prior information in various physical, engineering, and financial phenomena is closely associated with multivariate t distributions) and generate meaningful posterior distributions. This diversity and the apparent

ease of applications require careful analysis of the properties of the distribution in order to avoid pitfalls and misrepresentation.

The compilation of this book was a somewhat daunting task (as our Contents indicates). Indeed, the scope of the multivariate t distributions is unsurpassed, and, although there are books dealing with multivariate continuous distributions and review articles in the *Encyclopedia of Statistical Sciences and Biostatistics*, the material presented in these sources is quite limited.

Our goal was to collect and present in an organized and user-friendly manner all of the relevant information available in the literature worthy of publication. It is our hope that the readers – both novices and experts – will find the book useful. Our thanks are due to numerous authors who generously supplied us with their contributions and to Lauren Cowles, Elise Oranges and Lara Zoble at Cambridge University Press for their guidance. We also wish to thank Anusha Thiyagarajah for help with editing.

Samuel Kotz
Saralees Nadarajah

1
Introduction

1.1 Definition

There exist quite a few forms of multivariate t distributions, which will be discussed in subsequent chapters. In this chapter, however, we shall describe the most common and natural form. It directly generalizes the univariate Student's t distribution in the same manner that the multivariate normal distribution generalizes the univariate normal distribution.

A p-dimensional random vector $\mathbf{X} = (X_1, \ldots, X_p)^T$ is said to have the p-variate t distribution with degrees of freedom ν, mean vector $\boldsymbol{\mu}$, and correlation matrix $\mathbf{R}$ (and with $\boldsymbol{\Sigma}$ denoting the corresponding covariance matrix) if its joint probability density function (pdf) is given by

$$f(\mathbf{x}) = \frac{\Gamma((\nu+p)/2)}{(\pi\nu)^{p/2}\Gamma(\nu/2)\,|\mathbf{R}|^{1/2}}\left[1+\frac{1}{\nu}(\mathbf{x}-\boldsymbol{\mu})^T\mathbf{R}^{-1}(\mathbf{x}-\boldsymbol{\mu})\right]^{-(\nu+p)/2}. \tag{1.1}$$

The degrees of freedom parameter ν is also referred to as the shape parameter, because the peakedness of (1.1) may be diminished, preserved, or increased by varying ν (see Section 1.4). The distribution is said to be central if $\boldsymbol{\mu} = \mathbf{0}$; otherwise, it is said to be noncentral.

Note that if $p = 1$, $\boldsymbol{\mu} = 0$, and $\mathbf{R} = 1$, then (1.1) is the pdf of the univariate Student's t distribution with degrees of freedom ν. These univariate marginals have increasingly heavy tails as ν decreases toward unity. With or without moments, the marginals become successively less peaked about $0 \in \Re$ as $\nu \downarrow 1$.

If $p = 2$, then (1.1) is a slight modification of the bivariate surface of Pearson (1923). If $\nu = 1$, then (1.1) is the p-variate Cauchy distribution. If $(\nu+p)/2 = m$, an integer, then (1.1) is the p-variate Pearson type VII distribution. The limiting form of (1.1) as $\nu \to \infty$ is the joint pdf of the p-variate normal distribution with mean vector $\boldsymbol{\mu}$ and covariance matrix $\boldsymbol{\Sigma}$. Hence, (1.1) can be viewed as an approximation of the multivariate normal distribution. The particular case of (1.1) for $\boldsymbol{\mu} = \mathbf{0}$ and $\mathbf{R} = \mathbf{I}_p$ is a mixture of the normal density with zero means and covariance matrix $v\mathbf{I}_p$ - in the scale parameter v. The class of elliptically contoured distributions (see, for example, Fang et al., 1990) contain (1.1) as a particular case. Also (1.1) has the attractive property of being Schur-concave when elements of $\mathbf{R}$ satisfy $r_{ij} = \rho$, $i \neq j$ (see Marshall and Olkin, 1974). Namely, if $\mathbf{a}$ and $\mathbf{b}$ are two p-variate vectors with components ordered to achieve $a_1 \geq a_2 \geq \cdots \geq a_p$ and $b_1 \geq b_2 \geq \cdots \geq b_p$, and if this ordering implies $\sum_{i=1}^{k} a_i \leq \sum_{i=1}^{k} b_i$ for $k = 1, 2, \ldots, p-1$ and $\sum_{i=1}^{p} a_i \leq \sum_{i=1}^{p} b_i$, then (1.1) satisfies $f(\mathbf{a}) \geq f(\mathbf{b})$.

In Bayesian analyses, (1.1) arises as: (1) the posterior distribution of the mean of a multivariate normal distribution (Geisser and Cornfield, 1963; see also Stone, 1964); (2) the marginal posterior distribution of the regression coefficient vector of the traditional multivariate regression model (Tiao and Zellner, 1964); (3) the marginal prior distribution of the mean of a multinormal process (Ando and Kaufman, 1965); (4) the marginal posterior distribution of the mean and the predictive distribution of a future observation of the multivariate normal structural model (Fraser and Haq, 1969); (5) an approximation to posterior distributions arising in location-scale regression models (Sweeting, 1984, 1987); and (6) the prior distribution for set estimation of a multivariate normal mean (DasGupta et al., 1995). Additional applications of (1.1) can be seen in the numerous books dealing with the Bayesian aspects of multivariate analysis.

1.2 Representations

If $\mathbf{X}$ has the p-variate t distribution with degrees of freedom ν, mean vector $\boldsymbol{\mu}$, and correlation matrix $\mathbf{R}$, then it can be represented as

- If $\mathbf{Y}$ is a p-variate normal random vector with mean $\mathbf{0}$ and covariance matrix $\boldsymbol{\Sigma}$, and if $\nu S^2/\sigma^2$ is the chi-squared random variable with degrees of freedom ν, independent of $\mathbf{Y}$, then

$$\mathbf{X} = S^{-1}\mathbf{Y} + \boldsymbol{\mu}. \tag{1.2}$$

  This implies that $\mathbf{X} \mid S = s$ has the p-variate normal distribution with mean vector $\boldsymbol{\mu}$ and covariance matrix $(1/s^2)\boldsymbol{\Sigma}$.

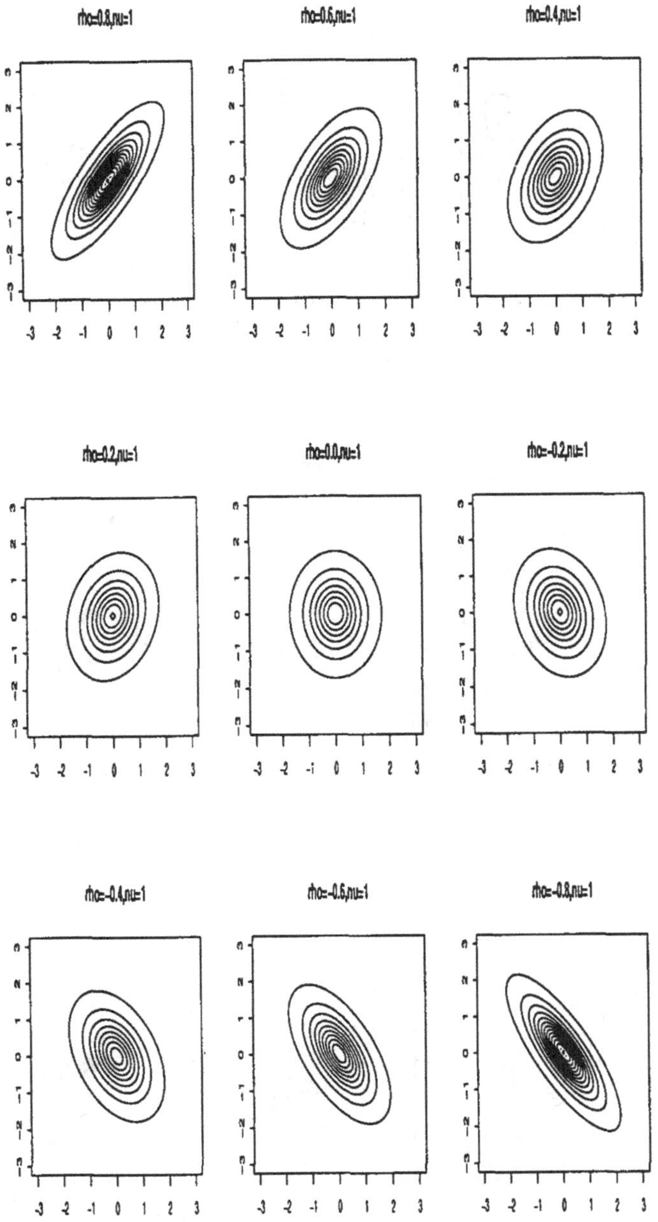

Fig. 1.1. Joint contours of (1.1) with degrees of freedom $\nu = 1$, zero means, and correlation coefficient $\rho = 0.8, 0.6, \ldots, -0.6, -0.8$

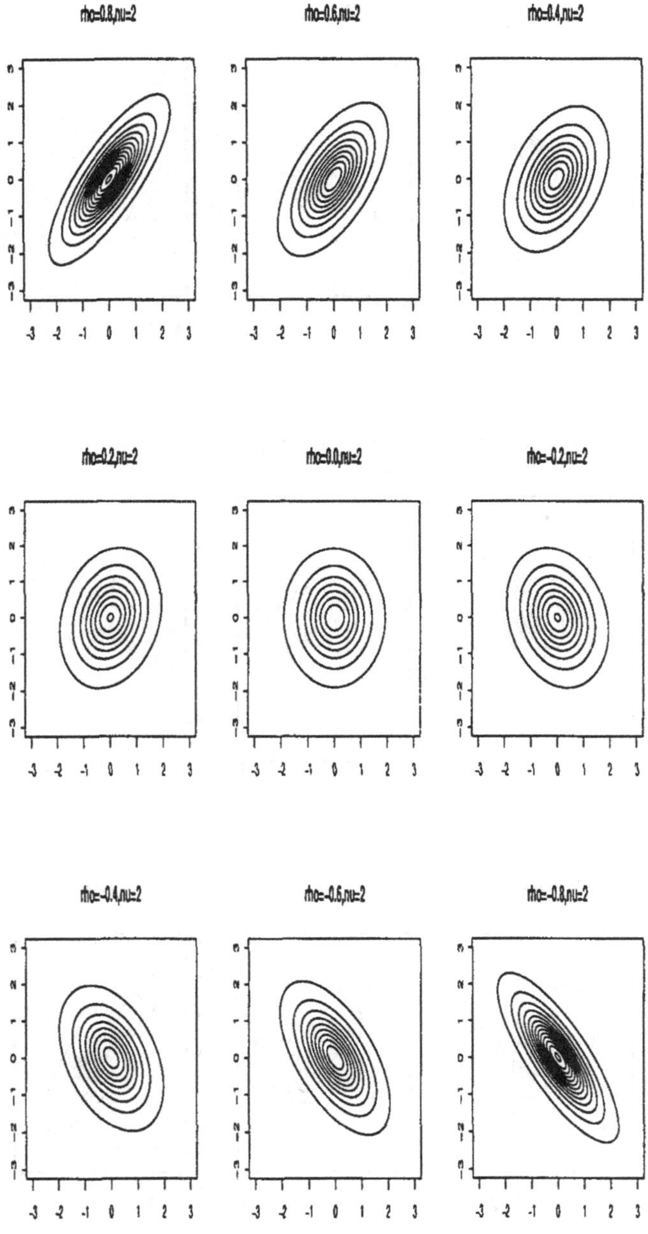

Fig. 1.2. Joint contours of (1.1) with degrees of freedom $\nu = 2$, zero means, and correlation coefficient $\rho = 0.8, 0.6, \ldots, -0.6, -0.8$

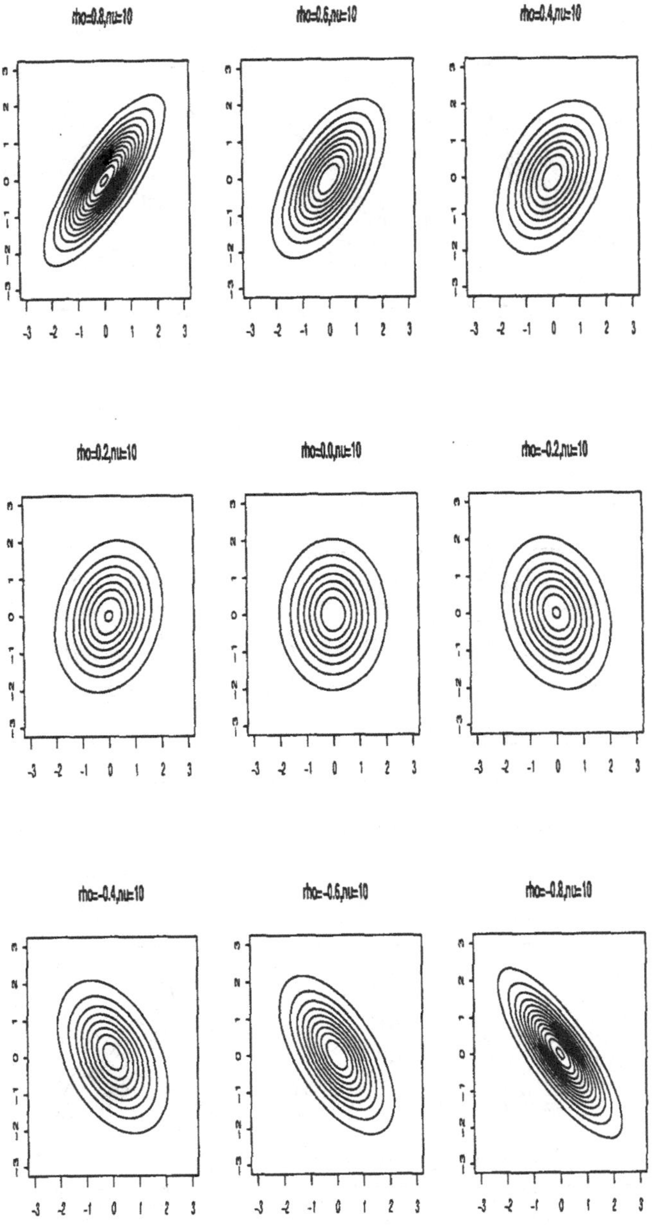

Fig. 1.3. Joint contours of (1.1) with degrees of freedom $\nu = 10$, zero means, and correlation coefficient $\rho = 0.8, 0.6, \ldots, -0.6, -0.8$

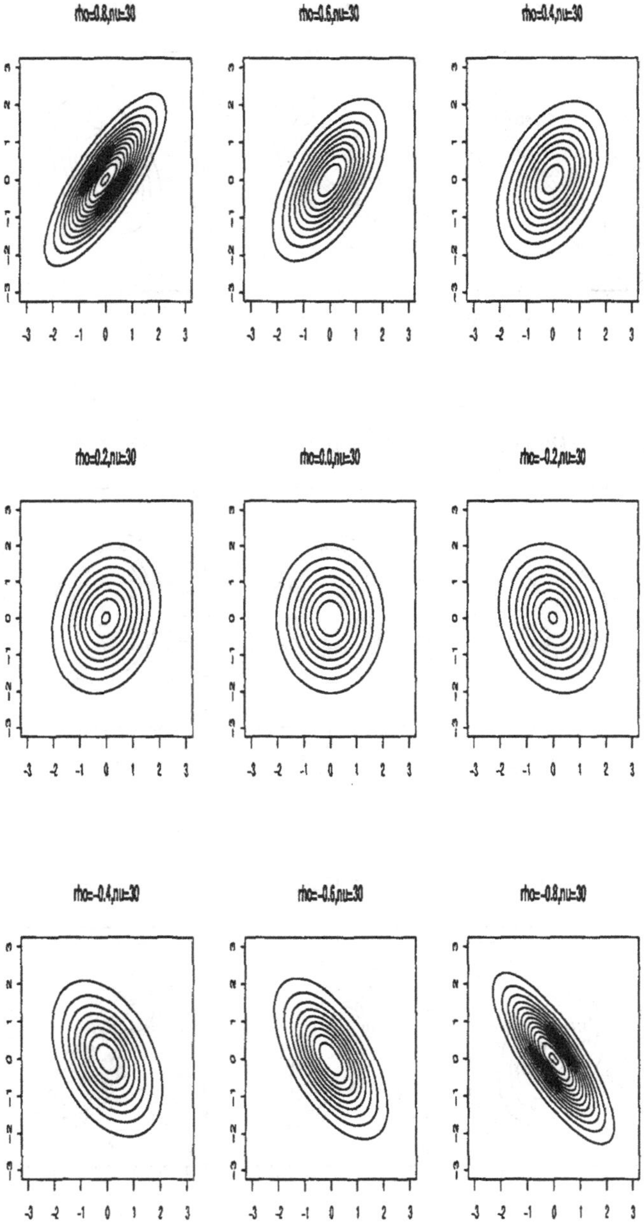

Fig. 1.4. Joint contours of (1.1) with degrees of freedom $\nu = 30$, zero means, and correlation coefficient $\rho = 0.8, 0.6, \ldots, -0.6, -0.8$

- If $\mathbf{V}^{1/2}$ is the symmetric square root of $\mathbf{V}$, that is,

$$\mathbf{V}^{1/2}\mathbf{V}^{1/2} = \mathbf{V} \sim \mathcal{W}_p\left(\mathbf{R}^{-1}, \nu + p - 1\right), \tag{1.3}$$

where $\mathcal{W}_p(\mathbf{\Sigma}, n)$ denotes the p-variate Wishart distribution with degrees of freedom n and covariance matrix $\mathbf{\Sigma}$, and if $\mathbf{Y}$ has the p-variate normal distribution with zero means and covariance matrix $\nu \mathbf{I}_p$ ($\mathbf{I}_p$ is the p-dimensional identity matrix), independent of $\mathbf{V}$, then

$$\mathbf{X} = \left(\mathbf{V}^{1/2}\right)^{-1} \mathbf{Y} + \boldsymbol{\mu} \tag{1.4}$$

(Ando and Kaufman, 1965). This implies that $\mathbf{X} \mid \mathbf{V}$ has the p-variate normal distribution with mean vector $\boldsymbol{\mu}$ and covariance matrix $\nu \mathbf{V}^{-1}$.

1.3 Characterizations

From representation (1.2) it easily follows for any $\mathbf{a} \neq \mathbf{0}$ that $\mathbf{X}$ has the joint pdf (1.1) if and only if

$$\begin{aligned} & \mathbf{X} \mid S^2 = s^2 \sim N\left(\boldsymbol{\mu}, s^{-2}\mathbf{\Sigma}\right) \\ \Leftrightarrow\ & \left(\mathbf{a}^T\mathbf{\Sigma}\mathbf{a}\right)^{-1/2} \mathbf{a}^T\left(\mathbf{X} - \boldsymbol{\mu}\right) \mid S^2 = s^2 \sim N\left(0, s^{-2}\right) \\ \Leftrightarrow\ & \left(\mathbf{a}^T\mathbf{\Sigma}\mathbf{a}\right)^{-1/2} \mathbf{a}^T\left(\mathbf{X} - \boldsymbol{\mu}\right) \sim t_\nu, \end{aligned}$$

and this is one of the earliest characterization results given in Cornish (1962). This result can also be obtained by using the representation (1.4): $\mathbf{X}$ has the joint pdf (1.1) if and only if

$$\begin{aligned} & \mathbf{X} \mid \mathbf{V} \sim N\left(\boldsymbol{\mu}, \nu\mathbf{V}^{-1}\right) \\ \Leftrightarrow\ & \left(\mathbf{a}^T\mathbf{\Sigma}\mathbf{a}\right)^{-1/2} \mathbf{a}^T\left(\mathbf{X} - \boldsymbol{\mu}\right) \mid \mathbf{V} \sim N\left(0, \nu\left(\mathbf{a}^T\mathbf{V}^{-1}\mathbf{a}\right) \Big/ \left(\mathbf{a}^T\mathbf{\Sigma}\mathbf{a}\right)\right) \\ \Leftrightarrow\ & \left(\mathbf{a}^T\mathbf{\Sigma}\mathbf{a}\right)^{-1/2} \mathbf{a}^T\left(\mathbf{X} - \boldsymbol{\mu}\right) \sim t_\nu, \end{aligned}$$

as noted by Lin (1972).

Lin (1972) obtained two further characterizations using the representation (1.2). Let $\nu S^2 \sim \chi^2_\nu$ and let $X_1, X_2, \ldots, X_p$ be conditionally independent continuous random variables symmetrically distributed with $E(X_k \mid S^2 = s^2) = \mu_k$ and $Var(X_k \mid S^2 = s^2) = \sigma_k^2/s^2 < \infty$ for $k = 1, \ldots, p$. Then the following characterizations are valid

- $(X_1, X_2, \ldots, X_p)^T$ has the joint pdf (1.1) with mean vector $\boldsymbol{\mu}$, covariance matrix $\mathbf{D}$, and degrees of freedom ν if and only if

$$\sum_{k=1}^{p} \frac{\left(X_k - \mu_k\right)^2}{p\sigma_k^2} \sim F_{p,\nu},$$

where $\mathbf{D}$ is a $p \times p$ diagonal matrix with its kth diagonal element equal to σ_k^2.

- In the special case $\sigma_k^2 = \sigma^2$ for all k and the conditional pdf of $X_k \mid S^2 = s^2$ is positive and differentiable for all $x \in \Re$, $(X_1, X_2, \ldots, X_p)^T$ has the joint pdf (1.1) with zero means, covariance matrix $\sigma^2 \mathbf{I}_p$, and degrees of freedom ν if and only if the joint pdf of $X_1, X_2, \ldots, X_p$ is a function of $x_1^2 + x_2^2 + \cdots + x_p^2$ only.

1.4 A Closure Property

Consider Studentizing transformations $T : \Re^n \to \Re^k$, depending on matrices $\mathbf{A}(n \times k)$, $\mathbf{B}(n \times \nu)$ and $\mathbf{\Omega}(n \times n)$, given by

$$T(\mathbf{X}) = \frac{\sqrt{\nu}\mathbf{A}^T\mathbf{X}}{\| \mathbf{B}^T\mathbf{X} \|} \tag{1.5}$$

such that $\mathbf{A}^T \mathbf{\Omega} \mathbf{B} = \mathbf{0}$. Jensen (1994) established that the class of multivariate t distributions is closed under the transform $T(\cdot)$. Specifically, assume $\mathbf{A}^T\mathbf{A} = \mathbf{I}_k$, $\mathbf{B}^T\mathbf{B} = \mathbf{I}_\nu$, and $\mathbf{X}$ is distributed according to (1.1) with zero means, correlation matrix $\mathbf{I}_n$, and degrees of freedom m. Under these assumptions, Jensen showed that $T(\mathbf{X})$ is also distributed according to (1.1) with zero means, correlation matrix $\mathbf{I}_k$, and degrees of freedom ν.

Jensen (1994) also studied the concentration properties of (1.1) via peakedness by varying its parameters. If $\mathbf{X}$ is multivariate normal, then the transformation $\mathbf{X} \to T(\mathbf{X})$ diminishes the peakedness. If, on the other hand, $\mathbf{X}$ is distributed according to (1.1) with mean vector $\mu \mathbf{1}_n$, covariance matrix $\sigma^2 \mathbf{I}_n$, and degrees of freedom m, then the transformation is peakedness-enhancing for all $m < \nu$. If $m > \nu > 2$, then the transformation serves to increase variances. For any $m > \nu > 0$ the marginal distributions are less peaked after $T(\mathbf{X})$ than before in the sense of Birnbaum (1948). If $m = \nu$, then the marginals are identical before and after $T(\mathbf{X})$, thus exhibiting identical tail behavior. If $\nu > m$ then marginals are more peaked (in the sense of Birnbaum, 1948) after applying $T(\mathbf{X})$ than before; and if $\nu > m > 2$, then $T(\mathbf{X})$ serves as a variance-diminishing transformation.

1.5 A Consistency Property

A random vector $\mathbf{X} = (X_1, \ldots, X_p)^T$ is said to have the spherical distribution if its joint pdf can be written in the form

$$g\left(\sum_{i=1}^{p} x_i^2 \,\middle|\, p\right),$$

where $g(\cdot)$ is referred to as the density generator. The p-variate t pdf (1.1) with $\boldsymbol{\mu} = \mathbf{0}$ and $\boldsymbol{\Sigma} = \mathbf{I}_p$ is spherical because in this case,

$$g(u) = \frac{\Gamma\left((\nu + p)/2\right)}{(\pi\nu)^{p/2}\Gamma\left(\nu/2\right)} \left(1 + \frac{u}{\nu}\right)^{-(p+\nu)/2}.$$

Other examples of spherical distributions include the multivariate normal and the multivariate power exponential. A spherical distribution is said to possess the consistency property if

$$\int_{-\infty}^{\infty} g\left(\sum_{i=1}^{p+1} x_i^2 \,\middle|\, p\right) dx_{p+1} = g\left(\sum_{i=1}^{p} x_i^2 \,\middle|\, p\right) \tag{1.6}$$

for any integer p and almost all $\mathbf{x} \in \Re^p$. This consistency property ensures that any marginal distribution of $\mathbf{X}$ also belongs to the same spherical family. Kano (1994) provided several necessary and sufficient conditions for a spherical distribution to satisfy (1.6). One of the them is that g must be a mixture of normal distributions; specifically, there exists a random variable $Z > 0$, unrelated to p, such that, for any p,

$$f(u \mid p) = \int \left(\frac{z}{2\pi}\right)^{p/2} \exp\left(-\frac{uz}{2}\right) F(dz),$$

where $F(\cdot)$ denotes the cumulative distribution function (cdf) of Z. Since the multivariate t is a mixture of normal distributions (see (1.2)), it follows that it must have the consistency property. Other distributions that have the consistency property include the multivariate normal and the multivariate Cauchy. Distributions that do not share this property include the multivariate logistic, multivariate Pearson type II, multivariate Pearson type VII , and the multivariate Bessel.

1.6 Density Expansions

Fisher (1925) and later Dickey (1967a) provided expansions of the pdf

$$f(x) = \frac{\Gamma((\nu + 1)/2)}{\sqrt{\pi\nu}\Gamma\left(\nu/2\right)} \left\{1 + \frac{x^2}{\nu}\right\}^{-(\nu+1)/2}$$

of the univariate Student's t distribution. The expansion in the latter paper involves Appell's polynomials, and hence recurrence schemes are available for its coefficients. Specifically,

$$f(x) = \sqrt{\frac{1+\nu}{2\pi\nu}} \exp\left(-\frac{1+\nu}{2\nu}x^2\right) \sum_{k=0}^{\infty} Q_k \left(-\frac{1+\nu}{2\nu}x^2\right)(1+\nu)^{-k}, \tag{1.7}$$

where

$$Q_k(t) = P_k(t) - \frac{1}{\sqrt{\pi}} \sum_{l=0}^{k-1} Q_l(t) P_{k-l}(\Gamma). \tag{1.8}$$

Here, $P_k(t)$ are polynomials (in powers of t) satisfying

$$\sum_k P_k(t)(1+\nu)^{-k} = \left(1 - \frac{2t}{1+\nu}\right)^{-(1+\nu)/2} \exp(-t)$$

and $P_k(\Gamma)$ denotes the polynomial $P_k(t)$ with the powers t^r replaced by $\Gamma(r+1/2)$. Dickey (1967a) also provided an analog of (1.7) for the multivariate t pdf (1.1). It takes the same form as (1.7) with x^2 replaced by $(\mathbf{x}-\boldsymbol{\mu})^T \mathbf{R}^{-1}(\mathbf{x}-\boldsymbol{\mu})$, $\nu+1$ replaced by $\nu+p$, and with (1.8) replaced by

$$Q_k(t) = P_k(t) - \frac{1}{\Gamma(p/2)} \sum_{l=0}^{k-1} Q_l(t) P_{k-l}(\Gamma_p),$$

where Γ_p indicates the substitution of $\Gamma(r+p/2)$ for t^r.

1.7 Moments

Since $\mathbf{Y}$ and S in (1.2) are independent, the conditional distribution of (X_i, X_j), given $S = s$, is bivariate normal with means (μ_i, μ_j), common variance σ^2/s^2, and correlation coefficient r_{ij}. Thus,

$$\begin{aligned} E(X_i) &= E[E(X_i | S = s)] \\ &= E(\mu_i) \\ &= \mu_i. \end{aligned}$$

To find the second moments, consider the classical identity

$$\begin{aligned} Cov(X_i, X_j) = {} & E[Cov(X_i, X_j) | S = s] \\ & + Cov[E(X_i | S = s) \, E(X_j | S = s)] \end{aligned}$$

for all $i, j = 1, \ldots, p$. Clearly, one has

$$E\left[Cov\left(X_i, X_j\right) \mid S = s\right] = \sigma^2 r_{ij} E\left(\frac{1}{S^2}\right)$$

and

$$Cov\left[E\left(X_i \mid S = s\right) E\left(X_j \mid S = s\right)\right] = 0.$$

If $\nu > 2$, then $E(1/S^2)$ exists and is equal to $\nu/\{\sigma^2(\nu - 2)\}$. Thus, by choosing $i = j$ and $i < j$, respectively, one obtains

$$Var\left(X_i\right) = \frac{\nu}{\nu - 2}$$

and

$$Cov\left(X_i, X_j\right) = \frac{\nu}{\nu - 2} r_{ij}.$$

Hence the matrix $\mathbf{R}$ is indeed the correlation matrix as stated in definition (1.1).

In the case where $\boldsymbol{\mu} = \mathbf{0}$, the product moments of $\mathbf{X}$ are easily found by exploiting the independence of $\mathbf{Y}$ and S in (1.2). One obtains

$$\begin{aligned} \mu_{r_1, r_2, \ldots, r_p} &= E\left[\prod_{j=1}^{p} X_j^{r_j}\right] \\ &= E\left[S^{-r}\left(\prod_{j=1}^{p} Y_j^{r_j}\right)\right] \\ &= \sigma^{-r} \nu^{r/2} E\left[\prod_{j=1}^{p} Y_j^{r_j}\right] E\left[\chi_\nu^{-r}\right], \end{aligned}$$

provided that $r = r_1 + r_2 + \cdots + r_p < \nu/2$. In the special case where $Y_1, \ldots, Y_p$ are mutually independent, one obtains

$$\mu_{r_1, r_2, \ldots, r_p} = \sigma^{-r} \nu^{r/2} E\left[\chi_\nu^{-r}\right] \prod_{j=1}^{p} E\left[Y_j^{r_j}\right].$$

If anyone of the r_j's is odd, then the moment is zero. If all of them are even, then

$$\mu_{r_1, r_2, \ldots, r_p} = \frac{\nu^{r/2} \prod_{j=1}^{p} \{1 \cdot 3 \cdot 5 \cdots (2r_j - 1)\}}{(\nu - 2)(\nu - 4) \cdots (\nu - r)}, \qquad \nu > r.$$

In particular,

$$\mu_{2,0,\ldots,0} = \frac{\nu}{\nu-2}, \qquad \nu > 2,$$

$$\mu_{4,0,\ldots,0} = \frac{3\nu^2}{(\nu-2)(\nu-4)}, \qquad \nu > 4,$$

$$\mu_{2,2,0,\ldots,0} = \frac{\nu^2}{(\nu-2)(\nu-4)}, \qquad \nu > 4,$$

and

$$\mu_{2,2,2,0,\ldots,0} = \frac{\nu^3}{(\nu-2)(\nu-4)(\nu-6)}, \qquad \nu > 6.$$

1.8 Maximums

Of special interest are the moments of $Z = \max(X_1, \ldots, X_p)$ when $\mathbf{X}^T = (X_1, \ldots, X_p)$ has the t pdf (1.1) with the mean vector $\boldsymbol{\mu}$ and covariance matrix $\boldsymbol{\Sigma}$. These moments have applications in decision theory, particularly in the selection and estimation of the maximum of a set of parameters. It also has applications in forecasting. The problem of finding the moments of Z has been considered by Raiffa and Schlaifer (1961), Afonja (1972), and Cain (1996).

Raiffa and Schlaifer (1961) provided an expression for $E(Z-\theta)$ for the case where $p = 3$ and $\boldsymbol{\mu} = \theta \mathbf{1}_p$ (where $\mathbf{1}_p$ denotes a vector of 1's). Afonja (1972) generalized this for the general case of unequal means, variances, and correlations. We mention later a particular case of this result for $\boldsymbol{\mu} = \theta \mathbf{1}_p$. Let $\phi_p(\mathbf{y}; \mathbf{R})$ denote a p-dimensional normal pdf with zero means, unit variances, and correlation matrix $\mathbf{R}$. Also let $\mathbf{R}_i$ denote a $p \times p$ matrix with its (j, j')th element equal to $r_{i,jj'}$, where $r_{i,jj'} (j, j' \neq i)$ is the correlation between $(X_i, -X_j)$ and $(X_i, -X_{j'})$ and $r_{i,ij} = \operatorname{corr}(X_i, X_i - X_j)$. Then the kth moment of Z is given by

$$E\left(Z^k\right) = \frac{1}{\Gamma(\nu/2)} \sum_{i=1}^{p} \sum_{j=0}^{k} \binom{k}{j} \theta^{k-j} \left(\frac{\nu^2 \sigma_{ii}}{2(\nu-2)}\right)^{j/2} \Gamma\left(\frac{\nu-j}{2}\right) \mu_j\left(y_i\right), \tag{1.9}$$

where

$$\mu_j\left(y_i\right) = \int_0^{\infty} \int_0^{\infty} \cdots \int_{-\infty}^{\infty} \cdots \int_0^{\infty} \int_0^{\infty} y_i^j \phi_p\left(\mathbf{y}; \mathbf{R}_i\right) dy_p dy_{p-1} \cdots dy_i \cdots dy_2 dy_1 \tag{1.10}$$

is the marginal moment (up to a constant) of truncated normal variates. The mean and variance can be derived easily from this formula. For example,

$$E(Z) = \theta + \{E(W) - \theta\} \Gamma\left(\frac{\nu-1}{2}\right) / \Gamma\left(\frac{\nu}{2}\right),$$

where $W = \max(Y_1, \ldots, Y_p)$ for a p-variate normal random vector $\mathbf{Y}^T = (Y_1, \ldots, Y_p)$ with means equal to θ and covariance matrix $(\nu/(\nu-2))\mathbf{\Sigma}$. Afonja (1972) showed further that

$$E(W) = \theta + \sqrt{\frac{\nu}{\nu-2}} \sum_{i=1}^{p} \sqrt{\sigma_{ii}} \mu_1(y_i),$$

where $\mu_1(y_i)$ is given by (1.10) for $j = 1$.

More recently, Cain (1996) considered two forecasts F_1 and F_2 of a future variable Y where the forecast errors $X_1 = F_1 - Y$ and $X_2 = F_2 - Y$ are assumed to have the bivariate t distribution with means (μ_1, μ_2), variances (σ_1^2, σ_2^2), correlation coefficient ρ, and degrees of freedom $\nu > 2$. Cain was interested in the maximum $Z = \max(X_1, X_2)$ of the two forecast errors and whether this nonlinear function could be useful as a component of a linear combination forecast. It was shown that the pdf of Z can be written as the sum

$$f(z) = f_1(z) + f_2(z),$$

where

$$f_j(z) = \frac{1}{\sigma_j}\sqrt{\frac{\nu}{\nu-2}} t_\nu\left(\sqrt{\frac{\nu}{\nu-2}} \frac{z-\mu_i}{\sigma_i}\right) \times T_{1+\nu}\left(\frac{1+\nu\left[\frac{z-\mu_k}{\sigma_k} - \rho\frac{z-\mu_i}{\sigma_i}\right]}{\sqrt{1-\rho^2}\sqrt{\nu-2+\left(\frac{z-\mu_i}{\sigma_i}\right)^2}}\right)$$

for $k = 3 - j$, $j = 1, 2$. Here, t_ν and T_ν are, respectively, the pdf and the cdf of the Student's t distribution with degrees of freedom ν. Integration by parts yields that

$$E(Z) = \mu_1 \int_{-\infty}^{\infty} f_1(z)dz + \mu_2 \int_{-\infty}^{\infty} f_2(z)dz + \tau t_{\nu-2}\left(\frac{\mu_1-\mu_2}{\tau}\right),$$

$$Var(Z) = \sigma_1^2 \int_{-\infty}^{\infty} f_1(z)dz + \sigma_2^2 \int_{-\infty}^{\infty} f_2(z)dz$$

$$
\begin{aligned}
&+ (\mu_1 - \mu_2)^2 \int_{-\infty}^{\infty} f_1(z)dz \int_{-\infty}^{\infty} f_2(z)dz \\
&+\tau (\mu_1 - \mu_2) t_{\nu-2} \left(\frac{\mu_1 - \mu_2}{\tau}\right) \int_{-\infty}^{\infty} f_2(z)dz \\
&-\tau (\mu_1 - \mu_2) t_{\nu-2} \left(\frac{\mu_1 - \mu_2}{\tau}\right) \int_{-\infty}^{\infty} f_1(z)dz \\
&+\frac{(\mu_1 - \mu_2)\left(\sigma_2^2 - \sigma_1^2\right)}{\tau(\nu-2)} t_{\nu-2} \left(\frac{\mu_1 - \mu_2}{\tau}\right) \\
&-\tau^2 t_{\nu-2}^2 \left(\frac{\mu_1 - \mu_2}{\tau}\right),
\end{aligned}
$$

and

$$
\begin{aligned}
Cov\,(Z, X_1) &= \sigma_1^2 \int_{-\infty}^{\infty} f_1(z)dz + \rho\sigma_1\sigma_2 \int_{-\infty}^{\infty} f_2(z)dz \\
&\quad + \frac{(\mu_1 - \mu_2)\left(\sigma_1^2 - \rho\sigma_1\sigma_2\right)}{\tau(\nu-2)} t_{\nu-2} \left(\frac{\mu_1 - \mu_2}{\tau}\right),
\end{aligned}
$$

where $\tau = \sqrt{\sigma_1^2 + \sigma_2^2 - 2\rho\sigma_1\sigma_2}$. The two integrals in the above expressions can be evaluated as

$$
\int_{-\infty}^{\infty} f_1(z)dz = T_\nu \left(\frac{\mu_1 - \mu_2}{\tau} \sqrt{\frac{\nu}{\nu - 2}}\right)
$$

and

$$
\int_{-\infty}^{\infty} f_2(z)dz = 1 - T_\nu \left(\frac{\mu_1 - \mu_2}{\tau} \sqrt{\frac{\nu}{\nu - 2}}\right).
$$

The expression for $Cov(Z, X_2)$ can be obtained by switching the subscripts 1 and 2. As $\nu \to \infty$, the above expressions can be reduced by replacing $t_\nu(\cdot)$ and $T_\nu(\cdot)$ by $\phi(\cdot)$ and $\Phi(\cdot)$, respectively. On the other extreme, as $\nu \to 2^+$, the expressions could be reduced by using the fact that

$$
\lim_{\nu \to 2^+} \frac{|x|}{\nu - 2} t_{\nu-2}(x) = \begin{cases} 0, & \text{if } x = 0, \\ 1/2, & \text{if } x \neq 0, \end{cases}
$$

and

$$
\lim_{\nu \to 2^+} T_{\nu-2} \left(\sqrt{\frac{\nu}{\nu - 2}} x\right) = \begin{cases} 1, & \text{if } x > 0, \\ 1/2, & \text{if } x = 0, \\ 0, & \text{if } x < 0. \end{cases}
$$

This suggests that the results for the maximum of bivariate t distributed errors may be materially different from those for bivariate normal errors.

Cain (1996) also investigated to see whether the maximum Z can provide information additional to that of F_1 and F_2 in forecasting Y via a linear combination of the form $F = \alpha + \beta_1 F_1 + \beta_2 F_2 + \gamma M$ with $\beta_1 + \beta_2 + \gamma = 1$. Cain showed that the mean squared error of F is minimized when $\gamma = 0$ and hence that M is linearly dominated by F_1 and F_2. Similar calculations reveal that the mean forecast $(F_1 + F_2)/2$ dominates M if and only if either $\mu_1 = \mu_2$ or $\sigma_1 = \sigma_2$. Evidently further investigations are in order (to consider, for example, the case of more than two forecasts).

1.9 Distribution of a Linear Function

If $\mathbf{X}$ has the p-variate t distribution with degrees of freedom ν, mean vector $\boldsymbol{\mu}$, and correlation matrix $\mathbf{R}$, then, for any nonsingular scalar matrix $\mathbf{C}$ and for any $\mathbf{a}$, $\mathbf{CX}+\mathbf{a}$ has the p-variate t distribution with degrees of freedom ν, mean vector $\mathbf{C}\boldsymbol{\mu}+\mathbf{a}$, and correlation matrix $\mathbf{CRC}^T$. This result is of importance in applications and is similar to the corresponding result for the multivariate normal distribution.

1.10 Marginal Distributions

Let $\mathbf{X}$ possess the p-variate t distribution with degrees of freedom ν, mean vector $\boldsymbol{\mu}$, and correlation matrix $\mathbf{R}$. Consider the partitions

$$\mathbf{X} = \begin{pmatrix} \mathbf{X}_1 \\ \mathbf{X}_2 \end{pmatrix}, \tag{1.11}$$

$$\boldsymbol{\mu} = \begin{pmatrix} \boldsymbol{\mu}_1 \\ \boldsymbol{\mu}_2 \end{pmatrix}, \tag{1.12}$$

and

$$\mathbf{R} = \begin{pmatrix} \mathbf{R}_{11} & \mathbf{R}_{12} \\ \mathbf{R}_{21} & \mathbf{R}_{22} \end{pmatrix}, \tag{1.13}$$

where $\mathbf{X}_1$ is $p_1 \times 1$ and $\mathbf{R}_{11}$ is $p_1 \times p_1$. Then $\mathbf{X}_1$ has the p_1-variate t distribution with degrees of freedom ν, mean vector $\boldsymbol{\mu}_1$, correlation matrix $\mathbf{R}_{11}$, and with the joint pdf given by

$$f(\mathbf{x}_1) = \frac{\Gamma((\nu + p_1)/2)}{(\pi\nu)^{p_1/2}\Gamma(\nu/2)\left|\mathbf{R}_{11}\right|^{1/2}} \times \left[1 + \frac{1}{\nu}(\mathbf{x}_1 - \boldsymbol{\mu}_1)^T \mathbf{R}_{11}^{-1}(\mathbf{x}_1 - \boldsymbol{\mu}_1)\right]^{-(\nu+p_1)/2}.$$

Moreover, $\mathbf{X}_2$ also has the $(p-p_1)$-variate t distribution with degrees of freedom ν, mean vector $\boldsymbol{\mu}_2$, correlation matrix $\mathbf{R}_{22}$, and with the joint pdf given by

$$\begin{aligned} f(\mathbf{x}_2) &= \frac{\Gamma((\nu+p-p_1)/2)}{(\pi\nu)^{p_1/2}\Gamma(\nu/2)\left|\mathbf{R}_{22}\right|^{1/2}} \\ &\quad \times \left[1+\frac{1}{\nu}(\mathbf{x}_2-\boldsymbol{\mu}_2)^T\mathbf{R}_{22}^{-1}(\mathbf{x}_2-\boldsymbol{\mu}_2)\right]^{-(\nu+p-p_1)/2}. \end{aligned}$$

1.11 Conditional Distributions

Several interesting properties have been obtained for conditional pdfs of the multivariate t distribution. If $\mathbf{X}$ has the central p-variate t distribution with degrees of freedom ν and correlation matrix $\mathbf{R}$, it then follows from Section 1.10 that the conditional pdf of $\mathbf{X}_2$ given $\mathbf{X}_1$ is given by

$$\begin{aligned} f(\mathbf{x}_2 \mid \mathbf{x}_1) &= \frac{\Gamma((\nu+p)/2)}{(\nu\pi)^{p_1/2}\Gamma((\nu+p_1)/2)} \frac{\left|\mathbf{R}_{11}\right|^{1/2}}{\left|\mathbf{R}\right|^{1/2}} \\ &\quad \times \frac{\left[1+(1/\nu)\mathbf{x}_1^T\mathbf{R}_{11}^{-1}\mathbf{x}_1\right]^{(\nu+p_1)/2}}{\left[1+(1/\nu)\mathbf{x}^T\mathbf{R}^{-1}\mathbf{x}\right]^{(\nu+p)/2}}. \end{aligned} \tag{1.14}$$

Since

$$|\mathbf{R}| = |\mathbf{R}_{11}|\left|\mathbf{R}_{22}-\mathbf{R}_{21}\mathbf{R}_{11}^{-1}\mathbf{R}_{12}\right|$$

and

$$\mathbf{x}^T\mathbf{R}^{-1}\mathbf{x} = \mathbf{x}_1^T\mathbf{R}_{11}^{-1}\mathbf{x}_1 + \mathbf{x}_{2\cdot1}^T\mathbf{R}_{22\cdot1}^{-1}\mathbf{x}_{2\cdot1},$$

where

$$\mathbf{x}_{2\cdot1} = \mathbf{x}_2 - \mathbf{R}_{21}\mathbf{R}_{11}^{-1}\mathbf{x}_1$$

and

$$\mathbf{R}_{22\cdot1} = \mathbf{R}_{22} - \mathbf{R}_{21}\mathbf{R}_{11}^{-1}\mathbf{R}_{12},$$

one can rewrite (1.14) as

$$\begin{aligned} f(\mathbf{x}_2 \mid \mathbf{x}_1) &= \frac{\Gamma((\nu+p)/2)}{\{(\nu+p_1)\pi\}^{(p-p_1)/2}\Gamma((\nu+p_1)/2)\left|\mathbf{R}_{22\cdot1}\right|^{1/2}} \\ &\quad \times \left[1+\frac{1}{\nu+p_1}\frac{((\nu+p_1)/\nu)\mathbf{x}_{2\cdot1}^T\mathbf{R}_{22\cdot1}^{-1}\mathbf{x}_{2\cdot1}}{1+(1/\nu)\mathbf{x}_1^T\mathbf{R}_{11}^{-1}\mathbf{x}_1}\right]^{-(\nu+p)/2} \\ &\quad \times \left[\frac{(\nu+p_1)/\nu}{1+(1/\nu)\mathbf{x}_1^T\mathbf{R}_{11}^{-1}\mathbf{x}_1}\right]^{(p-p_1)/2}. \end{aligned} \tag{1.15}$$

Landenna and Ferrari (1988) noted that this conditional pdf is not a $(p-p_1)$-variate t unless the values of $\mathbf{x}_1$ are ± 1. For example, consider the special case of (1.15) for $\mathbf{R} = \mathbf{I}_p$. In this case, (1.15) becomes

$$f(\mathbf{x}_2 \mid \mathbf{x}_1) = \frac{\Gamma((\nu+p)/2)}{\pi^{(p-p_1)/2}\Gamma((\nu+p_1)/2)\left(\nu+\sum_{j=1}^{p_1} x_j^2\right)^{(p-p_1)/2}} \times \left[1+\frac{1}{\nu+\sum_{j=1}^{p_1} x_j^2}\sum_{j=p_1+1}^{p} x_j^2\right]^{-(\nu+p)/2}. \tag{1.16}$$

When $x_j = \pm 1$, $j = 1, 2, \ldots, p_1$, (1.16) reduces to

$$f(\mathbf{x}_2 \mid \mathbf{x}_1) = \frac{\Gamma((\nu+p)/2)}{\pi^{(p-p_1)/2}\Gamma((\nu+p_1)/2)(\nu+p_1)^{(p-p_1)/2}} \times \left[1+\frac{1}{\nu+p_1}\sum_{j=p_1+1}^{p} x_j^2\right]^{-(\nu+p)/2},$$

which is the joint pdf of a central $(p-p_1)$-variate t distribution with degrees of freedom $(\nu+p_1)$ and correlation matrix $\mathbf{I}_{p-p_1}$. Landenna and Ferrari (1988) also described the manner in which the probabilities of the conditional pdf (1.15) can be expressed in terms of the probabilities of $\mathbf{x}_2$ conditioned on $\mathbf{x}_1$ taking the values ± 1.

The form of the conditional pdf (1.15) also suggests that

$$\mathbf{Y}_1 = \mathbf{X}_1 \tag{1.17}$$

and

$$\mathbf{Y}_2 = \sqrt{\frac{\nu+p_1}{\nu}}\left(1+\frac{1}{\nu}\mathbf{X}_1^T\mathbf{R}_{11}^{-1}\mathbf{X}_1\right)^{-1/2}\left(\mathbf{X}_2-\mathbf{R}_{21}\mathbf{R}_{11}^{-1}\mathbf{X}_1\right) \tag{1.18}$$

are independent, that $\mathbf{Y}_1$ has the central p_1-variate t distribution with degrees of freedom ν and correlation matrix $\mathbf{R}_{11}$, and that $\mathbf{Y}_2$ has the central $(p-p_1)$-variate t distribution with degrees of freedom $\nu+p_1$ and correlation matrix $\mathbf{R}_{22\cdot 1}$. From this observation, it follows easily that the conditional expectation of $\mathbf{X}_2$ given $\mathbf{X}_1$ is linear and that $E(\mathbf{X}_2 \mid \mathbf{X}_1 = \mathbf{x}_1) = \mathbf{R}_{21}\mathbf{R}_{11}^{-1}\mathbf{x}_1$. In particular,

$$E(X_p \mid X_1 = x_1, \ldots, X_{p-1} = x_{p-1}) = \frac{1}{r_{pp}^*}\sum_{j=0}^{p-1} r_{jp}^* x_j$$

and

$$
\begin{aligned}
& Var\,(X_p\,|X_1 = x_1, \ldots, X_{p-1} = x_{p-1}) \\
= \; & \frac{1}{r^*_{pp}} \frac{\nu}{\nu + p - 3} \left[1 + \frac{1}{\nu} \sum_{j,k=0}^{p-1} \left\{ r^*_{jk} - \frac{r^*_{jp} r^*_{kp}}{r^*_{pp}} \right\} x_j x_k \right], \quad (1.19)
\end{aligned}
$$

where r^*_{jk} is the (j,k)th element of $\mathbf{R}^{-1}$ (Bennett, 1961). It is illuminating to compare the conditional variance (1.19) with the value $1/r^*_{pp}$ corresponding to the conditional variance of the multivariate normal distribution.

Siotani (1976) generalized the result of (1.17)–(1.18) by splitting $\mathbf{X}$ into more than two sets of variates. Let

$$
\mathbf{X} = \begin{pmatrix} \mathbf{X}_1 \\ \mathbf{X}_2 \\ \vdots \\ \mathbf{X}_k \end{pmatrix} \quad (1.20)
$$

and

$$
\mathbf{R} = \begin{pmatrix} \mathbf{R}_{11} & \mathbf{R}_{12} & \cdots & \mathbf{R}_{1k} \\ \mathbf{R}_{21} & \mathbf{R}_{22} & \cdots & \mathbf{R}_{2k} \\ \vdots & \vdots & \ddots & \vdots \\ \mathbf{R}_{k1} & \mathbf{R}_{k2} & \cdots & \mathbf{R}_{kk} \end{pmatrix},
$$

where $\mathbf{X}_l$ is $p_1 \times 1$ for $l = 1, 2, \ldots, k$ and $\mathbf{R}_{lm}$ is $p_l \times p_m$ for $l = 1, 2, \ldots, k$, $m = 1, 2, \ldots, k$. Clearly $p_1 + p_2 + \cdots + p_k = p$. Introducing the notations

$$
q_l = p_1 + p_2 + \cdots + p_l, \quad (1.21)
$$

$$
\mathbf{X}_{(l)} = \begin{pmatrix} \mathbf{X}_1 \\ \mathbf{X}_2 \\ \vdots \\ \mathbf{X}_l \end{pmatrix}, \quad (1.22)
$$

$$
\mathbf{R}_{(l)} = \begin{pmatrix} \mathbf{R}_{11} & \mathbf{R}_{12} & \cdots & \mathbf{R}_{1l} \\ \mathbf{R}_{21} & \mathbf{R}_{22} & \cdots & \mathbf{R}_{2l} \\ \vdots & \vdots & \ddots & \vdots \\ \mathbf{R}_{l1} & \mathbf{R}_{l2} & \cdots & \mathbf{R}_{ll} \end{pmatrix}, \quad (1.23)
$$

$$\mathbf{R}_{(l)}^{(l+1)} = \begin{pmatrix} \mathbf{R}_{1,l+1} \\ \mathbf{R}_{2,l+1} \\ \vdots \\ \mathbf{R}_{l,l+1} \end{pmatrix}, \tag{1.24}$$

and

$$\mathbf{R}_{l+1,l+1\cdot(l)} = \mathbf{R}_{l+1,l+1} - \mathbf{R}_{(l)}^{(l+1)^T} \mathbf{R}_{(l)}^{-1} \mathbf{R}_{(l)}^{(l+1)}, \tag{1.25}$$

Siotani showed that

$$\mathbf{Y}_1 = \mathbf{X}_1$$

and

$$\begin{aligned} \mathbf{Y}_{l+1} &= \sqrt{\frac{\nu + q_l}{\nu}} \left(1 + \frac{1}{\nu} \mathbf{X}_{(l)}^T \mathbf{R}_l^{-1} \mathbf{X}_{(l)}\right)^{-1/2} \\ &\quad \times \left(\mathbf{X}_{(l+1)} - \mathbf{R}_{(l)}^{(l+1)^T} \mathbf{R}_{(l)}^{-1} \mathbf{X}_{(l)}\right) \end{aligned}$$

for $l = 1, \ldots, k-1$ are independent, that $\mathbf{Y}_1$ has the central p_1-variate t distribution with degrees of freedom ν and correlation matrix $\mathbf{R}_{11}$, and that $\mathbf{Y}_{l+1}$ has the central p_{l+1}-variate t distribution with degrees of freedom $(\nu + q_l)$ and correlation matrix $\mathbf{R}_{l+1,l+1\cdot(l)}$ for $l = 1, \ldots, k-1$. In the special case for $\mathbf{R} = \mathbf{I}_p$, the $\mathbf{Y}$'s can be written as

$$\mathbf{Y}_1 = \mathbf{X}_1$$

and

$$\mathbf{Y}_{l+1} = \sqrt{\frac{\nu + q_l}{\nu}} \left(1 + \frac{1}{\nu} \sum_{m=1}^{l} \mathbf{X}_m^T \mathbf{X}_m\right)^{-1/2} \mathbf{X}_{l+1}.$$

1.12 Quadratic Forms

If $\mathbf{X}$ has the p-variate t distribution with degrees of freedom ν, mean vector $\boldsymbol{\mu}$, and correlation matrix $\mathbf{R}$, then $\mathbf{X}^T\mathbf{R}^{-1}\mathbf{X}/p$ has the noncentral F distribution with degrees of freedom p and ν and noncentrality parameter $\boldsymbol{\mu}^T\mathbf{R}^{-1}\boldsymbol{\mu}/p$. See Hsu (1990) for a particular case of this result. When $\boldsymbol{\mu} = \mathbf{0}$, the distribution is central F and so $\mathbf{X}^T\mathbf{R}^{-1}\mathbf{X}/(p + \mathbf{X}^T\mathbf{R}^{-1}\mathbf{X})$ has the $Beta(p/2, \nu/2)$ distribution. There are a number of problems related to quadratic forms of multivariate t that are worthy of further investigation.

1.13 F Matrix

Consider two independent random samples $\mathbf{x}_1^{(1)}, \ldots, \mathbf{x}_{n_1}^{(1)}$ and $\mathbf{x}_1^{(2)}, \ldots, \mathbf{x}_{n_2}^{(2)}$ from two different elliptical distributions (which contain multivariate t as a particular case – as already mentioned in Section 1.1). Let

$$\mathbf{S}_i = \sum_{k=1}^{n_i} \mathbf{x}_k^{(i)} \mathbf{x}_k^{(i)T}$$

for $i = 1, 2$. Then $\mathbf{F} = (\mathbf{S}_1/n_1)/(\mathbf{S}_2/n_2)$ is the multivariate F matrix. Hayakawa (1989) studied the asymptotic behavior of the determinant, latent roots, latent vectors, and the trace of the F matrix for an elliptical population. These results are useful in the study of the robustness of the statistics derived for testing several hypotheses about parameters of a normal population with the elliptical distribution introduced as the alternative population. Hayakawa (1989) illustrated the usefulness of the results through a multivariate t-population.

1.14 Association

The well known definition states that the random variables $X_1, \ldots, X_p$ are said to be associated if

$$Cov(f(X_1, \ldots, X_p), g(X_1, \ldots, X_p)) \geq 0$$

for all nondecreasing functions f, g (Esary et al., 1967). Association implies positive quadrant dependence, that is, that $\Pr\{\cap(X_i \leq x_i)\} \geq \prod_{i=1}^{p} \Pr(X_i \leq x_i)$ for all real numbers $x_1, \ldots, x_p$ (Lehmann, 1966). Jogdeo (1977) and Abdel-Hameed and Sampson (1978) established that the components of a multivariate t random vector are associated under certain conditions on correlations. More generally, the following result holds. Let $\mathbf{Z}$ be a p-variate vector with independent and real components, each having a symmetric unimodal distribution. Suppose $\mathbf{Y} = \mathbf{Z} + \mathbf{U}$, where $\mathbf{U}$ is independent of $\mathbf{Z}$ and either

(i) $\mathbf{U} = (\alpha_1 V, \ldots, \alpha_k V, \alpha_{k+1} W, \ldots, \alpha_n W)$, where (V, W) has a bivariate normal distribution centered at $\mathbf{0}$,

(ii) or $\mathbf{U} = \boldsymbol{\alpha} W$, where $\boldsymbol{\alpha}$ is an arbitrary but fixed p-variate vector and W is an arbitrary real random variable.

For $(n+1)$ independent and identically distributed (iid) copies $\mathbf{Y}_i^T = (Y_{i1}, \ldots, Y_{ip})$, $i = 0, 1, \ldots, n$ of $\mathbf{Y}$ define X_j^2, $j = 1, \ldots, p$ by

$$X_j^2 = \frac{nY_{0j}^2}{\sum_{i=1}^n Y_{ij}^2}.$$

Then the variables X_j^2 (or, equivalently, $| X_j |$), $j = 1, \ldots, p$ are associated.

Now, redefine $\mathbf{Y}$ as a p-variate normal random vector with zero means and covariance matrix specified by $\boldsymbol{\Sigma} = \{r_{ij}\sigma_i\sigma_j\}$. Let S_k^2 and S_k^{*2} be independent chi-squared random variables with degrees of freedom n and q_k, respectively, for $k = 1, \ldots, p$. Also assume that $\mathbf{X}$, S_k^2, and S_k^{*2} are mutually independent. Then, as a consequence of the above general result, one could provide the following assertions about bivariate and trivariate t vectors

- For $p = 2$, the random variables

  $$(X_1, X_2) = \left(\frac{| Y_1 |}{\sqrt{S_1^2 + S_1^{*2}}}, \frac{| Y_2 |}{\sqrt{S_2^2 + S_2^{*2}}} \right)$$

  are associated.
- For $p = 3$, if $\prod_{i<j} \operatorname{sign}(\lambda_{ij}) \le 0$, where $\boldsymbol{\Lambda} = \{\lambda_{ij}\} = \boldsymbol{\Sigma}^{-1}$, then the random variables

  $$(X_1, X_2, X_3) = \left(\frac{| Y_1 |}{\sqrt{S_1^2 + S_1^{*2}}}, \frac{| Y_2 |}{\sqrt{S_2^2 + S_2^{*2}}}, \frac{| Y_3 |}{\sqrt{S_3^2 + S_3^{*2}}} \right)$$

  are associated.

1.15 Entropy

The entropy of a continuous random vector $\mathbf{X}$ may be regarded as a descriptive quantity, just as the median, mode, variance, and the coefficient of skewness may be regarded as descriptive parameters. The entropy is a measure of the extent to which a multivariate distribution is concentrated on a few points or dispersed over many points. Thus, the entropy is a measure of dispersion, somewhat like the standard deviation in the univariate case.

Mathematically, the entropy of $\mathbf{X}$ is defined by

$$H(\mathbf{X}) = E[-\log f(\mathbf{X})]$$

$$= -\int f(\mathbf{x}) \log f(\mathbf{x})\, d\mathbf{x}. \tag{1.26}$$

Guerrero-Cusumano (1996a) derived the forms of this for the multivariate t distribution. For a central p-variate t, it turns out that

$$\begin{aligned} H(\mathbf{X};\mathbf{R}) &= \frac{1}{2}\log|\mathbf{R}| + \log\left[\frac{(\nu\pi)^{p/2}}{\Gamma(p/2)} B\left(\frac{p}{2},\frac{\nu}{2}\right)\right] \\ &\quad + \frac{\nu+p}{2}\left[\psi\left(\frac{\nu+p}{2}\right) - \psi\left(\frac{\nu}{2}\right)\right], \end{aligned} \tag{1.27}$$

where $\psi(t) = d\log\Gamma(t)/dt$ denotes the digamma function. Note that (1.27) can reexpressed as $H(\mathbf{X}) = 1/2 \mid \mathbf{R} \mid + \Phi(\nu, p)$, where $\Phi(\nu, p)$ is a constant that depends only on ν and p. Table 1 in Guerrero-Cusumano (1996a) tabulates $\Phi(\nu, p)$ for $\nu = 1(1)35$ and $p = 1(1)5$. The following is an abridged version of the table.

Constant Φ for $H(\mathbf{X}) = 1/2 \mid \mathbf{R} \mid + \Phi(\nu, p)$

ν	$p = 1$	$p = 2$	$p = 3$	$p = 4$	$p = 5$
1	2.53102	4.83788	7.06205	9.24381	11.3999
2	1.96028	3.83788	5.67306	7.48261	9.27502
3	1.77348	3.50454	5.20997	6.89826	8.57432
4	1.68176	3.33788	4.97687	6.60362	8.22121
5	1.62750	3.23788	4.83602	6.42500	8.00685
6	1.59172	3.17121	4.74153	6.30474	7.86226
7	1.56638	3.12359	4.67368	6.21809	7.75785
8	1.54750	3.08788	4.62257	6.15261	7.67878
9	1.53289	3.06010	4.58266	6.10135	7.61677
10	1.52126	3.03788	4.55062	6.06010	7.56678

The particular case of (1.27) for $\nu = 1$ gives the entropy for the multivariate Cauchy distribution

$$\begin{aligned} H(\mathbf{X};\mathbf{R}) &= \frac{1}{2}\log|\mathbf{R}| + \log\left[\frac{\pi^{p/2}}{\Gamma(p/2)} B\left(\frac{p}{2},\frac{1}{2}\right)\right] \\ &\quad + \frac{1+p}{2}\left[\psi\left(\frac{1+p}{2}\right) - \psi\left(\frac{1}{2}\right)\right]. \end{aligned}$$

As $\nu \to \infty$, (1.27) converges to the entropy of the normal distribution given by

$$H(\mathbf{X};\mathbf{R}) = \frac{p}{2}\log(2e\pi) + \frac{1}{2}\log|\mathbf{R}|. \tag{1.28}$$

The sampling properties of (1.27) will be discussed in Chapter 9.

For the noncentral p-variate t, (1.26) takes the general form

$$H(\mathbf{X};\mathbf{R}) = \frac{1}{2}\log|\mathbf{R}| + \log\left[\frac{(\nu\pi)^{p/2}}{\Gamma(p/2)}B\left(\frac{p}{2},\frac{\nu}{2}\right)\right] + \frac{\nu+p}{2}M(\nu,p,\Delta), \tag{1.29}$$

where $\Delta = \boldsymbol{\mu}^T\mathbf{R}^{-1}\boldsymbol{\mu}$ and $M(\nu,p,\Delta)$ is given by

$$M(\nu,p,\Delta) = \exp\left(-\frac{\Delta}{2}\right)\sum_{j=0}^{\infty}\frac{1}{j!}\left\{\psi\left(\frac{\nu+p+2j}{2}\right) - \psi\left(\frac{\nu}{2}\right)\right\}.$$

Setting $\nu = 1$ in (1.29), one can obtain the entropy of the noncentral p-variate Cauchy distribution. In the case $p = 1$, (1.29) coincides with the entropies for the univariate Student's t and Cauchy distributions given, for example, in Lazo and Rathie (1978).

Zografos (1999) provided a maximum entropy characterization of (1.1). The maximum entropy principle suggests to approximate the unknown pdf of $\mathbf{X}$ by the model that maximizes (1.26) subject to the constraints that define the class of pdfs considered. Jaynes (1957) asserted that the maximum entropy distribution, obtained by this constrained maximization problem, "is the only unbiased assignment we can make; to use any other would amount to an arbitrary assumption of information which by hypothesis we do not have." Zografos (1999) showed that (1.1) is the solution to maximizing $E[-\log f(\mathbf{X})]$ subject to the constraint

$$E\left[\log\left\{1 + \frac{1}{\nu}(\mathbf{X}-\boldsymbol{\mu})^T\mathbf{R}^{-1}(\mathbf{X}-\boldsymbol{\mu})\right\}\right] = w\left(\frac{p+\nu}{2};\frac{p}{2}\right),$$

where $w(x;a) = \psi(x) - \psi(x-a)$, $x > a$, and $\psi(\cdot)$ denotes the digamma function. For further discussion of maximum entropy methods, see Fry (2002).

1.16 Kullback-Leibler Number

The mutual information of a continuous random vector $\mathbf{X}$ with joint pdf $f(\mathbf{x})$ and marginal pdfs $f(x_i)$, $i = 1,\ldots,p$ is defined by

$$T(\mathbf{X}) = E\left[-\log\left\{\frac{f(\mathbf{X})}{f(x_1)\cdots f(x_p)}\right\}\right] \tag{1.30}$$

with the domain of variation given by $0 \le T(\mathbf{X}) < \infty$. (The reader should not confuse this with the transformation $T(\mathbf{X})$ given in (1.5).) The quantity (1.30) can be considered a measure of dependence (Joe, 1989). The larger the $T(\mathbf{X})$, the higher the dependence among the variables X_i, $i = 1, \ldots, p$. Naturally, $T(\mathbf{X}) = 0$ implies that the variables are independent; this latter statement follows from the fact that T is a special case of the Kullback-Leibler number, $KL(f, g)$ (Kullback, 1968). When the variables of $\mathbf{X}$ are multivariate normal with covariance matrix $\mathbf{\Sigma}$, it is easy to compute $T(\mathbf{X})$ as the difference between entropies given by (1.28); specifically,

$$T(\mathbf{X}; \mathbf{\Sigma}) = H(\mathbf{X}; \mathbf{\Sigma}) - H(\mathbf{X}; \mathbf{D}),$$

where $\mathbf{D}$ is a diagonal matrix corresponding to $\mathbf{\Sigma}$ with the elements $\sigma_{11}, \ldots, \sigma_{pp}$. This is due to the well known fact that uncorrelatedness implies independence in the normal case. This fact also implies that $T(\mathbf{X}; \mathbf{I}) = 0$. In general, for any member of an elliptical family of distributions, this is not true; in other words, uncorrelatedness does not imply that $T(\mathbf{X}) = 0$. The mutual information attempts to summarize in a single number the whole dependence structure of the multivariate distribution of $\mathbf{X}$.

Guerrero-Cusumano (1996b) derived the form of (1.30) for the multivariate t distribution. For a central p-variate t, it turns out that

$$T(\mathbf{X}) = \Omega - \frac{1}{2} \log |\mathbf{R}|, \tag{1.31}$$

where Ω is given by

$$\begin{aligned} \Omega = {} & \log \left\{ \frac{\Gamma(p/2)}{\pi^{p/2}} \frac{B^p\left(\frac{1+\nu}{2}, \frac{1}{2}\right)}{B\left(\frac{p+\nu}{2}, \frac{p}{2}\right)} \right\} + \frac{p(1+\nu)}{2} \left\{ \psi\left(\frac{1+\nu}{2}\right) - \psi\left(\frac{\nu}{2}\right) \right\} \\ & - \frac{p+\nu}{2} \left\{ \psi\left(\frac{p+\nu}{2}\right) - \psi\left(\frac{\nu}{2}\right) \right\}. \end{aligned} \tag{1.32}$$

It is easy to see that $\Omega \to 0$ as $\nu \to \infty$. The mutual information for the multivariate normal distribution with correlation matrix $\mathbf{R}$ is given by $-(1/2) \log |\mathbf{R}|$ (Kullback, 1968). The particular case of (1.31) for $\nu = 1$ gives the mutual information for the multivariate Cauchy distribution with Ω taking the simpler form

$$\Omega = \log \left\{ \frac{8^p}{\pi^{p/2}} \frac{\Gamma\left(p + \frac{1}{2}\right)}{\Gamma\left(\frac{1+p}{2}\right)} \right\} - \frac{1+p}{2} \left\{ \psi\left(\frac{1+p}{2}\right) - \psi\left(\frac{1}{2}\right) \right\}.$$

Table 1 in Guerrero-Cusumano (1996b) provides values of (1.32) for a range of ν and p. The following is an abridged version.

Constant Ω for $T(\mathbf{X}) = \Omega - (1/2)\log|\mathbf{R}|$

ν	$p=1$	$p=2$	$p=3$	$p=4$	$p=5$
1	0	0.4196180	0.949615	1.530690	2.141170
2	0	0.2927000	0.705474	1.184010	1.704100
3	0	0.2254360	0.565424	0.975130	1.431820
4	0	0.1835450	0.473177	0.832265	1.240460
5	0	0.1548760	0.407380	0.727338	1.096790
6	0	0.1339950	0.357917	0.646600	0.984235
7	0	0.1180970	0.319304	0.582368	0.893344
8	0	0.1055830	0.288289	0.529959	0.818244
9	0	0.0954730	0.262813	0.486337	0.755056
10	0	0.0871342	0.241503	0.449434	0.701101

Figures 1.5 and 1.6 graph $T(\mathbf{X})$ in (1.31) for $p = 2$ and $p = 4$, respectively. The correlation matrix $\mathbf{R}$ is taken to have the equicorrelation structure $r_{ij} = \rho$, $i \neq j$. It is interesting to see the "dale-shaped" three-dimensional plot. The figures show that, as one moves toward the center of the "dale," the dependence among the variables decreases, and, as one moves away from the center, the dependence increases.

For the normal case, Linfoot (1957) and Joe (1989) suggested a parameterization for $T(\mathbf{X})$ to make it comparable to a correlation coefficient. They defined the induced correlation coefficient based on the mutual information as

$$\rho_I = \sqrt{1 - \exp\{-2T(\mathbf{X})\}}. \qquad (1.33)$$

Guerrero-Cusumano (1998) suggested a similar measure for the multivariate t distribution referred to as the *dependence coefficient.* It is given by

$$\rho_I = \sqrt{1 - |\mathbf{R}| \exp(-2\Omega)}. \qquad (1.34)$$

The dependence coefficient is a quantification of dependence among the p variables of $\mathbf{X}$. This follows from the fact that independence implies

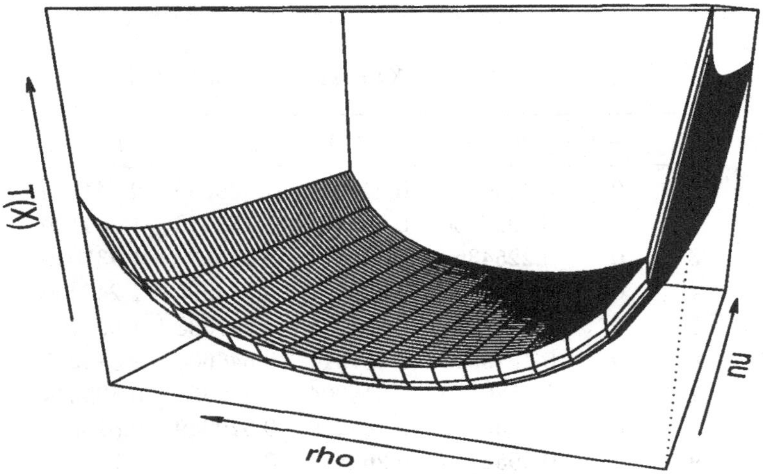

Fig. 1.5. Mutual information, (1.31), for $p = 2$

$\rho_I = 0$ and that $T(\mathbf{X}) = \infty$ implies $\rho_I = 1$. When $\nu \to \infty$, (1.34) coincides with (1.33).

The sampling properties of (1.31) will be discussed in Chapter 9.

1.17 Rényi Information

Since the concept of Rényi information is not widely available in the literature, we provide here a brief discussion of the concept. Rényi information of order λ for a continuous random variable with pdf f is defined as

$$\mathcal{I}_R(\lambda) \quad := \quad \frac{1}{1-\lambda} \log \left(\int f^{\lambda}(x) dx \right) \tag{1.35}$$

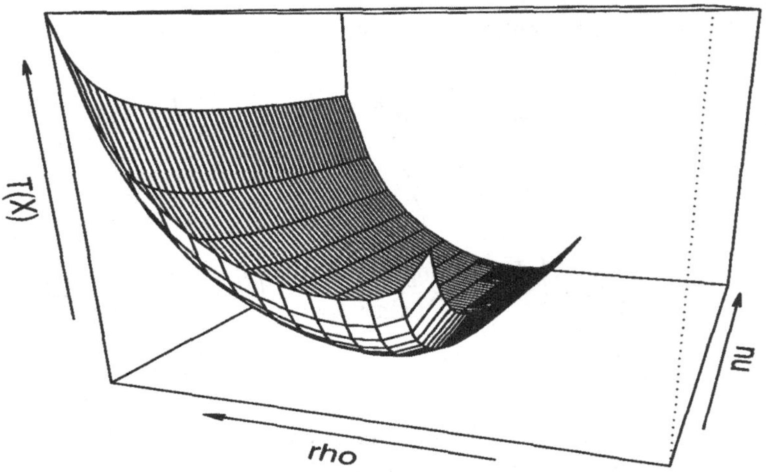

Fig. 1.6. Mutual information, (1.31), for $p = 4$

for $\lambda \neq 1$. Its value for $\lambda = 1$ is taken as the limit

$$\begin{aligned} \mathcal{I}_R(1) &:= \lim_{\lambda \to 1} \mathcal{I}_R(\lambda) \\ &= -\int f(x) \log(f(x))\, dx \\ &= -E[\log f(X)], \end{aligned}$$

which is the well known Shannon entropy. Rényi's (1959, 1960, 1961) generalization of the Shannon entropy allows for "different averaging of probabilities" via λ. Sometimes (1.35) is also referred to as the spectrum of Rényi information. Rényi information finds its applications as a measure of complexity in areas of physics, information theory, and engineering to describe many nonlinear dynamical or chaotic systems (Kurths et al., 1995) and in statistics as certain appropriately scaled test statistics (Rényi distances or relative Rényi information) for testing hypotheses in parametric models (Morales et al., 1997). The gradient $\mathcal{I}'_R(\lambda) = d\mathcal{I}_R(\lambda)/d\lambda$ also conveys useful information. In fact, a direct calculation based on (1.35) – assuming that the integral $\int f^\lambda(x)dx$ is well defined and differentiation operations are legitimate – shows that

$$\begin{aligned}\mathcal{I}'_R(1) &= \lim_{\lambda\to 1}\left[(1-\lambda)\frac{\int f^\lambda(x)\log f(x)dx}{\int f^\lambda(x)dx} + \log\left(\int f^\lambda(x)dx\right)\right] \Bigg/ (1-\lambda)^2 \\ &= -\frac{1}{2}\lim_{\lambda\to 1}\left\{\frac{\int f^\lambda(x)\log^2 f(x)dx}{\int f^\lambda(x)dx} - \left(\frac{\int f^\lambda(x)\log f(x)dx}{\int f^\lambda(x)dx}\right)^2\right\} \\ &= -\frac{1}{2}Var\left[\log f(X)\right].\end{aligned}$$

In other words, the gradient of Rényi information at $\lambda = 1$ is simply the negative half of the variance of the log-likelihood compared to the entropy as the negative of the expected log-likelihood. Thus, the variance of the log-likelihood $\mathcal{I}_f := 2\mathcal{I}'_R(1)$ measures the intrinsic shape of the distribution. This can be seen by observing that $\mathcal{I}_f$, where $f(x) = (1/\sigma)g((x-\mu)/\sigma)$. In fact, according to Bickel and Lehmann (1975), it can serve as a measure of the shape of a distribution. In the case where $f(x)$ has a finite fourth moment, it plays a similar role as a kurtosis measure in comparing the shapes of various frequently used densities and measuring the heaviness of tails, but it measures more than what kurtosis measures.

Rényi information of order λ for a p-variate random vector with joint pdf $\mathbf{x}$ is defined as

$$\mathcal{I}_R(\lambda) \quad := \quad \frac{1}{1-\lambda}\log\left(\int f^\lambda\left(x_1,\ldots,x_p\right)dx_1\cdots dx_p\right). \qquad (1.36)$$

The gradient $\mathcal{I}'_R(\lambda)$ and the measure $\mathcal{I}_f$ are defined similarly.

Song (2001) provided a comprehensive account of $\mathcal{I}_R(\lambda)$, $\mathcal{I}'_R(\lambda)$, and $\mathcal{I}_f$ for well known univariate and multivariate distributions. For the univariate Student's t distribution with degrees of freedom ν, it can be shown for $\lambda > 1/(1+\nu)$ that

$$\mathcal{I}_R(\lambda) \quad = \quad \frac{1}{1-\lambda}\log\left\{\frac{B\left((\nu\lambda+\lambda-1)/2, 1/2\right)}{B^\lambda\left(\nu/2, 1/2\right)}\right\} + \frac{1}{2}\log(\nu),$$

$$
\begin{aligned}
\mathcal{I}'_R(\lambda) &= \left[\log\left\{\frac{B\left((\nu\lambda+\lambda-1)/2,1/2\right)}{B\left(\nu/2,1/2\right)}\right\}\right. \\
&\quad +\frac{(1-\lambda)(1+\nu)}{2}\psi\left(\frac{\nu\lambda+\lambda-1}{2}\right) \\
&\quad \left.-\frac{(1-\lambda)(1+\nu)}{2}\psi\left(\frac{(1+\nu)\lambda}{2}\right)\right]\Big/(1-\lambda)^2,
\end{aligned}
$$

and

$$
\mathcal{I}_f(\nu) = \frac{(1+\nu)^2}{4}\left\{\psi'\left(\frac{\nu}{2}\right)-\psi'\left(\frac{1+\nu}{2}\right)\right\}.
$$

Using tables in Abramowitz and Stegun (1965), one obtains the particular values

$$
\begin{aligned}
\mathcal{I}_f(1) &= \frac{\pi^2}{3}, \\
\mathcal{I}_f(2) &= 9-\frac{3\pi^2}{4}, \\
\mathcal{I}_f(3) &= \frac{4\pi^2}{3}-12, \\
\mathcal{I}_f(4) &= \frac{775}{36}-\frac{25\pi^2}{12}, \\
\mathcal{I}_f(5) &= 3\pi^2-\frac{115}{4}.
\end{aligned}
$$

It is interesting to note that the measure $\mathcal{I}_f(\nu)$ decreases as ν increases, which makes sense since the tails become lighter as ν increases. In fact, it can be shown, using asymptotic formulas for the trigamma function, that $\lim_{\nu\to\infty}\mathcal{I}_f(\nu)=1/2$, which corresponds to the measure $\mathcal{I}_f(\nu)$ for the normal distribution.

For the central p-variate t distribution with correlation matrix $\mathbf{R}$ and degrees of freedom ν, it can be shown for $\lambda > p/(p+\nu)$ that

$$
\begin{aligned}
\mathcal{I}_R(\lambda) &= \frac{1}{1-\lambda}\log\left\{\frac{B\left((\nu\lambda+p\lambda-p)/2,p/2\right)}{B^{\lambda}\left(\nu/2,p/2\right)}\right\}+\frac{1}{2}\log\left\{(\nu\pi)^p\mid\mathbf{R}\mid\right\} \\
&\quad -\log\Gamma\left(\frac{p}{2}\right),
\end{aligned}
$$

$$
\begin{aligned}
\mathcal{I}'_R(\lambda) &= \left[\log\left\{\frac{B\left((\nu\lambda+p\lambda-1)/2,p/2\right)}{B\left(\nu/2,p/2\right)}\right\}\right. \\
&\quad +\frac{(1-\lambda)(p+\nu)}{2}\psi\left(\frac{\nu\lambda+p\lambda-p}{2}\right)
\end{aligned}
$$

$$-\frac{(1-\lambda)(p+\nu)}{2}\psi\left(\frac{(p+\nu)\lambda}{2}\right)\Bigg]\Bigg/(1-\lambda)^2,$$

and

$$\mathcal{I}_f(\nu) = \frac{(p+\nu)^2}{4}\left\{\psi'\left(\frac{\nu}{2}\right)-\psi'\left(\frac{p+\nu}{2}\right)\right\}.$$

For $p=1$, these expressions reduce to those derived for the Student's t distribution.

1.18 Identities

In one of the earliest papers on the subject, Dickey (1965, 1968) provided two multidimensional-integral identities involving the multivariate t distribution. This first identity expresses a moment of a product of multivariate t densities of the form (1.1) as an integral of dimension 1 less than the number of factors. Consider the product

$$g(\mathbf{x}) = \prod_{k=1}^{K}\left[1+(\mathbf{x}-\boldsymbol{\mu}_k)^T\mathbf{R}_k(\mathbf{x}-\boldsymbol{\mu}_k)\right]^{-\nu_k/2}, \qquad (1.37)$$

where each $\mathbf{R}_k \geq 0$ and $\nu_k > 0$, and so each term may not have a finite integral. The identity seeks an expression for the complete p-dimensional integral of $s \cdot g$, where $s(\mathbf{x})$ is a polynomial in the coordinates of $\mathbf{x}$. Let $\mathbf{Y}$ be a p-variate normal random vector with the covariance matrix and mean vector given by

$$\mathbf{D}_u^{-1} = \left(\sum_{k=1}^{K}u_k\mathbf{R}_k\right)^{-1}$$

and

$$\bar{\boldsymbol{\mu}}_u = \mathbf{D}_u^{-1}\sum_{k=1}^{K}u_k\mathbf{R}_k\boldsymbol{\mu}_k,$$

respectively. For given constants $c_k > 0$, $k=1,\ldots,K$, let $u_\cdot = \sum_{k=1}^{K}c_k u_k$ and $u_k = v_k u_\cdot$. Then the quantity defined by $N_{s|u} = E(s(\mathbf{Y}))$ can be expanded as a polynomial in $1/u_\cdot$ as

$$N_{s|u} = \sum_j h_j(v_1,\ldots,v_K)\,u_\cdot^{-j}.$$

Given this terminology, the identity can now be expressed as

$$\begin{aligned}
&\int_{\Re^p} s(\mathbf{x})\, g(\mathbf{x})\, d\mathbf{x} \\
&= \frac{K_0}{c_K} \sum_j 2^{-j} \Gamma\left(\frac{\nu_. - p}{2} - j\right) \int_\sigma |\mathbf{D}_v|^{-1/2} h_j(v_1, \ldots, v_K) \\
&\quad \times \left(\prod_{k=1}^K v_k^{\nu_k/2-1}\right) W_v^{j-(\nu_.-p)/2} dv_1 \cdots dv_{K-1}, \qquad (1.38)
\end{aligned}$$

where

$$K_0 = \pi^{p/2} \Big/ \prod_{k=1}^K \Gamma(\nu_k/2),$$

$$\nu_. = \sum_{k=1}^K \nu_k,$$

$$\mathbf{D}_v = \sum_{k=1}^K v_k \mathbf{R}_k,$$

$$W_v = \sum_{k=1}^K v_k \left\{1 + \boldsymbol{\mu}_k^T \mathbf{R}_k \boldsymbol{\mu}_k\right\} - \left(\sum_{k=1}^K v_k \mathbf{R}_k \boldsymbol{\mu}_k\right)^T \mathbf{D}_v^{-1} \left(\sum_{k=1}^K v_k \mathbf{R}_k \boldsymbol{\mu}_k\right),$$

and σ is the simplex

$$\sigma = \left\{(v_1, \ldots, v_K) : \sum_{k=1}^K c_k v_k = 1, \quad v_k > 0\right\}.$$

This identity has applications to inference concerning the location parameters of a multivariate normal distribution. In the particular case $K = 2$, $\mathbf{R}_k = \gamma_k \mathbf{I}_p$, and $s \equiv 1$, (1.38) reduces to

$$\begin{aligned}
\int_{\Re^p} g(\mathbf{x})\, d\mathbf{x} &= C \gamma_2^{(\nu_.-p)/2} B\left(\frac{\nu_1}{2}, \frac{\nu_2}{2}\right) \\
&\quad \times F_1\left(\frac{\nu_1}{2}; \frac{\nu_. - p}{2}, \frac{\nu_. - p}{2}, \frac{\nu_.}{2}; z_1, z_2\right), \qquad (1.39)
\end{aligned}$$

where

$$C = \frac{\Gamma((\nu_. - p)/2)}{\Gamma(\nu_1/2)\Gamma(\nu_2/2)} \frac{\pi^{p/2}}{\gamma_1^{\nu_1/2} \gamma_2^{\nu_2/2}},$$

F_1 is Appell's hypergeometric function of two variables defined by

$$\begin{aligned} & F_1(\alpha; \beta, \beta'; \gamma; x, y) \\ = & \frac{1}{B(\alpha, \gamma - \alpha)} \int_0^1 t^{\alpha-1}(1-t)^{\gamma-\alpha-1}(1-tx)^{-\beta}(1-ty)^{-\beta'} dt \end{aligned} \tag{1.40}$$

(see, for example, Erdélyi et al., 1953), and z_1 and z_2 are the two real roots of the equation

$$z^2 + \left(\gamma_2 \parallel \boldsymbol{\mu}_2 - \boldsymbol{\mu}_1 \parallel^2 + \frac{\gamma_2}{\gamma_1} - 1\right) z - \gamma_2 \parallel \boldsymbol{\mu}_2 - \boldsymbol{\mu}_1 \parallel^2 = 0.$$

The integral (1.39) is proportional to a multivariate generalization of the Behren-Fisher density. For an asymptotic expansion of (1.37) in powers of ν_k, see Dickey (1967a).

The second identity given by Dickey (1968) – see also Dickey (1966b) – expresses the density of a linear combination of independently distributed multivariate t vectors as an integral of dimension 1 less than the number of summands. Consider the r-variate vector $\boldsymbol{\delta}$ formed by the linear combination

$$\boldsymbol{\delta} = \sum_{k=1}^{K} \mathbf{B}_k \mathbf{X}_k,$$

where $\mathbf{X}_k$ are independent q_k-variate standard t random vectors with zero means, covariance matrix $\mathbf{I}_{q_k}$, and degrees of freedom ν_k. Dickey (1968) showed that $\boldsymbol{\delta}$ has the representation

$$\boldsymbol{\delta} = \sqrt{\sum_{k=1}^{K} \nu_k U_k^{-1} \mathbf{B}_k \mathbf{B}_k^T} \mathbf{Y},$$

where U_k are independent chi-squared random variables with degrees of freedom ν_k and $\mathbf{Y}$ is an independent r-variate standard normal vector. As a consequence, $\boldsymbol{\delta}$ has the further representation

$$\boldsymbol{\delta} = \sqrt{\sum_{k=1}^{K} \nu_k V_k^{-1} (\nu_k/\nu.) \mathbf{B}_k \mathbf{B}_k^T} \mathbf{W},$$

where $\nu. = \sum_{k=1}^{K} \nu_k$, $V_k = U_k / \sum_{j=1}^{K} U_j$ and $\mathbf{W}$ is an independent r-variate standard t vector with degrees of freedom $\nu.$. If the matrix $\sum \mathbf{B}_k \mathbf{B}_k^T$ is nonsingular, the distribution of $\boldsymbol{\delta}$ is nondegenerate with the joint pdf

$$f(\boldsymbol{\delta}) = C\int_{\sigma}\left(\prod_{k=1}^{K} v_k^{\nu_k/2-1}\right)\left\{1+\boldsymbol{\delta}^T\left(\sum_{k=1}^{K}(\nu_k/v_k)\,\mathbf{B}_k\mathbf{B}_k^T\right)^{-1}\boldsymbol{\delta}\right\}^{-(\nu.+r)/2}$$
$$\Big/\sqrt{\sum_{k=1}^{K}(\nu_k/v_k)\,\mathbf{B}_k\mathbf{B}_k^T dv_1\cdots dv_{K-1}}, \qquad (1.41)$$

where

$$C = \frac{\Gamma((\nu.+r)/2)}{\pi^{r/2}\Gamma(\nu_1/2)\cdots\Gamma(\nu_K/2)}$$

and as above

$$\sigma = \left\{(v_1,\ldots,v_K) : \sum_{k=1}^{K} c_k v_k = 1, \quad v_k > 0\right\}.$$

This identity has applications in Behrens-Fisher problems. The version of (1.41) for $K = 2$ and $\mathbf{B}_k = \beta_k$ is

$$f(\boldsymbol{\delta}) = CB\left(\frac{\nu_1+p}{2}, \frac{\nu_2+p}{2}\right)$$
$$\times F_1\left(\frac{\nu_1+p}{2}; \frac{\nu.+p}{2}, \frac{\nu.+p}{2}; \frac{\nu.}{2}+p; z_1, z_2\right),$$

where

$$C = \frac{\Gamma((\nu.+p)/2)}{\pi^{p/2}\Gamma(\nu_1/2)\cdots\Gamma(\nu_2/2)}\left(\nu_1\beta_1^2\right)^{\nu_1/2}\left(\nu_2\beta_2^2\right)^{-(\nu_1+p)/2},$$

F_1 is Appell's hypergeometric function as defined in (1.40), and z_1 and z_2 are the two real roots of the equation

$$z^2 + \left(\frac{\|\boldsymbol{\delta}\|^2}{\nu_2\beta_2^2} + \frac{\nu_1\beta_1^2}{\nu_2\beta_2^2} - 1\right)z - \frac{\|\boldsymbol{\delta}\|^2}{\nu_2\beta_2^2} = 0.$$

This special case is essentially equivalent to the two-factor version of (1.38). Moreover, (1.41) is a generalization of Ruben's (1960) integral representation (in the univariate case) for the usual Behrens-Fisher densities.

1.19 Some Special Cases

A number of special cases of (1.1) have been studied in the literature with great detail. Cornish (1954), in his early paper, considered the special case of (1.1) when $\boldsymbol{\mu} = \mathbf{0}$ and $\mathbf{R}$ is given by the equicorrelation matrix

$$\mathbf{R} = \begin{pmatrix} 1 & -1/p & \dots & -1/p \\ -1/p & 1 & \dots & -1/p \\ \vdots & \vdots & \dots & \vdots \\ -1/p & -1/p & \dots & 1 \end{pmatrix}.$$

The following interesting properties were established

- $\mathbf{X}^T\mathbf{R}^{-1}\mathbf{X}$ has the noncentral F distribution with degrees of freedom p and ν.
- $\mathbf{X}^T\mathbf{R}^{-1}\mathbf{X}$ has the Fisher's z distribution with degrees of freedom $p-q$ and ν – when $\mathbf{X}$ is subject to the linearly independent homogeneous conditions represented by the equation $\mathbf{SX} = \mathbf{0}$, where $\mathbf{S}$ is of order $q \times p$ and rank $q < p$.
- The cdf of the quadratic form $Q = \mathbf{X}^T\mathbf{A}\mathbf{X}$ when $\mathbf{A}$ is of rank $q \leq p$ is given by

$$\frac{\Gamma((\nu+q)/2)}{(\pi\nu)^{q/2}\Gamma(\nu/2)} \int \cdots \int \left(1 + \frac{\mathbf{x}_1^T\mathbf{x}_1}{\nu}\right)^{-(\nu+q)/2} d\mathbf{x}_1,$$

where $\mathbf{x}_1^T = (x_1, \dots, x_q)$ and the domain of integration is defined by

$$\sum_{i=1}^{q} \lambda_i x_i^2 \geq Q,$$

where λ_i are the roots of the equation $\mid \lambda\mathbf{R}^{-1} - \mathbf{A} \mid= 0$ or, alternatively, the latent roots of the matrix $\mathbf{RA}$. Consequently, the distribution of $\mathbf{X}^T\mathbf{A}\mathbf{X}$ is Fisher's z with degrees of freedom q and ν if and only if the nonzero latent roots of $\mathbf{RA}$ are all equal to unity.
- If the distribution of $\mathbf{X}$ is partitioned as in (1.11)–(1.13), then

$$E(\mathbf{X}_1 \mid \mathbf{X}_2) = -\mathbf{R}_{11}^{-1}\mathbf{R}_{22}\mathbf{x}_2,$$

and

$$Var(\mathbf{X}_1 \mid \mathbf{X}_2) = \frac{\nu + \mathbf{x}_2^T\left(\mathbf{R}_{22} - \mathbf{R}_{21}\mathbf{R}_{11}^{-1}\mathbf{R}_{12}\right)\mathbf{x}_2}{\nu + p - p_1 - 2}\mathbf{R}_{11}^{-1}.$$

In the particular case $p_1 = 1$,

$$E(X_1 \mid \mathbf{X}_2) = -\frac{1}{2}\sum_{j=2}^{p} x_j,$$

$$
\begin{aligned}
Var\,(X_1 \mid \mathbf{X}_2) &= \frac{(p+1)\nu}{2p(\nu+p-3)} + \frac{3}{4(\nu+p-3)} \sum_{j=2}^{p} x_j \\
&\quad + \frac{p+1}{2(\nu+p-3)} \sum_{j<k} x_j x_k, \\
E\,[Var\,(X_1 \mid \mathbf{X}_2)] &= \frac{\nu}{\nu-2}\frac{p+1}{2p}, \\
Var\,(\mathbf{X}_2) &= \frac{\nu}{\nu-2}\mathbf{R}_{22}, \\
Cov\,(X_1, X_i) &= -\frac{\nu}{p(\nu-2)}, \qquad i = 2, \ldots, p.
\end{aligned}
$$

Furthermore, the residual variance of X_1 with respect to $\mathbf{X}_2$ is

$$
\frac{\nu}{\nu-2}\frac{p+1}{2p},
$$

and the partial correlation coefficient of X_1 with respect to $\mathbf{X}_2$ is $-1/2$.

Patil and Kovner (1968) provided a detailed study of the trivariate t density

$$
\begin{aligned}
f\,(x_1, x_2, x_3) &= \frac{\Gamma\,((n+3)/2)}{(n\pi)^{3/2}\sqrt{1-\rho^2}\,\Gamma\,(n/2)} \\
&\quad \times \left(1 + \frac{1}{n}\frac{x_1^2 - 2\rho x_1 + x_2^2}{1-\rho^2} + x_3^2\right)^{-(n+3)/2}.
\end{aligned}
$$

Among other results, Taylor series expansions – in powers of $1/n$ – of the density and associated probabilities in rectangles were given.

2

The Characteristic Function

The characteristic function (cf) of the univariate Student's t distribution for odd degrees of freedom was derived by Fisher and Healy (1956). Ifram (1970) gave a general result for all degrees of freedom, but Pestana (1977) pointed out that this result is not quite correct. More recent derivations are presented in Drier and Kotz (2002). Here we discuss two independent results on the characteristic function for the multivariate t distribution. The first one, due to Sutradhar (1986, 1988a), provides a series representation for the cf while the other, due to Joarder and Ali (1996), derives an expression in terms of the MacDonald function. The expressions given are rather complicated; thus, further research and possible simplifications may be desirable.

2.1 Sutradhar's Approach

Let $\mathbf{X}$ be distributed according to

$$f(\mathbf{x}) = \frac{\Gamma((\nu+p)/2)}{(\pi\nu)^{p/2}\Gamma(\nu/2)\,|\mathbf{R}|^{1/2}}\left[1+\frac{1}{\nu}(\mathbf{x}-\boldsymbol{\mu})^T\mathbf{R}^{-1}(\mathbf{x}-\boldsymbol{\mu})\right]^{-(\nu+p)/2}. \tag{2.1}$$

Consider the transformation $\mathbf{Y} = \mathbf{R}^{-1/2}(\mathbf{X}-\boldsymbol{\mu})$. It then follows that the joint pdf of $\mathbf{Y}$ is

$$f_{\mathbf{Y}}(\mathbf{y}) = \nu^{\nu/2}\frac{\Gamma((\nu+p)/2)}{\Gamma(\nu/2)}\left(\nu+\sum_{k=1}^{p} y_k^2\right)^{-(\nu+p)/2}.$$

The cf of $\mathbf{Y}$ is

$$\phi_Y(\mathbf{t};\nu)$$

$$= \frac{\Gamma((\nu+p)/2)}{\nu^{-\nu/2}\Gamma(\nu/2)} \int \exp\left(i\mathbf{t}^T\mathbf{y}\right)\left(\nu + \sum_{k=1}^{p} y_k^2\right)^{-(\nu+p)/2} dy_1 dy_2 \cdots dy_p.$$

To evaluate this integral, Sutradhar (1986) makes the orthogonal transformation $\mathbf{Y} = \mathbf{\Gamma Z}$, where the first column of the $p \times p$ matrix $\mathbf{\Gamma}$ is the vector

$$\left(\frac{t_1}{\parallel \mathbf{t} \parallel}, \frac{t_2}{\parallel \mathbf{t} \parallel}, \ldots, \frac{t_p}{\parallel \mathbf{t} \parallel}\right)$$

with $\parallel \mathbf{t} \parallel = \sqrt{\mathbf{t}^T\mathbf{t}}$. It follows that the cf of $\mathbf{Z}$ is given by

$$\begin{aligned}\phi_{\mathbf{Z}}(\mathbf{t};\nu) &= \nu^{\nu/2}\frac{\Gamma((\nu+p)/2)}{\Gamma(\nu/2)} \int \exp\left(i \parallel \mathbf{t} \parallel z_1\right) dz_1 \\ &\quad \times \int\int\cdots\int \left(c_p + z_p^2\right)^{-(\nu+p)/2} dz_2 \cdots dz_p, \quad (2.2)\end{aligned}$$

where $z_k \in \Re$, $k = 1, \ldots, p$ and $c_p = \nu + \sum_{k=1}^{p-1} z_k^2$. Successive integration of (2.2) with respect to $z_p, z_{p-1}, \ldots, z_2$ yields

$$\phi_{\mathbf{Z}}(\mathbf{t};\nu) = \nu^{\nu/2}\frac{\Gamma((\nu+1)/2)}{\sqrt{\nu}\Gamma(\nu/2)} J_1, \quad (2.3)$$

where

$$J_1 = \int_{-\infty}^{\infty} \exp\left(i \parallel \mathbf{t} \parallel w\right)\left(w^2 + \nu\right)^{-(\nu+1)/2} dz. \quad (2.4)$$

Note that J_1 is an improper integral along the real axis, where w denotes a complex variable. For odd ν, the integrand has poles of integer order while, for fractional and even ν, the poles are of fractional order. Sutradhar (1986) evaluated J_1 separately for the three cases: odd ν; even ν; and fractional ν, using the relations that

$$\phi_{\mathbf{X}}(\mathbf{t};\nu) = \exp\left(i\mathbf{t}^T\boldsymbol{\mu}\right)\phi_{\mathbf{Y}}\left(\mathbf{R}^{1/2}\mathbf{t};\nu\right) \quad (2.5)$$

and

$$\phi_{\mathbf{Y}}(\mathbf{t};\nu) = \phi_{\mathbf{Z}}(\mathbf{t};\nu) \quad (2.6)$$

to obtain the following expressions for $\phi_{\mathbf{X}}$.

For the case of odd ν

$$\begin{aligned}\phi_{\mathbf{X}}(\mathbf{t};\nu) &= \frac{\sqrt{\pi}\Gamma((\nu+1)/2)\exp\left(i\mathbf{t}^T\boldsymbol{\mu} - \sqrt{\nu\mathbf{t}^T\mathbf{R}\mathbf{t}}\right)}{2^{\nu-1}\Gamma(\nu/2)} \\ &\quad \times \sum_{k=1}^{m}\binom{2m-k-1}{m-k}\frac{\left(2\sqrt{\nu\mathbf{t}^T\mathbf{R}\mathbf{t}}\right)^{k-1}}{(k-1)!},\end{aligned}$$

where $m = (\nu + 1)/2$. For the case of even ν

$$\begin{aligned}\phi_{\mathbf{X}}(\mathbf{t};\nu) &= \frac{(-1)^{m+1}\Gamma\left((\nu+1)/2\right)\exp\left(i\mathbf{t}^T\boldsymbol{\mu}\right)}{\sqrt{\pi}\Gamma\left(\nu/2\right)\prod_{k=1}^{m}(m-k+1/2)} \\ &\quad \times \sum_{n=0}^{\infty}\left[\frac{1}{(n!)^2}\left(\frac{\sqrt{\nu\mathbf{t}^T\mathbf{R}\mathbf{t}}}{2}\right)^{2n}\left\{\sum_{j=0}^{m-1}\left(\prod_{k=0,k\neq j}^{m-1}(n-k)\right)\right.\right. \\ &\quad \left.\left. + \prod_{j=0}^{m-1}(n-j)\left(\log\frac{\nu\mathbf{t}^T\mathbf{R}\mathbf{t}}{4} - \frac{\Gamma'(n+1)}{\Gamma(n+1)}\right)\right\}\right], \qquad (2.7)\end{aligned}$$

where $m = \nu/2$. Finally, when ν is of fractional order

$$\begin{aligned}\phi_{\mathbf{X}}(\mathbf{t};\nu) &= \frac{\pi(-1)^m\nu^{(\nu/2)-m}\Gamma\left((\nu+1)/2\right)\exp\left(i\mathbf{t}^T\boldsymbol{\mu}\right)}{2^{\xi}\Gamma\left(\nu/2\right)\sin(\xi\pi)\Gamma\left(\xi+1/2\right)\prod_{k=1}^{m}\left(\frac{\nu+1}{2}-k\right)} \\ &\quad \times \sum_{n=0}^{\infty}\left[\frac{1}{n!}\left(\frac{\sqrt{\nu\mathbf{t}^T\mathbf{R}\mathbf{t}}}{2}\right)^{2n}\left\{\frac{2^{\xi}\prod_{k=0}^{m-1}(n-\xi-k)}{\nu^{\xi}\Gamma(n+1-\xi)}\right.\right. \\ &\quad \left.\left. - \frac{\left(\mathbf{t}^T\mathbf{R}\mathbf{t}\right)^{\xi}\prod_{k=0}^{m-1}(n-k)}{2^{\xi}\Gamma\left(n+1+\xi\right)}\right\}\right], \qquad (2.8)\end{aligned}$$

where $m = [(\nu+1)/2]$ is the integer part of $(\nu+1)/2$ and $\xi = (\nu/2) - m$ is such that $0 <| \xi |< 1/2$.

Both (2.7) and (2.8) involve infinite series. Checking the convergence of these series is an open problem. For another series representation for the cf of the multivariate t, see Javier and Srivastava (1988).

2.2 Joarder and Ali's Approach

An integral representation of the MacDonald function is

$$K_\alpha(r) = \left(\frac{2}{r}\right)^{\alpha}\frac{\Gamma(\alpha+1)}{\sqrt{\pi}}\int_0^\infty\left(1+u^2\right)^{-(\alpha+1/2)}\cos(ru)du, \quad (2.9)$$

where $r > 0$ and $\alpha > -1/2$ (see, for example, Watson, 1958, page 172). A series representation of $K_\alpha(r)$ for $r > 0$ and $\alpha \geq 0$ is given by

$$\begin{aligned}K_\alpha(r) &= 2^{\alpha-1}\sum_{j=0}^{\alpha-1}\frac{(-1)^j(\alpha-j-1)!}{j!4^j}r^{2j-\alpha} \\ &\quad +(-1)^\alpha 2^{-(1+\alpha)}\sum_{j=0}^{\infty}\frac{\psi(1+j)}{j!(j+a)4^j}r^{2j+\alpha}\end{aligned}$$

$$
-(-1)^{\alpha}2^{-(1+\alpha)}\sum_{j=0}^{\infty}\frac{\psi(1+\alpha+j)}{j!(j+a)4^j}r^{2j+\alpha}
$$

$$
-(-1)^{\alpha}2^{-(1+\alpha)}\sum_{j=0}^{\infty}\frac{\log\left(r^2/4\right)}{j!(j+a)4^j}r^{2j+\alpha} \qquad (2.10)
$$

(see, for example, Lebedev, 1965, page 107, 110). A series representation of $K_\alpha(r)$ for $r > 0$ and nonintegral positive values of α is

$$
K_\alpha(r) = 2^{\alpha-1}\sum_{j=0}^{\infty}\frac{r^{2j-\alpha}}{j!(1-\alpha)_j 4^j} + 2^{-(1+\alpha)}\sum_{j=0}^{\infty}\frac{r^{2j+\alpha}}{j!(1+\alpha)_j 4^j} \qquad (2.11)
$$

(see, for example, Spainer and Oldham, 1987, Chapter 51), where $(c)_j = c(c+1)\cdots(c+j-1)$ denotes the ascending factorial. Using (2.9), Joarder and Ali (1996) rewrote the integral (2.4) as

$$
\begin{aligned}
J_1 &= \nu^{-(\nu+1)/2}\int_{-\infty}^{\infty}\{\cos\left(\parallel \mathbf{t} \parallel w\right) + i\sin\left(\parallel \mathbf{t} \parallel w\right)\}\left(1+\frac{w^2}{\nu}\right)^{-\frac{\nu+1}{2}} dz \\
&= 2\nu^{-(\nu+1)/2}\int_{0}^{\infty}\cos\left(\parallel \mathbf{t} \parallel w\right)\left(1+\frac{w^2}{\nu}\right)^{-\frac{\nu+1}{2}} dz \\
&= \frac{\sqrt{\pi}\parallel \mathbf{t} \parallel^{\nu/2}}{2^{\nu/2-1}\Gamma((\nu+1)/2)}K_{\nu/2}\left(\sqrt{\nu}\parallel \mathbf{t} \parallel\right).
\end{aligned}
$$

Thus, using (2.3), one obtains

$$
\phi_{\mathbf{Z}}(\mathbf{t}) = \frac{\parallel \sqrt{\nu}\mathbf{t} \parallel^{\nu/2}}{2^{\nu/2-1}\Gamma(\nu/2)}K_{\nu/2}\left(\parallel \sqrt{\nu}\mathbf{t} \parallel\right).
$$

Hence, using the relationships (2.5) and (2.6), one arrives at the expression for the cf given by Joarder and Ali (1996)

$$
\phi_{\mathbf{X}}(\mathbf{t}) = \exp\left(i\mathbf{t}^T\boldsymbol{\mu}\right)\frac{\parallel \sqrt{\nu\mathbf{R}}\mathbf{t} \parallel^{\nu/2}}{2^{\nu/2-1}\Gamma(\nu/2)}K_{\nu/2}\left(\parallel \sqrt{\nu\mathbf{R}}\mathbf{t} \parallel\right). \qquad (2.12)
$$

Joarder and Ali also provided expansions of this cf using the series representations (2.10) and (2.11). For positive and even ν, applying (2.10), one obtains

$$
\begin{aligned}
\phi_{\mathbf{X}}(\mathbf{t}) = \exp\left(i\mathbf{t}^T\boldsymbol{\mu}\right)\Bigg\{&\sum_{j=0}^{\nu/2-1} C_1(j)\parallel \sqrt{\nu\mathbf{R}}\mathbf{t} \parallel^{2j} \\
&+\sum_{j=0}^{\infty} C_2(j)\parallel \sqrt{\nu\mathbf{R}}\mathbf{t} \parallel^{\nu+2j}
\end{aligned}
$$

$$- \sum_{j=0}^{\infty} C_3(j) \parallel \sqrt{\nu} \mathbf{R} \mathbf{t} \parallel^{\nu+2j} \log \left(\parallel \sqrt{\nu} \mathbf{R} \mathbf{t} \parallel \right) \Bigg\},$$

where

$$C_1(j) = \frac{(-1)^j (\nu/2 - j - 1)!}{(\nu/2 - 1)! j! 4^j},$$

$$C_2(j) = \frac{\psi(1+j) + \psi(1+\nu/2+j) + \log 4}{2^\nu (\nu/2-1)! j! (\nu/2+j!) 4^j},$$

and

$$C_3(j) = \frac{2^{1-\nu} 4^{-j}}{(\nu/2-1)! (\nu/2+j)!}.$$

For positive and odd or fractional ν, applying (2.11), the cf (2.12) becomes

$$\phi_{\mathbf{X}}(\mathbf{t}) = \exp\left(i\mathbf{t}^T \boldsymbol{\mu}\right) \sum_{j=0}^{\infty} \left\{ D_1(j) \parallel \sqrt{\nu} \mathbf{R} \mathbf{t} \parallel^{2j} + D_2(j) \parallel \sqrt{\nu} \mathbf{R} \mathbf{t} \parallel^{\nu+2j} \right\},$$

where

$$D_1(j) = \frac{4^{-j}}{j! (1-\nu/2)_j}$$

and

$$D_2(j) = \frac{2^{-\nu} 4^{-j} \Gamma(-\nu/2)}{\Gamma(\nu/2) (1+\nu/2)_j}.$$

Since the univariate Student's t, multivariate Cauchy, and Pearson's type VII are all particular cases of (2.1), the corresponding cfs in terms of the MacDonald function can be obtained from (2.12). They are as follows

- For the univariate Student's t distribution with the pdf

$$f(x) = \frac{\Gamma((1+\nu)/2)}{\sqrt{\nu\pi}\Gamma(\nu/2)} \left(1 + \frac{x^2}{\nu}\right)^{-(1+\nu)/2}$$

(where $x \in \Re$ and $\nu > 0$), the cf in terms of the MacDonald function is

$$\phi_X(t) = \frac{\nu^{\nu/4} \mid t \mid^{\nu/2}}{2^{\nu/2-1} \Gamma(\nu/2)} K_{\nu/2} \left(\sqrt{\nu} \mid t \mid\right)$$

(compare with Dreier and Kotz, 2002).

- For the p-variate Cauchy distribution with the joint pdf

$$f(\mathbf{x}) = \frac{\Gamma((1+p)/2)}{(\pi)^{(1+p)/2}|\mathbf{R}|^{1/2}}\left[1+(\mathbf{x}-\boldsymbol{\mu})^T\mathbf{R}^{-1}(\mathbf{x}-\boldsymbol{\mu})\right]^{-(1+p)/2}$$

(where $x \in \Re^p$), the cf is

$$\begin{aligned}\phi_{\mathbf{X}}(\mathbf{t}) &= \exp\left(i\mathbf{t}^T\boldsymbol{\mu}\right)\sqrt{2/\pi} \parallel \sqrt{\mathbf{R}}\mathbf{t} \parallel^{1/2} K_{1/2}\left(\parallel \sqrt{\mathbf{R}}\mathbf{t} \parallel\right) \\ &= \exp\left(i\mathbf{t}^T\boldsymbol{\mu} - \parallel \sqrt{\mathbf{R}}\mathbf{t} \parallel\right),\end{aligned}$$

which follows by using the result that $K_{1/2}(r) = \sqrt{\pi/(2r)}\exp(-r)$ (see, for example, Tranter, 1968, page 19).

- For the p-variate Pearson type VII distribution with the joint pdf

$$f(\mathbf{x}) = \frac{\Gamma(m)}{\sqrt{\nu}\pi^{p/2}\Gamma(m-p/2)|\mathbf{R}|^{1/2}}\left[1+\frac{1}{\nu}(\mathbf{x}-\boldsymbol{\mu})^T\mathbf{R}^{-1}(\mathbf{x}-\boldsymbol{\mu})\right]^{-m}$$

(where $x \in \Re^p$, $m \geq p/2$ and $\nu > 0$), the cf becomes

$$\phi_{\mathbf{X}}(\mathbf{t}) = \exp\left(i\mathbf{t}^T\boldsymbol{\mu}\right)\frac{\parallel \sqrt{\nu\mathbf{R}}\mathbf{t} \parallel^{m-p/2}}{2^{m-p/2-1}\Gamma(m-p/2)}K_{m-p/2}\left(\parallel \sqrt{\nu\mathbf{R}}\mathbf{t} \parallel\right).$$

2.3 Lévy Representation

Infinite divisibility of the univariate Student's t distribution was first proved by Grosswald (1976) – see also Kelker (1971) for a partial result. Later, Halgreen (1979) established the Lévy representation of its cf. For a multivariate t, Takano (1994) provided the first proof of infinite divisibility and the corresponding Lévy representation. Consider the standard case $\boldsymbol{\mu} = \mathbf{0}$ and $\mathbf{R} = \mathbf{I}_p$. In this case, after suitable transformation, the joint pdf (2.1) can be written in the form

$$f(\mathbf{x}) = \frac{\Gamma(m+p/2)}{\pi^{p/2}\Gamma(m)}\left(1+\parallel \mathbf{x} \parallel^2\right)^{-(m+p/2)}.$$

The corresponding cf is

$$\phi(\mathbf{t}) = \frac{2^{1-m}}{\Gamma(m)} \parallel \mathbf{t} \parallel^m K_m\left(\parallel \mathbf{t} \parallel\right). \tag{2.13}$$

Takano (1994) derived the Lévy representation of (2.13) in the form

$$\phi(\mathbf{t}) = \exp\left[\int_{\Re^p}\frac{2}{\parallel \mathbf{x} \parallel^p}\left\{\exp\left(i\mathbf{t}^T\mathbf{x}\right)-1-\frac{i\mathbf{t}^T\mathbf{x}}{1+\parallel \mathbf{x} \parallel^2}\right\}\right.$$

$$\times \left\{ \int_0^\infty g_m(2w) L_{p/2}\left(\sqrt{2w} \parallel \mathbf{x} \parallel\right) dw \right\} dx_1 \cdots dx_p \Bigg],$$

where

$$g_\alpha(x) = \frac{2}{\pi^2 x} \Big/ \left\{ J_\alpha^2\left(\sqrt{x}\right) + Y_\alpha^2\left(\sqrt{x}\right) \right\},$$

$$L_\alpha(x) = (2\pi)^{-\alpha} x^\alpha K_\alpha(x),$$

$$J_\alpha(x) = \frac{x^\alpha}{\sqrt{\pi} 2^\alpha \Gamma(\alpha + 1/2)} \int_{-1}^{1} \left(1 - z^2\right)^{\alpha - 1/2} \exp(ixz)\, dz,$$

and

$$Y_\alpha(x) = \frac{1}{\sin(\alpha\pi)} \left\{ \cos(\alpha\pi) J_\alpha(x) - J_{-\alpha}(x) \right\}.$$

Note that $J_\alpha(\cdot)$ and $Y_\alpha(\cdot)$ are the Bessel functions of the first and second kinds, respectively, of order m.

Now consider (2.13) itself as a joint pdf

$$f_m(\mathbf{x}) = C \parallel \mathbf{x} \parallel^m K_m(\parallel \mathbf{x} \parallel), \qquad (2.14)$$

where the normalizing constant C is

$$C = \frac{2^{1-m-p/2}}{(2\pi)^{p/2} \Gamma(m + p/2)}.$$

Using properties of the MacDonald function $K_\alpha(\cdot)$, (2.14) can be reduced to the simpler forms

$$f_{1/2}(\mathbf{x}) = C\sqrt{\frac{\pi}{2}} \exp(- \parallel \mathbf{x} \parallel),$$

$$f_{n+1/2}(\mathbf{x}) = C\sqrt{\frac{\pi}{2}} \exp(- \parallel \mathbf{x} \parallel) \sum_{k=0}^{n} \frac{(n+k)!}{2^k k!(n-k)!} \parallel \mathbf{x} \parallel^{n-k},$$

and

$$\begin{aligned} f_n(\mathbf{x}) = {} & C 2^{n-1} \sum_{k=0}^{n-1} \frac{(-1)^k (n-k-1)!}{4^k k!} \parallel \mathbf{x} \parallel^{2k} \\ & + C(-1)^{n+1} \sum_{k=0}^{\infty} \frac{1}{k!(n+k)!} \left(\frac{\parallel \mathbf{x} \parallel}{2} \right)^{2(n+k)} \\ & \times \left\{ \log\left(\frac{\parallel \mathbf{x} \parallel}{2} \right) - \frac{1}{2}\psi(1+k) - \frac{1}{2}\psi(1+n+k) \right\}. \end{aligned}$$

Takano (1994) established further that the joint pdf (2.14) is also infinitely divisible and that its cf

$$\phi(\mathbf{t}) = (1+ \| \mathbf{t} \|)^{-(m+p/2)}$$

admits the Lévy representation

$$\phi(\mathbf{t}) = \exp\left[(2m+p)\int_{\Re^p}\left\{\exp\left(i\mathbf{t}^T\mathbf{x}\right)-1\right\}\frac{L_{p/2}(\| \mathbf{x} \|)}{\| \mathbf{x} \|^p}dx_1\cdots dx_p\right],$$

where $L_{p/2}(\cdot)$ is as defined above.

3

Linear Combinations, Products, and Ratios

3.1 Linear Combinations

The distribution of linear combinations of independent t random variables has been studied by numerous authors, among them Fisher (1935), Chapman (1950), Fisher and Healy (1956), Ruben (1960), Patil (1965), Ghosh (1975), and Walker and Saw (1978). Johnson and Weerahandi (1988) tackled the distribution of linear combinations for multivariate t random vectors. Their results are included here for completeness and to motivate further multivariate extensions. We hope that our readers will benefit from studying this material, which contains fruitful ideas and also refers to the original papers for further details.

Chapman (1950) considered the difference $D = X_1 - X_2$ of two independent t random variables X_j with common degree of freedom ν. If ν is odd, then it is known that the characteristic function of $Y_j = X_j/\sqrt{\nu}$ is

$$\begin{aligned} \phi_\nu(t) &= E\left[\exp\left(itY_j\right)\right] \\ &= \frac{\sqrt{\pi}\exp\left(-\mid t \mid\right)}{2^{\nu-1}\Gamma(\nu/2)} \sum_{k=0}^{(\nu-1)/2} \frac{((\nu-1)/2+k)!}{\nu!((\nu-1)/2-k)!} \mid 2t \mid^{\frac{\nu-1}{2}-k}. \end{aligned} \tag{3.1}$$

Using this representation, Chapman provided the following general expression for the pdf of $W = D/\sqrt{\nu}$

$$f(w) = \frac{1}{2\pi}\int_{-\infty}^{\infty} \exp\left\{-i(\nu+1)t\right\} \phi_\nu^2(t)dt,$$

which may be integrated to obtain the pdf of D in a closed form for values such as $\nu = 1, 3, 5$, and so on. Chapman tabulated the distribution of D for $\nu = 1, 3, 5, 7, 9, 11$.

Fisher and Healy (1956) considered the mixture $D = a_1X_1 + a_2X_2$ of two independent t random variables X_j with degrees of freedom ν_j when $a_j > 0$, $j = 1, 2$. It is obvious that the characteristic function of D is the product

$$\phi_{\nu_1}(a_1 \mid t \mid) \phi_{\nu_2}(a_2 \mid t \mid),$$

where $\phi_\nu(\cdot)$ is as defined in (3.1). Since the product is simply a polynomial in t of degree $(\nu_1 + \nu_2)/2 - 1$, it can be expanded in a finite series of terms of $\phi_m(\cdot)$ in which the highest value of $m = \nu_1 + \nu_2 - 1$. For example, in the special case $\nu_1 = 3$ and $\nu_2 = 5$, one can write

$$\begin{aligned}
& \exp\left\{(\sin\theta + \cos\theta)\, t\right\} \phi_3(\sin\theta t)\, \phi_5(\cos\theta t) \\
= & \frac{25\sqrt{3}\tan\theta}{(\sqrt{5} + \sqrt{3}\tan\theta)^3} \phi_7 \left[\left(\sqrt{\frac{3}{7}}\sin\theta + \sqrt{\frac{5}{7}}\cos\theta\right) t\right] \\
& + \frac{5\sqrt{5} - 10\sqrt{3}\tan\theta + 9\sqrt{5}\tan^2\theta}{(\sqrt{5} + \sqrt{3}\tan\theta)^3} \phi_5\left[\left(\sqrt{3}\sin\theta + \sqrt{5}\cos\theta\right) t\right] \\
& + \left(\frac{\sqrt{3}\tan\theta}{\sqrt{5} + \sqrt{3}\tan\theta}\right)^3 \phi_3\left[\left(\sqrt{3}\sin\theta + \sqrt{5}\cos\theta\right) t\right];
\end{aligned}$$

from this one can easily deduce the pdf of D.

Ruben (1960) provided results on the distribution of $D = X_1 \sin\theta - X_2 \cos\theta$, when X_j are independent t random variables with degrees of freedom ν_j and θ is a fixed angle between 0 and 90 degrees. This statistic was originally proposed by Fisher (1935) as the basis for testing or estimating the difference in means of two unconnected and totally unknown normal populations, the "fiducial distribution" of the difference between the latter quantity and the corresponding sample mean difference, when suitably standardized, being supposed to be that of the statistic D. Ruben obtained the pdf of D in the integral form

$$\begin{aligned}
f(d) = & \int_0^1 \frac{1}{\sqrt{\nu_1 + \nu_2} B((\nu_1 + \nu_2)/2, 1/2)} \left\{1 + \frac{d^2\phi^2(s)}{\nu_1 + \nu_2}\right\}^{-(\nu_1+\nu_2+1)/2} \\
& \times \psi(s) \frac{s^{(\nu_1/2)-1}(1 - s)^{(\nu_2/2)-1}}{B(\nu_1/2, \nu_2/2)} ds, \qquad (3.2)
\end{aligned}$$

where

$$\psi(s) = \sqrt{\frac{(\nu_1 + \nu_2)s(1 - s)}{\nu_1(1 - s)\sin^2\theta + \nu_2 s \cos^2\theta}}.$$

It follows directly from (3.2) that D may be expressed in the form

$$D = \frac{X}{\psi(S)},$$

where X is a Student's t random variable with degrees of freedom $\nu_1 + \nu_2$ and S is a Beta variable with parameters $\nu_1/2$ and $\nu_2/2$, that is, a variable with pdf given by the second term under the integral sign in (3.2), with the first term under the integral sign representing the conditional pdf of $X/\psi(S)$ for fixed S.

In the special case $\nu_1 = \nu_2 = \nu$ and $\theta = 45$ degrees, (3.2) reduces to

$$f(d) = \sqrt{\frac{2}{\nu}} \frac{\Gamma(\nu + 1/2)\Gamma((\nu+1)/2)}{\Gamma^2(\nu/2)\Gamma(\nu/2+1)} \, {}_2F_1\left(\nu + \frac{1}{2}, \frac{\nu+1}{2}; \frac{\nu}{2} + 1; -\frac{d^2}{2\nu}\right),$$

where ${}_2F_1$ denotes the Gauss hypergeometric function. By using the appropriate four of the group of 24 transformations of the hypergeometric function ${}_2F_1$ (see, for example, Whittaker and Watson, 1952, page 284), the above pdf may be expressed in the following three additional ways

$$f(d) = \sqrt{\frac{2}{\nu}} \frac{\Gamma(\nu + 1/2)\Gamma((\nu+1)/2)}{\Gamma^2(\nu/2)\Gamma(\nu/2+1)} \left(1 + \frac{d^2}{2\nu}\right)^{-\nu} \times {}_2F_1\left(\frac{1-\nu}{2}, \frac{1}{2}; \frac{\nu+2}{2}; -\frac{d^2}{2\nu}\right), \quad (3.3)$$

$$f(d) = \sqrt{\frac{2}{\nu}} \frac{\Gamma(\nu + 1/2)\Gamma((\nu+1)/2)}{\Gamma^2(\nu/2)\Gamma(\nu/2+1)} \left(1 + \frac{d^2}{2\nu}\right)^{-(\nu+1/2)} \times {}_2F_1\left(\nu + \frac{1}{2}, \frac{1}{2}; \frac{\nu+2}{2}; \frac{d^2}{d^2+2\nu}\right), \quad (3.4)$$

and

$$f(d) = \sqrt{\frac{2}{\nu}} \frac{\Gamma(\nu + 1/2)\Gamma((\nu+1)/2)}{\Gamma^2(\nu/2)\Gamma(\nu/2+1)} \left(1 + \frac{d^2}{2\nu}\right)^{-(\nu+1)/2} \times {}_2F_1\left(\frac{1+\nu}{2}, \frac{1-\nu}{2}; \frac{\nu+2}{2}; \frac{d^2}{d^2+2\nu}\right). \quad (3.5)$$

Note that (3.3) and (3.5) may each be expanded as a terminating series (refer to the definition of the Gauss hypergeometric function) when ν is odd, and also that (3.4) is expressed as the product of the pdf of a t random variable with degrees of freedom 2ν and the Gauss hypergeometric function.

In the special case $\nu_1 = \infty$ and $\nu_2 = \nu$, (3.2) reduces to

$$f(d) = \frac{\Gamma(\nu)\Gamma((\nu+1)/2)}{\sqrt{2\pi}\sin\theta\Gamma(\nu/2)\Gamma(\nu+1/2)} {}_1F_1\left(\frac{\nu+1}{2}; \nu+\frac{1}{2}; -\frac{d^2}{2\sin^2\theta}\right),$$

where ${}_1F_1$ is the confluent hypergeometric function. Using Kummer's first transformation for the confluent hypergeometric function, one can obtain the alternative form

$$\begin{aligned} f(d) &= \frac{\Gamma(\nu)\Gamma((\nu+1)/2)}{\sqrt{2\pi}\sin\theta\Gamma(\nu/2)\Gamma(\nu+1/2)} \exp\left(-\frac{d^2}{2\sin^2\theta}\right) \\ &\quad \times {}_1F_1\left(\frac{\nu}{2}; \nu+\frac{1}{2}; \frac{d^2}{2\sin^2\theta}\right). \end{aligned}$$

Ruben (1960) also provided expressions for the cdf of D, but these were infinite series involving incomplete gamma function ratios. For tables of percentage points of D, see Sukhatme (1938), Fisher and Yates (1943), and Weir (1966).

Ghosh (1975) provided explicit expressions for the cdf in terms of simple hypergeometric functions when $D = X_1 - X_2$ and X_j are independent Student's t random variables with common degree of freedom $\nu \leq 4$. In particular, Ghosh obtained the following expressions for $\Pr(0 < D < d)$

$$\frac{1}{\pi}\arctan\left(\frac{d}{2}\right),$$

$$\frac{1}{4\sqrt{q}}\left\{(1+q)E(q) - (1-q)K(q)\right\},$$

$$\frac{2\sqrt{3}d\left(18+d^2\right)}{\pi\left(12+d^2\right)^2} + \frac{1}{\pi}\arctan\left(\frac{d}{2\sqrt{3}}\right),$$

and

$$\begin{aligned} \frac{1}{64p\sqrt{p}}\Big\{ &\left(8p^4 - 31p^3 + 48p^2 + 5p + 2\right)E(p) \\ &- \left(4p^4 - 16p^3 + 6p^2 + 4p + 2\right)K(p)\Big\}, \end{aligned}$$

for $\nu = 1$, $\nu = 2$, $\nu = 3$, and $\nu = 4$, respectively, where $p = d^2/(16+d^2)$, $q = d^2/(8+d^2)$,

$$E(x) = \int_0^{\pi/2} \sqrt{1 - x\sin^2 s}\, ds$$

is the complete elliptical integral of the second kind, and

$$K(x) = \int_0^{\pi/2} \frac{ds}{\sqrt{1 - x\sin^2 s}}$$

is the complete elliptical integral of the first kind. Similar expressions for $\Pr(0 < D < d)$ as a finite sum of terms can be obtained for any positive integer ν. Ghosh also provided a tabulation of the numerical values of $\Pr(D < d)$ for $\nu = 1(1)20$ and $d = 0.0(0.5)10.0$.

Walker and Saw (1978) expressed the cdf of the linear combinations of any number of Student's t random variables with odd degrees of freedom as a mixture of t distributions. Define the linear combination as

$$D = \sum_{k=1}^{n} \frac{a_k}{\sqrt{\nu_k}} X_k,$$

where $a_k \geq 0$, $a_1 + \cdots + a_n = 1$ and X_k are independent Student's t random variables with degrees of freedom $\nu_k = 2m_k + 1$, $m_k = 0, 1, 2, \ldots$. Construct a matrix $\mathbf{Q}$ whose element in row i and column j is the coefficient of $\exp(- \mid t \mid) \mid t \mid^j$ in $\phi_{2i+1}(t)$ (see equation (3.1)), that is,

$$Q_{ij} = \begin{cases} \frac{2^j i!(2i-j)!}{(2i)!j!(i-j)!} & \text{if } j = 0, 1, 2, \ldots, i,\ i = 0, 1, 2, \ldots, \\ 0 & \text{if } j > i. \end{cases}$$

The characteristic function of D when $\nu_k = 2m_k + 1$ can be written as

$$\phi(t) = E\left[\exp\left(itD\right)\right] = \prod_{k=1}^{n} \phi_{\nu_k}\left(a_k t\right),$$

and since $\exp(\mid t \mid)\phi(t)$ is a polynomial in $\mid t \mid$, one may obtain a vector $\boldsymbol{\lambda}$ such that

$$\phi(t) = \exp(- \mid t \mid) \sum_{k=0}^{\infty} \lambda_k \mid t \mid^k .$$

Walker and Saw (1978) showed that the cdf of D can expressed as the weighted sum

$$\Pr(D \leq d) = \sum_{k=0}^{S} \eta_k H_k(d),$$

where

$$\eta^T = \boldsymbol{\lambda}^T \mathbf{Q}^{-1},$$

$$S = \sum_{k=1}^{n} m_k,$$

and

$$H_k(d) = \Pr\left[X_{2k+1} \leq \sqrt{2k+1}d\right].$$

This result can be used to calculate probabilities of D utiliizing only tables of the Student's t distribution.

The general distribution of $D = X_1 - X_2$ is very complicated when the X_j are independent Student's t random variables with $\nu_1 \neq \nu_2$. It is therefore natural to ask whether a reasonably good approximation can be found. Chapman (1950) suggested the simple approximation

$$\Pr(D < d) \approx \Phi\left[d \Big/ \sqrt{\frac{\nu_1}{\nu_1 - 2} + \frac{\nu_2}{\nu_2 - 2}}\right],$$

where $\Phi(\cdot)$ is the cdf of the standard normal distribution. This idea is, of course, prompted by the fact that $X_1\sqrt{\nu_1 - 2}/\sqrt{\nu_1}$ and $X_2\sqrt{\nu_2 - 2}/\sqrt{\nu_2}$ are both asymptotically standard normal random variables, approaching normality more rapidly than X_1 and X_2 do. However, a few calculations show that this approximation is quite unsatisfactory even for moderately large values of $\nu_1 \geq \nu_2 > 2$. Based on a t-approximation proposed by Patil (1965), an improved approximation is

$$\Pr(D < d) \approx T_\nu\left(\frac{hd}{\sqrt{2}}\right),$$

where T_ν is the cdf of the Student's t distribution of ν degrees of freedom,

$$\nu = 4 + \frac{\left(\frac{\cos^2 \nu_2}{\nu_2 - 2} + \frac{\sin^2 \nu_1}{\nu_1 - 2}\right)^2}{\frac{\cos^4 \nu_2^2}{(\nu_2-2)^2(\nu_2-4)} + \frac{\sin^4 \nu_1^2}{(\nu_1-2)^2(\nu_1-4)}}$$

(where $\nu_1 \geq \nu_2 > 4$), and

$$h^2 = \frac{\nu}{\nu - 2} \Big/ \left(\frac{\cos^2 \nu_2}{\nu_2 - 2} + \frac{\sin^2 \nu_1}{\nu_1 - 2}\right),$$

where $\nu_1 \geq \nu_2 > 4$. Ghosh (1975) derived another approximation that requires only tables of the normal distribution

$$\Pr(D < d) = \Phi\left(\frac{d}{\sqrt{2}}\right) - \frac{d\phi(d/\sqrt{2})}{32\sqrt{2}}\left\{\frac{Q_1(d)}{\nu_2} + \frac{Q_2(d)}{\nu_2^2} + \frac{Q_3(d)}{\nu_2^3}\right.$$

$$+O\left(\frac{1}{\nu_2^4}\right)\Bigg\},$$

where

$$Q_1(d) = (1+\lambda)\left(d^2+10\right),$$

$$Q_2(d) = \frac{1+\lambda^2}{384}\left(3d^6+98d^4+620d^2+168\right) + \frac{\lambda}{64}\left(d^6-10d^4+36d^2-456\right),$$

$$Q_3(d) = \frac{1+\lambda^2}{24576}\left(d^{10}+66d^8+1016d^6-1296d^4-65328d^2-141408\right) + \frac{\lambda(1+\lambda)}{24576}\left(3d^{10}-58d^8-280d^6+6864d^4-70032d^2\right) + \frac{1277\lambda(1+\lambda)}{256},$$

and $\lambda = \nu_2/\nu_1$. Ghosh showed evidence to suggest that this is far more accurate than Patil's approximation.

Johnson and Weerahandi (1988) considered linear combinations of t random vectors in a Bayesian context. Suppose $\mathbf{y}_{11}, \ldots, \mathbf{y}_{1m_1}$ and $\mathbf{y}_{21}, \ldots, \mathbf{y}_{2m_2}$ are independent samples from two p-variate normal populations $N(\boldsymbol{\mu}_1, \boldsymbol{\Sigma}_1)$ and $N(\boldsymbol{\mu}_2, \boldsymbol{\Sigma}_2)$, respectively, where the population covariance matrices are unknown and not necessarily equal. Let $\bar{\mathbf{y}}_i$ and $\mathbf{S}_i$ denote the corresponding sample mean vectors and sample covariance matrices. Johnson and Weerahandi considered the distribution of the quadratic form

$$Q = (\boldsymbol{\delta}-\mathbf{d})^T \mathbf{V}^{-1} (\boldsymbol{\delta}-\mathbf{d}),$$

where $\boldsymbol{\delta} = \boldsymbol{\mu}_1 - \boldsymbol{\mu}_2$, $\mathbf{d} = \bar{\mathbf{y}}_1 - \bar{\mathbf{y}}_2$, and $\mathbf{V}$ is any $p \times p$ positive definite matrix. Note that $\boldsymbol{\mu}_i - \bar{\mathbf{y}}_i$ have the central p-variate t distribution with covariance matrix $\mathbf{S}_i/(m_i - p)$ and degrees of freedom $m_i - p$. Under the diffuse prior distribution

$$p(\boldsymbol{\mu}_1, \boldsymbol{\Sigma}_1, \boldsymbol{\mu}_2, \boldsymbol{\Sigma}_2) = \sqrt{|\boldsymbol{\Sigma}_1|\,|\boldsymbol{\Sigma}_1|}, \tag{3.6}$$

the posterior cdf of Q can be expressed as

$$F(q) = \sum_{j=0}^{\infty} E(w_j)\, F_{p+2j,n}\left[\frac{nq}{\theta(p+2j)}\right], \tag{3.7}$$

where $n = m_1 + m_2 - 2p$ and the expectation is taken with respect to the beta random variable B, which is distributed as $Be((m_1 - p)/2, (m_2 - p)/2)$. The w_j are defined in terms of θ (an arbitrary constant) and λ_j by the recursive relation

$$w_r = \frac{1}{2r}\sum_{j=0}^{r-1} H_{r-j} w_j,$$

where

$$w_0 = \prod_{j=1}^{p}\sqrt{\frac{\theta}{\lambda_j}},$$

$$H_r = \sum_{j=1}^{p}\left(1-\frac{\theta}{\lambda_j}\right)^r,$$

and λ_j are the ordered eigenvalues $\lambda_1 \leq \cdots \leq \lambda_p$ of the matrix

$$\frac{1}{m_1 B}\mathbf{V}^{-1/2}\mathbf{S}_1\mathbf{V}^{-1/2} + \frac{1}{m_2(1-B)}\mathbf{V}^{-1/2}\mathbf{S}_2\mathbf{V}^{-1/2}.$$

In the particular case $\mathbf{V} = c\mathbf{S}_1 + (1-c)\mathbf{S}_2$, the λ_j can be conveniently obtained by using the relation

$$\lambda_j = \frac{m_2(1-B) + m_1 B\xi_j}{m_1 m_2 B(1-B)\{c + (1-c)\xi_j\}},$$

where $\xi_1, \ldots, \xi_p$ are the eigenvalues of $\mathbf{S}_1^{-1}\mathbf{S}_2$. In the univariate case, (3.7) reduces to give the posterior cdf of $Y = (\mu_1 - \mu_2) - (\bar{x}_1 - \bar{x}_2)$ as

$$F(y) = E\left[T_{m_1+m_2-2}\left(\frac{y\sqrt{m_1 - m_2 - 2}}{\sqrt{\frac{m_1-1}{m_1}\frac{s_1^2}{B} + \frac{m_2-1}{m_2}\frac{s_2^2}{1-B}}}\right)\right],$$

where s_1^2 and s_2^2 are the sample variances and the expectation is taken with respect to B, which is distributed as $Be((m_1 - 1)/2, (m_1 - 1)/2)$. The result (3.7) can also be used to deduce the pdf of $U = (\mathbf{T}_1 \pm \mathbf{T}_2)^T(\mathbf{T}_1 \pm \mathbf{T}_2)$, where the $\mathbf{T}_i$ are independent p-variate random vectors having the t distribution with covariance matrix $(a_i/m_i)\mathbf{I}_p$ and degrees of freedom m_i. It turns out that the cdf of U is

$$F(u) = E\left[F_{p,m_1+m_2}\left(\frac{u\,(m_1+m_2)}{p}\frac{B(1-B)}{a_1 + B\,(a_2 - a_1)}\right)\right],$$

where the expectation is taken with respect to B, which is distributed as $Be(m_1/2, m_2/2)$. Johnson and Weerahandi also established several interesting bounds on (3.7), one upper bound being

$$F(q) \quad < \quad E\left[F_{p,n}\left(\frac{nq}{p}\left|\frac{1}{m_1 B}\bar{\mathbf{S}}_1 + \frac{1}{m_2(1-B)}\bar{\mathbf{S}}_2\right|^{-1/p}\right)\right],$$

where $\bar{\mathbf{S}}_i = \mathbf{V}^{-1/2}\mathbf{S}_i\mathbf{V}^{-1/2}$. Furthermore, it was shown that similar results hold if the diffuse prior in (3.6) is replaced by the natural conjugate prior distribution.

3.2 Products

The distribution of products of Student's t random variables has been studied by Harter (1951) and Wallgren (1980).

Products of Student's t random variables arise naturally in classification problems. In many educational and industrial problems it may be necessary to classify persons or objects into one of two categories - those fit and those unfit for a particular purpose. In formulating a classification problem, assume that for p tests one knows the scores of N_1 individuals known to belong to population Π_1 and of N_2 individuals known to belong to population Π_2, along with those of the individual under consideration, a member of the population Π, where it is known a priori that Π is identical with either Π_1 or Π_2. Assume further that the distribution of the test scores of the individuals making up Π_1 and Π_2 are two p-dimensional normal distributions, which possess the same covariance matrix but are independent of each other. In order to classify the individual in question into either Π_1 or Π_2, Wald (1944) introduced the statistic V given by

$$V \quad = \quad \sum_{i=1}^{p}\sum_{j=1}^{p} s^{ij} Y_{i,n+1} Y_{j,n+2}, \tag{3.8}$$

where $n = N_1 + N_2 - 2$, s^{ij} is the (i,j)th element of the inverse of the matrix $\mathbf{S}$ defined by

$$s_{ij} \quad = \quad \frac{1}{n}\sum_{k=1}^{n} Y_{i,k} Y_{j,k},$$

and $Y_{i,k}$ $(i = 1, \ldots, p;\ k = 1, \ldots, n+2)$ are iid normally distributed random variables with unit variances and expected values $E(Y_{i,k}) = 0$, $k = 1, \ldots, n$, $E(Y_{i,n+1}) = \mu_1$, and $E(Y_{i,n+2}) = \mu_2$.

In the particular case $p = 1$, (3.8) can be written as

$$V = \frac{Y_{1,n+1} Y_{1,n+2}}{Z}, \tag{3.9}$$

where

$$Z = \frac{1}{n} \sum_{k=1}^{n} Y_{1,k}^2.$$

Thus, one sees that the V in (3.9) is a product of two independent Student's t variates. Harter (1951) derived the exact distribution of this V. In particular, he showed that the pdf of $\mid V \mid$ is given by

$$\begin{aligned} f(\mid v \mid) &= \frac{1}{\pi \Gamma(n/2)} \sum_{j=0}^{\infty} (-1)^j n^{n/2+j} \mid v \mid^{-(1+j+n/2)} \\ & \quad \times \Gamma^2 \left(\frac{2+2j+n}{4} \right) \Big/ \Gamma(1+j) \end{aligned}$$

if $\mid v \mid > n/2$, and by

$$\begin{aligned} f(\mid v \mid) &= \frac{\exp\left(-\mu_2^2/2\right)}{\pi \Gamma(n/2)} \sum_{j=0}^{\infty} \frac{(-1)^j n^{n/2+j}}{\Gamma(1+j)} \mid v \mid^{-(1+j+n/2)} \\ & \quad \times \Gamma^2 \left(\frac{2+2j+n}{4} \right) \sum_{k=0}^{\infty} \frac{\left(2\mu_2^2\right)^k}{(2k)!} \Gamma\left(k + \frac{2+n+2j}{4}\right) \end{aligned}$$

if $\mid v \mid \leq n/2$.

Wallgren (1980) studied the product of two correlated Student's t variates. This was motivated by hypotheses testing problems that assume that the relationship between two regression lines $y = \alpha_1 + \beta_1 x$ and $y = \alpha_2 + \beta_2 x$ hold for all real x. Although it is often true that such a relationship holds for all real x, there are instances in which the relationship may hold true only for x's in a given interval, say $[c_1, c_2]$. Wallgren (1980) showed that the statistic for testing this hypothesis is a product of two correlated t-variates; specifically, it takes the form $W = XY/S^2$, where (X, Y) has a bivariate normal distribution with means (μ_1, μ_2), common variance σ^2, and the correlation matrix

$$\mathbf{R} = \begin{pmatrix} 1 & \rho \\ \rho & 1 \end{pmatrix}. \tag{3.10}$$

Moreover, nS^2/σ^2 is independent of (X, Y) and has a chi-squared distribution with degrees of freedom ν.

The limiting distribution of $W = XY/S^2$ as $\nu \to \infty$ is $Z = XY/\sigma^2$. The distribution of Z has been studied by Aroian et al. (1978). The study of the distribution of W is, therefore, a generalization of the study by Aroian et al. since σ^2 is unknown. .

In the central case $\mu_1 = 0$ and $\mu_2 = 0$, Wallgren (1980) showed that the cdf $F(w) = \Pr(W \leq w)$ of W is given by

$$F(w) = 1 - \int_0^{\alpha} Q_\nu(\theta; \rho, w)\, d\theta$$

for $-1 < \rho \leq 0$, $w \geq 0$;

$$F(w) = \int_{\alpha-\pi}^{0} Q_\nu(\theta; \rho, w)\, d\theta$$

for $-1 < \rho \leq 0$, $w \leq 0$;

$$F(w) = 1 - \int_0^{\pi+\alpha} Q_\nu(\theta; \rho, w)\, d\theta$$

for $1 > \rho \geq 0$, $w \geq 0$; and

$$F(w) = \int_{\alpha}^{0} Q_\nu(\theta; \rho, w)\, d\theta$$

for $1 > \rho \geq 0$, $w \leq 0$, where

$$\pi Q_\nu(\theta; \rho, w) = \left(\frac{\nu \sin\theta \sin(\theta + A)}{w + \nu \sin\theta \sin(\theta + A)}\right)^{\nu/2},$$

$$\alpha = \arctan\left(-\frac{\sqrt{1-\rho^2}}{\rho}\right),$$

and angle A is defined by $\sin A = \sqrt{1-\rho^2}$, $\cos A = \rho$ for $0 < A < \pi$. The corresponding pdf $f(w)$ can be obtained by differentiation. The pdf has a singularity at $w = 0$ and, considered as a function of ρ and w, $f(w;\rho) = f(-w;-\rho)$. The limit of $f(w)$ as $\rho \to 1^-$ is the $F_{1,\nu}$ density. Moreover, if $\rho_1 < \rho_2$, then $F(w;\rho_1) > F(w;\rho_2)$ for any w.

In the noncentral case, the cdf $F(w)$ is given by

$$F(w) = \int_0^{\infty} \int_{-\lambda_2}^{\infty} \frac{h(s) \exp(-v^2/2)}{\sqrt{2\pi}} \times \Phi\left(\frac{ws^2/(v+\lambda_2) - \lambda_1 - \rho v}{\sqrt{1-\rho^2}}\right) dv ds$$

or, equivalently,

$$F(w) = \int_0^{\infty} \int_{-\infty}^{-\lambda_2} \frac{h(s) \exp\left(-v^2/2\right)}{\sqrt{2\pi}} \times \Phi\left(\frac{-ws^2/(v+\lambda_2)+\lambda_1+\rho v}{\sqrt{1-\rho^2}}\right) dvds,$$

where $\lambda_i = \mu_i/\sigma$, $i = 1, 2$ are the noncentrality parameters and

$$h(s) = \frac{\nu^{\nu/2} s^{\nu-1} \exp\left(-\nu s^2/2\right)}{2^{(\nu-2)/2}\Gamma(\nu/2)}.$$

The two double integrals above are bounded above by $\Phi(\lambda_2)$ and $\Phi(-\lambda_2)$, respectively. As a function of λ_1 and λ_2, F has the following properties:

$$F(w; \lambda_1, \lambda_2) = F(w; \lambda_2, \lambda_1),$$

$$F(w; -\lambda_1, -\lambda_2) = F(w; \lambda_1, \lambda_2),$$

$$F(w; \lambda_1, -\lambda_2) = F(w; -\lambda_1, \lambda_2),$$

$$\lim_{\lambda_2 \to \infty} F(w; c\lambda_2, \lambda_2) = 0 \text{ if } c > 0,$$

$$\lim_{\lambda_2 \to \infty} F(w; c\lambda_2, \lambda_2) = 1/2 \text{ if } c = 0,$$

$$\lim_{\lambda_2 \to \infty} F(w; c\lambda_2, \lambda_2) = 1 \text{ if } c < 0,$$

$$\lim_{\lambda_1 \to \infty} F(w; \lambda_1, B) = \Phi(-B),$$

and

$$\lim_{\lambda_1 \to -\infty} F(w; \lambda_1, B) = \Phi(B).$$

Also, for $\lambda_1 \geq 0$, $\lambda_2 \geq 0$, $w > 0$, and $-1 < \rho < 0$, $F(w; \lambda_1, \lambda_2)$ is a strictly decreasing function of λ_1 and λ_2. Thus, the maximum of $F(w)$ over the region $\lambda_1 \geq 0$, $\lambda_2 \geq 0$ occurs at $\lambda_1 = \lambda_2 = 0$.

Since (X, Y) and S^2 are independent, the first two moments of W are given by

$$E(W) = E(XY) E\left(1/S^2\right)$$

and

$$E\left(W^2\right) = E\left(X^2Y^2\right) E\left(1/S^4\right).$$

It is known (see, for example, Craig, 1936) that

$$E(XY) = \sigma^2 (\lambda_1 \lambda_2 + \rho),$$

$$E(X^2 Y^2) = \sigma^4 \left(\lambda_1^2 + \lambda_2^2 + 4\rho\lambda_1\lambda_2 + \lambda_1^2\lambda_2^2 + 1 + 2\rho^2\right),$$

and

$$E\left[\left(1/S^2\right)^i\right] = \frac{\nu^i \Gamma(\nu/2 - i)}{(2\sigma^2)^i \Gamma(\nu/2)}$$

for $\nu > 2^i$.

3.3 Ratios

For a bivariate t random vector (X_1, X_2) with degrees of freedom ν, the mean vector (m_x, m_y) and correlation matrix $\mathbf{I}_2$ define the ratio $W = X_1/X_2$. The distribution of this ratio is of interest in problems in econometrics and ranking and selection. Press (1969) derived the pdf of W as

$$f(w) = \frac{k_1}{1+w^2}\left[1 + \frac{k_2 q}{(q^*)^{\nu+1}}\left\{2T_{\nu+1}\left(\frac{q\sqrt{\nu+1}}{q^*}\right) - 1\right\}\right],$$
$$-\infty < w < \infty, \qquad (3.11)$$

where $T_{\nu+1}$ denotes the cdf of the univariate Student's t distribution with degrees of freedom $\nu + 1$ and k_1, k_2, q, and q_* are constants defined by

$$k_1 = \frac{1}{\pi}\left(1 + \frac{m_x^2 + m_y^2}{\nu}\right)^{-\nu/2},$$

$$k_2 = \frac{\sqrt{\pi}\nu^{(\nu+2)/2}\Gamma((\nu+1)/2)}{2\Gamma((\nu+2)/2)}\left(1 + \frac{m_x^2 + m_y^2}{\nu}\right)^{-\nu/2},$$

$$q = -\frac{m_x w + m_y}{\sqrt{1+w^2}},$$

and

$$q^* = \sqrt{m_x^2 + m_y^2 + \nu - q^2},$$

respectively. In the special case $m_x = m_y = 0$, (3.11) is the pdf of a Cauchy distribution. The asymptotic distribution of W as $\nu \to \infty$ is that of the ratio of independent normal random variables. This fact can be verified as follows. Let $f_\infty(w) = \lim_{\nu\to\infty} f(w)$. Then it is easy to check that

$$\lim_{\nu\to\infty} \frac{k_2}{(q^*)^{\nu+1}} = \frac{1}{2\phi(q)},$$

$$\lim_{\nu\to\infty} \frac{\sqrt{\nu+1}}{q^*} = 1,$$

$$\lim_{\nu\to\infty} F_{\nu+1}(x) = \Phi(x),$$

and

$$\lim_{\nu\to\infty} k_1 = \frac{1}{\pi} \exp\left\{-\frac{1}{2}\left(m_x^2 + m_y^2\right)\right\},$$

where ϕ and Φ are, respectively, the pdf and the cdf of the standard normal distribution; thus, the limiting density f_∞ is given by

$$f_\infty(w) = \frac{1}{\pi\left(1+w^2\right)}\left[1 + \frac{q\{2\Phi(q)-1\}}{2\phi(q)}\right] \exp\left\{-\frac{1}{2}\left(m_x^2 + m_y^2\right)\right\},$$

which is the same density found by Marsaglia (1965, page 196, equation (5)) for the ratio of independent normal random variables with means (m_x, m_y) and unit variances.

The density $f(w) = f(w; m_x, m_y)$ may be confined for positive values $m_x > 0$, $m_y > 0$ since (3.11) shows that

$$f(w; -m_x, m_y) = f(-w; m_x, m_y),$$

$$f(w; -m_x, -m_y) = f(w; m_x, m_y),$$

and

$$f(w; m_x, -m_y) = f(-w; m_x, m_y).$$

Figure 3.1 shows how variations in (m_x, m_y, ν) affect the shape of the density. It may be seen by reference to Marsaglia (1965) that the shapes are similar to that of the ratio of normal random variables, even for small values of ν.

The percentage points of W defined by $\Pr(W \le w) = p$ are tabulated in Press (1969) for cumulative probabilities $p = 0.01$, 0.05, 0.10, 0.90, 0.95, 0.99; $\nu = 1$, 2, 5, 10, 30; and for some 16 selected values of (m_x, m_y). Here we provide the tables of w for $(m_x, m_y) = (0, 0)$, $(m_x, m_y) = (1, 3)$, $(m_x, m_y) = (3, 1)$, and $(m_x, m_y) = (3, 3)$.

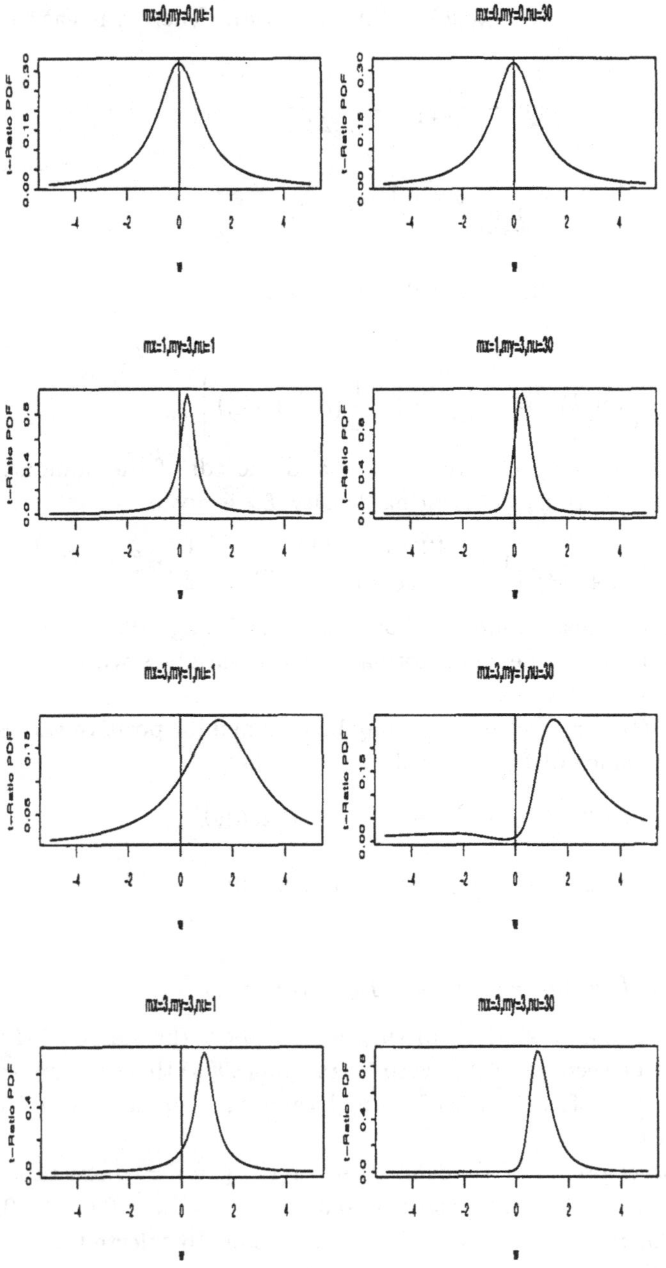

Fig. 3.1. Densities of the t-ratio distribution (3.11) for $(m_x, m_y, \nu) = (0, 0, 1)$, $(0, 0, 30)$, $(1, 3, 1)$, $(1, 3, 30)$, $(3, 1, 1)$, $(3, 1, 30)$, $(3, 3, 1)$, and $(3, 3, 30)$

Percentage points w for $(m_x, m_y) = (0, 0)$

ν	$p = 0.01$	$p = 0.05$	$p = 0.1$	$p = 0.9$	$p = 0.95$	$p = 0.99$
1	-31.820	-6.314	-3.078	3.078	6.314	31.820
2	-31.820	-6.314	-3.078	3.078	6.314	31.820
5	-31.820	-6.314	-3.078	3.078	6.314	31.820
10	-31.820	-6.314	-3.078	3.078	6.314	31.820
30	-31.820	-6.314	-3.078	3.078	6.314	31.820

Percentage points w for $(m_x, m_y) = (1, 3)$

ν	$p = 0.01$	$p = 0.05$	$p = 0.1$	$p = 0.9$	$p = 0.95$	$p = 0.99$
1	-10.254	-1.794	-0.721	1.321	2.394	10.853
2	-5.791	-0.902	-0.357	1.120	1.795	6.842
5	-1.331	-0.382	-0.166	1.018	1.486	6.464
10	-1.041	-0.312	-0.135	0.951	1.288	2.892
30	-0.681	-0.256	-0.109	0.922	1.211	2.350

Percentage points w for $(m_x, m_y) = (3, 1)$

ν	$p = 0.01$	$p = 0.05$	$p = 0.1$	$p = 0.9$	$p = 0.95$	$p = 0.99$
1	-51.268	-8.970	-3.604	6.604	11.970	54.267
2	-58.076	-10.187	-4.034	7.337	13.422	61.290
5	-64.675	-11.427	-4.524	8.007	14.774	67.984
10	-67.570	-11.984	-4.756	8.293	15.355	70.894
30	-69.742	-12.406	-4.934	8.505	15.788	73.074

Percentage points w for $(m_x, m_y) = (3, 3)$

ν	$p = 0.01$	$p = 0.05$	$p = 0.1$	$p = 0.9$	$p = 0.95$	$p = 0.99$
1	-12.970	-1.852	-0.442	2.242	3.652	14.770
2	-8.172	-0.474	0.183	2.141	3.205	11.028
5	-2.668	0.229	0.425	2.038	2.802	7.498
10	-0.118	0.334	0.477	1.986	2.627	6.019
30	0.130	0.384	0.505	1.943	2.497	4.997

Since it is evident from (3.11) that $f(w) \to 0$ rapidly as m_x and m_y become large, values of m_x, m_y greater than 3 are not considered.

The t-ratio distribution has one or two modes, depending upon the values of the parameters. The location of these modes are solutions of the equation

$$T_{\nu+1}\left(\frac{q}{q_*}\sqrt{\nu+1}\right) + A(w)\left(\frac{q}{q_*}\sqrt{\nu+3}\right) = B(w), \qquad (3.12)$$

where

$$A(w) = \frac{(\nu+1)q^2}{(q^*)^2}\frac{qw + m_x\sqrt{1+w^2}}{3qw + m_x\sqrt{1+w^2}},$$

$$\begin{aligned} B(w) = {} & \frac{1}{2} + \frac{A(w)}{2} - \frac{(\nu+2)\Gamma(\nu/2)}{2\sqrt{\pi}\Gamma((\nu+1)/2)}\frac{q^*w\left\{1+(q/q^*)^2\right\}^{-\nu/2}}{m_x\sqrt{1+w^2}+3qw} \\ & + \frac{\Gamma((\nu+2)/2)}{\sqrt{\pi}\Gamma((\nu+1)/2)}\frac{\sqrt{1+w^2}\left\{1+(q/q^*)^2\right\}^{-(\nu+2)/2}}{q^*\left\{m_x\sqrt{1+w^2}+3qw\right\}} \\ & \times\left\{w\sqrt{1+w^2} - qm_x\right\}, \end{aligned}$$

and q and q^* are as defined above. Note that since T_ν may be expressed in closed form in terms of elementary functions, (3.12) yields the modes in terms of elementary functions only.

From (3.11),

$$\lim_{w\to\infty} q = -m_x$$

and

$$\lim_{w\to\infty} q^* = \sqrt{m_y^2+\nu}.$$

Thus,

$$\lim_{w\to\infty} w^2 f(w) = \text{Constant}.$$

Hence the distribution of W can have no finite moments of order above zero.

Kappenman (1971) extended Press's ratio distribution for the multivariate case by considering the joint pdf of $\mathbf{W}^T = (X_2/X_1, X_3/X_1, \ldots, X_p/X_1)$, where $\mathbf{X} = (X_1, \ldots, X_p)^T$ is a p-variate t random vector with degrees of freedom ν, mean vector $\boldsymbol{\mu}$, and correlation matrix $\mathbf{R}$. The expressions for the joint pdf turned out to depend on whether p is odd or even. Introduce the following notation

$$\begin{aligned}
\mathbf{V}^T &= \left(1, \mathbf{W}^T\right), \\
M &= \mathbf{V}^T \mathbf{R}^{-1} \mathbf{V}, \\
K &= -2\mathbf{V}^T \mathbf{R}^{-1} \boldsymbol{\mu}, \\
L &= \nu + \boldsymbol{\mu}^T \mathbf{R}^{-1} \boldsymbol{\mu}, \\
a &= \frac{L}{M} - \frac{K^2}{4M^2},
\end{aligned}$$

and

$$b = -\frac{K}{2M}.$$

Then the expressions for $f(\mathbf{w})$ are

$$\begin{aligned}
f(\mathbf{w}) &= \frac{a \left|\mathbf{R}\right|^{-1/2} \nu^{\nu/2} \Gamma\left((\nu+p)/2\right)}{\pi^{p/2} (Ma)^{(\nu+p)/2} \Gamma(\nu/2)} \sum_{k=0}^{(p-1)/2} \binom{p-1}{p-1-2k} a^{2k} b^{p-1-2k} \\
&\quad \times \int_{-\infty}^{\infty} u^{2k} \left\{au^2 + 1\right\}^{-(\nu+p)/2} du
\end{aligned}$$

if p is odd;

$$\begin{aligned}
f(\mathbf{w}) &= \frac{2ab^{p-1} \left|\mathbf{R}\right|^{-1/2} \nu^{\nu/2} \Gamma\left((\nu+p)/2\right)}{\pi^{p/2} (Ma)^{(\nu+p)/2} \Gamma(\nu/2)} \\
&\quad \times \left[\sum_{k=0}^{p/2} \binom{p-1}{p-2k} \left(\frac{a}{b}\right)^{2k-1} \right. \\
&\quad \times \int_{-b/a}^{\infty} u^{2k-1} \left\{a^2u^2 + a\right\}^{-(\nu+p)/2} du \\
&\quad + \sum_{k=0}^{(p-2)/2} \binom{p-1}{p-2k-1} \left(\frac{a}{b}\right)^{2k} \\
&\quad \left. \times \int_{0}^{-b/a} u^{2k} \left\{a^2u^2 + a\right\}^{-(\nu+p)/2} du \right]
\end{aligned}$$

if p is even and $b < 0$; and

$$
\begin{aligned}
f(\mathbf{w}) &= \frac{2ab^{p-1}\left|\mathbf{R}\right|^{-1/2}\nu^{\nu/2}\Gamma\left((\nu+p)/2\right)}{\pi^{p/2}M^{(\nu+p)/2}\Gamma\left(\nu/2\right)} \\
&\quad \times \left[\sum_{k=0}^{p/2}\binom{p-1}{p-2k}\left(\frac{a}{b}\right)^{2k-1}\right. \\
&\quad \times \int_{b/a}^{\infty} u^{2k-1}\left\{a^2u^2+a\right\}^{-(\nu+p)/2} du \\
&\quad + \sum_{k=0}^{(p-2)/2}\binom{p-1}{p-2k-1}\left(\frac{a}{b}\right)^{2k} \\
&\quad \left. \times \int_{0}^{b/a} u^{2k}\left\{a^2u^2+a\right\}^{-(\nu+p)/2} du\right]
\end{aligned}
$$

if p is even and $b > 0$. The integrals in these expressions can easily be rewritten in terms of the gamma and incomplete beta functions; see Section 3 of Kappenman (1971) for details.

4

Bivariate Generalizations and Related Distributions

In this chapter, we shall survey a number of specific bivariate distributions that contain Student's t components.

4.1 Owen's Noncentral Bivariate t Distribution

Let Y_1 and Y_2 have the bivariate normal distribution with zero means, unit variances, and correlation 1. Let νS^2 have the chi-squared distribution with degrees of freedom ν and be independent of the X's. Then $X_1 = (Y_1 + \delta_1)/S$ and $X_2 = (Y_2 + \delta_2)/S$ have the noncentral univariate t distributions with degrees of freedom ν and noncentrality parameters δ_1 and δ_2, respectively. Owen (1965) studied the joint distribution of (X_1, X_2), a noncentral bivariate distribution.

The marginal cdf of X_j, $j = 1, 2$ may be written as

$$\Pr(X_j \leq y) = \frac{\sqrt{2\pi}}{\Gamma(\nu/2)2^{(\nu-2)/2}} \int_0^\infty x^{\nu-1}\phi(x)\Phi\left(\frac{yx}{\sqrt{\nu}} - \delta\right) dx, \tag{4.1}$$

where ϕ and Φ are, respectively, the pdf and the cdf of the standard normal distribution. Integrating by parts repeatedly, one obtains for odd values of ν

$$\begin{aligned} \Pr(X_j \leq y) &= \Phi\left(-\delta_j\sqrt{B}\right) + 2T\left(\delta_j\sqrt{B}, A\right) \\ &\quad +2\left[M_1 + M_3 + \cdots + M_{\nu-2}\right] \end{aligned}$$

and for even values of ν

$$\Pr(X_j \leq y) = \Phi(-\delta_j) + \sqrt{2\pi}\left[M_0 + M_2 + \cdots + M_{\nu-2}\right].$$

Here,

$$A = \frac{y}{\sqrt{\nu}},$$

$$B = \frac{\nu}{\nu + y^2},$$

$$T(h, a) = \frac{1}{2\pi} \int_0^a \frac{\exp\{-(1+x^2)h^2/2\}}{1+x^2} dx \tag{4.2}$$

(a function discussed and tabulated in Section 46.4 of Kotz et al., 2000), and the M's are defined recursively by

$$M_{-1} = 0,$$

$$M_0 = A\sqrt{B}\Phi'\left(\delta_j\sqrt{B}\right)\Phi\left(\delta_j A\sqrt{B}\right),$$

$$M_1 = B\left\{\delta_j A M_0 + \frac{A}{\sqrt{2\pi}}\Phi'\left(\delta_j\right)\right\},$$

$$M_2 = \frac{B}{2}\left\{\delta_j A M_1 + M_0\right\},$$

$$M_3 = \frac{2B}{3}\left\{\delta_j A M_2 + M_1\right\},$$

and

$$M_k = \frac{(k-1)B}{k}\left\{a_k \delta_j A M_{k-1} + M_{k-2}\right\}, \qquad k \geq 4,$$

where $a_k = 1/((k-2)a_{k-1})$, $k \geq 2$, and $a_2 = 1$. Two special cases of (4.1) are

$$\Pr(X_j \leq 0) = \Phi(-\delta_j)$$

and

$$\Pr(X_j \leq 1) = 1 - \Phi\left(\frac{\delta_j}{\sqrt{2}}\right), \qquad \nu = 1.$$

Also, if $\delta_j = 0$, then (4.1) is just the cdf of the Student's t distribution.

Owen (1965) expressed the joint cdf of (X_1, X_2) in terms of functions of the form

$$Q(y, \delta; 0, R) = \frac{\sqrt{2\pi}}{\Gamma(\nu/2)2^{(\nu-2)/2}} \int_0^R \int_0^\infty x^{\nu-1}\phi(x)\Phi\left(\frac{yx}{\sqrt{\nu}} - \delta\right) dx$$

and

$$Q(y,\delta;R,\infty) = \frac{\sqrt{2\pi}}{\Gamma(\nu/2)2^{(\nu-2)/2}} \int_0^R \int_0^\infty x^{\nu-1}\phi(x)\Phi\left(\frac{yx}{\sqrt{\nu}}-\delta\right) dx.$$

For example, if one assumes, without loss of generality, that $y_1 > y_2$, then the following relations hold

$$\Pr(X_1 \le y_1, X_2 \le y_2) = Q(y_1,\delta_1;0,R) + Q(y_2,\delta_2;R,\infty)$$

and

$$\Pr(X_1 \le y_1, X_2 \le y_2) = \Pr(X_2 \le y_2),$$

for $\delta_1 > \delta_2$ and $\delta_1 < \delta_2$, respectively. The formulas for $Q(y,\delta;0,R)$ and $Q(y,\delta;R,\infty)$ can be obtained by integration by parts. Since $Q(y,\delta;0,R)+Q(y,\delta;R,\infty) = \Pr(X_j \le y)$, it is sufficient to know the formula for one of the Q terms. Owen obtained the following formulas for $Q(y,\delta;0,R)$ for odd and even values of ν, respectively

$$\begin{aligned} Q(y,\delta;0,R) = {} & \Phi(R) - 2T(R,(AR-\delta)/R) \\ & -2T\left(\delta\sqrt{B},(\delta AB-R)/B\delta\right) + 2T\left(\delta\sqrt{B},A\right) \\ & +I\{\delta<0\} - 1 + 2\Big\{M_1^* + H_1 + M_3^* + H_3 \\ & + \cdots + M_{\nu-2}^* + H_{\nu-2}\Big\} \end{aligned}$$

and

$$\begin{aligned} Q(y,\delta;0,R) = {} & \Phi(-\delta) + \sqrt{2\pi}\Big\{M_0^* + H_0 + M_2^* + H_2 \\ & \cdots + M_{\nu-2}^* + H_{\nu-2}\Big\}. \end{aligned}$$

Here, $T(h,a)$ is as defined in (4.2) and the M^*'s are defined recursively by

$$M_0^* = A\sqrt{B}\phi\left(\delta\sqrt{B}\right)\left\{\Phi\left(\delta A\sqrt{B}\right) - \Phi\left((\delta AB-R)/\sqrt{B}\right)\right\},$$

$$\begin{aligned} M_1^* = {} & B\Big[\delta AM_0^* + A\phi\left(\delta\sqrt{B}\right)\Big\{\phi\left(\delta A\sqrt{B}\right) \\ & -\phi\left((\delta AB-R)/\sqrt{B}\right)\Big\}\Big], \end{aligned}$$

$$M_2^* = \frac{B}{2}\{\delta AM_1^* + M_0^*\} - L_1,$$

and

$$M_k^* = \frac{(k-1)B}{k}\left\{a_k\delta A M_{k-1}^* + M_{k-2}^*\right\} - L_{k-1}, \qquad k \geq 3,$$

and the H's are defined recursively by

$$H_0 = -\phi(R)\Phi(AR-\delta)$$

and

$$H_k = a_{k+2}RH_{k-1}, \qquad k \geq 1,$$

where

$$L_{k-1} = a_{k+2}RL_{k-2}, \qquad k \geq 3$$

with the initial value $L_1 = (ABR/2)\phi(R)\phi(AR-\delta)$ and

$$a_k = \frac{1}{(k-2)a_{k-1}}, \qquad k \geq 3$$

with the initial values $a_1 = a_2 = 1$.

4.2 Siddiqui's Noncentral Bivariate t Distribution

Siddiqui (1967) considered the joint distribution of Student's t variates when (Y_{1i}, Y_{2i}), $i = 1, \ldots, N$ is a random sample from the bivariate normal distribution with zero means, unit variances, and correlation coefficient ρ. Let

$$\bar{Y}_1 = \frac{1}{N}\sum_{i=1}^{N} Y_{1i},$$

$$\bar{Y}_2 = \frac{1}{N}\sum_{i=1}^{N} Y_{2i},$$

$$S_1^2 = \frac{1}{N}\sum_{i=1}^{N}\left(Y_{1i} - \bar{Y}_1\right)^2,$$

$$S_2^2 = \frac{1}{N}\sum_{i=1}^{N}\left(Y_{2i} - \bar{Y}_2\right)^2,$$

and

$$R = \frac{1}{NS_1S_2}\sum_{i=1}^{N}\left(Y_{1i}-\bar{Y}_1\right)\left(Y_{2i}-\bar{Y}_2\right).$$

The interest is in the joint distribution of the Student's t variates

$$(X_1, X_2) = \left(\frac{\bar{Y}_1}{\sqrt{N-1}S_1}, \frac{\bar{Y}_2}{\sqrt{N-1}S_2}\right). \tag{4.3}$$

It is well known that the joint pdf of $(\bar{Y}_1, \bar{Y}_2, S_1, S_2, R)$ is

$$\begin{aligned} & f\left(\bar{y}_1, \bar{y}_2, s_1, s_2, r\right) \\ = & \frac{N^N\left(s_1 s_2\right)^{N-2}\left(1-r^2\right)^{(N-4)/2}}{2\pi^2\Gamma(N-2)\left(1-\rho^2\right)^{N/2}} \exp\left[-\frac{N}{2\left(1-\rho^2\right)}\left\{\bar{x}^2+s_1^2\right.\right. \\ & \left.\left. +\bar{y}^2+s_2^2-2\rho\left(\bar{x}\bar{y}+rs_1s_2\right)\right\}\right] \end{aligned}$$

for $-\infty < \bar{y}_1 < \infty$, $-\infty < \bar{y}_2 < \infty$, $0 \leq s_1 < \infty$, $0 \leq s_2 < \infty$, and $-1 \leq r \leq 1$ (see, for example, Kendall and Stuart, 1958). After suitable transformation, Siddiqui obtained the joint pdf of (X_1, X_2, R) in the form

$$\begin{aligned} f\left(x_1, x_2, r\right) = & \frac{\Gamma(\nu+2)\left(1-\rho^2\right)^{(\nu+1)/2}\left(1-r^2\right)^{(\nu-2)/2}}{(2\pi)^{3/2}\Gamma(\nu+3/2)(1-b-cr)^{\nu+1/2}} \\ & \times\left\{\left(1+\frac{x_1^2}{\nu}\right)\left(1+\frac{x_2^2}{\nu}\right)\right\}^{-(\nu+1)/2} \\ & \times {}_2F_1\left(\frac{1}{2}, \frac{1}{2}; \nu+\frac{3}{2}; \frac{1+b+cr}{2}\right), \end{aligned} \tag{4.4}$$

where $\nu = N-1$, ${}_2F_1$ is the Gauss hypergeometric function,

$$b = \frac{\rho x_1 x_2}{\nu\sqrt{1+\frac{x_1^2}{\nu}}\sqrt{1+\frac{x_2^2}{\nu}}},$$

and

$$c = \frac{\rho}{\sqrt{1+\frac{x_1^2}{\nu}}\sqrt{1+\frac{x_2^2}{\nu}}}.$$

It is easily seen that the limit of this joint pdf as $\nu \to \infty$ is trivariate normal with independent components (X_1, X_2) and R. Integrating out

r in (4.4), Siddiqui showed further that the asymptotic joint pdf of (X_1, X_2) as $\nu \to \infty$ becomes

$$\begin{aligned} f(x_1,x_2) &\sim \frac{\Gamma(\nu+2)\left(1-\rho^2\right)^{(\nu+1)/2}}{2\pi\sqrt{\nu}\Gamma(\nu+3/2)}\left\{(1-b)^2-c^2\right\}^{1-\nu/2} \\ &\quad \times\left\{\left(1+\frac{x_1^2}{\nu}\right)\left(1+\frac{x_2^2}{\nu}\right)\right\}^{-(\nu+1)/2} \\ &\quad \times {}_2F_1\left(\frac{1}{2},\frac{1}{2};\nu+\frac{3}{2};\frac{1-b^2+c^2}{2(1-b)}\right). \end{aligned}$$

The exact joint pdf of (X_1, X_2) was also given for the two special cases $\nu = 1$ and $\nu = 3$. For $\nu = 1$, the joint pdf reduces to a bivariate Cauchy distribution

$$f(x_1,x_2) = \frac{\left(1-\rho^2\right)\csc^2\theta}{\pi^2\left(1+x_1^2\right)\left(1+x_2^2\right)}\left\{1+\left(\frac{\pi}{2}-\theta\right)\cot\theta\right\}, \qquad (4.5)$$

where

$$\cos\theta = \frac{2\rho\left(1-\rho^2\right)\left(1+y_1y_2\right)}{\sqrt{1+y_1^2}\sqrt{1+y_2^2}}.$$

In fact, if $\rho = 0$ in (4.5), then one arrives at a product of two independent Cauchy densities. For $\nu = 3$ the joint pdf is

$$f(x_1,x_2) = \frac{32\sqrt{2}\left(1-\rho^2\right)^2}{35\pi^2}\left(1+\frac{y_1^2}{3}\right)^{-2}\left(1+\frac{y_2^2}{3}\right)^{-2} I_3,$$

where

$$\begin{aligned} I_3 &= \sum_{k=0}^{\infty}\frac{\Gamma(9/2)\Gamma^2(k+1/2)}{\Gamma(k+9/2)\Gamma^2(1/2)k!}\sum_{l=0}^{k}\frac{1}{5-2l}\left(-\frac{1}{2}\right)^l\binom{k}{l} \\ &\quad \times\left\{(1-b-c)^{l-5/2}-(1-b+c)^{l-5/2}\right\}. \end{aligned}$$

4.3 Patil and Liao's Noncentral Bivariate t Distribution

Patil and Liao (1970) provided an extension of Siddiqui's work when (Y_{1i}, Y_{2i}), $i = 1, \ldots, N$ is a random sample from the bivariate normal distribution with zero means, common variance σ^2, and correlation coefficient ρ. Instead of considering the joint distribution of (4.3), they considered the joint distribution of

$$(X_1, X_2) = \left(\frac{\sqrt{N}\bar{Y}_1}{S}, \frac{\sqrt{N}\bar{Y}_2}{S}\right), \qquad (4.6)$$

where S is the pooled sample standard deviation defined by

$$S = \sqrt{\frac{\sum_{i=1}^{N}\left(Y_{1i}-\bar{Y}_1\right)^2+\sum_{i=1}^{N}\left(Y_{2i}-\bar{Y}_2\right)^2}{2N-1}}.$$

Exact expressions were given for the joint pdf and cdf of (X_1, X_2) and for the corresponding marginal distributions. For instance, if N is odd and equal to $2q+3$, then the joint pdf of (X_1, X_2) can be represented in the form

$$\begin{aligned} f(x_1, x_2) = & \frac{1}{2^{2q+5}\pi(q+1)\left(1-\rho^2\right)^{q+3/2}\Gamma^2(q+1)} \\ & \times \sum_{k=0}^{q}\binom{q}{k}(-1)^{q-k}\left(\frac{1-\rho^2}{\rho}\right)^{2q-k+1}\Gamma(2q-k+1) \\ & \times \left[\Gamma(k+2)\left\{\frac{1}{2(1+\rho)}+\frac{y_1^2-2\rho y_1 y_2+y_2^2}{8(q+1)\left(1-\rho^2\right)}\right\}^{-(k+2)}\right. \\ & -\sum_{l=0}^{2q-k}\left(\frac{\rho}{1-\rho^2}\right)^l \frac{\Gamma(k+l+2)}{l!}\left\{\frac{1}{2(1-\rho)}\right. \\ & \left.\left.+\frac{y_1^2-2\rho y_1 y_2+y_2^2}{8(q+1)\left(1-\rho^2\right)}\right\}^{-(k+l+2)}\right] \end{aligned}$$

while if N is even, then the joint pdf of (X_1, X_2) becomes

$$\begin{aligned} f(x_1, x_2) = & \frac{(1+\rho)^{(1-N)/2}}{\pi 2^{N+1}(N-1)(1-\rho)^{(N-1)/2}\sqrt{1-\rho^2}} \\ & \times \sum_{k=0}^{\infty}\binom{(1-N)/2}{k}(N+k-1)\left\{\frac{\rho}{(1+\rho)^2}\right\}^2 \\ & \times \left\{\frac{1}{2(1+\rho)}+\frac{y_1^2-2\rho y_1 y_2+y_2^2}{4(q-1)\left(1-\rho^2\right)}\right\}^{-(N+k)}. \end{aligned}$$

4.4 Krishnan's Noncentral Bivariate t Distribution

Krishnan (1972) provided another extension of Siddiqui's work when (Y_{1i}, Y_{2i}), $i = 1, \ldots, N$ is a random sample from the bivariate normal distribution with means (γ, δ), unit variances, and correlation coefficient ρ. She considered the joint distribution of

$$(X_1, X_2) = \left(\frac{\sqrt{N-1}\bar{Y}_1}{S_1}, \frac{\sqrt{N-1}\bar{Y}_2}{S_2}\right), \tag{4.7}$$

where S_1 and S_2 are correlated chi-squared random variables independent from $\bar{Y}_1$ and $\bar{Y}_2$, respectively. Series representations were derived for the joint pdf and the cdf of (X_1, X_2). One of the representations given for the joint pdf is

$$\begin{aligned} f(x_1, x_2) &= \frac{B}{4} \sum_{j=0}^{\infty} c_j \left\{ A^2 \left(1 + \frac{x_1^2}{N-1}\right) \left(1 + \frac{x_2^2}{N-1}\right) \right\}^{-(N+j)/2} \\ &\quad \times \sum_{m=0}^{\infty} d_{m,j}, \end{aligned} \tag{4.8}$$

where

$$A = \frac{N}{2(1-\rho^2)},$$

$$B = \frac{\sqrt{A^3 N^N 2^{4-N} \rho^{3-N}}}{\pi \Gamma((N-1)/2)} \exp\left\{-A\left(\gamma^2 + \delta^2 - 2\rho\gamma\delta\right)\right\},$$

$$c_j = \sum_{k=1}^{j} \frac{\nu_k (2\rho A x_1 x_2)^{j-k}}{(N-1)^{j-k}(j-k)!},$$

$$\nu_k = \begin{cases} (\rho A)^{(N-3)/2+k} \Big/ \{(k/2)!\Gamma(N+k-1)/2)\}, & \text{if } k \text{ is even}, \\ 0, & \text{if } k \text{ is odd}, \end{cases}$$

$$d_{m,j} = \sum_{i=0}^{m} \frac{(-2c)^i (-2d)^{m-i}}{i!(m-i)!} \Gamma\left(\frac{N+i+j}{2}\right) \Gamma\left(\frac{N+j+m-i}{2}\right),$$

$$c = \frac{\sqrt{A}(\rho\delta - \gamma) x_1}{\sqrt{N-1+x_1^2}},$$

and

$$d = \frac{\sqrt{A}(\rho\gamma - \delta) x_2}{\sqrt{N-1+x_2^2}}.$$

In the central case $\gamma = \delta = 0$, (4.8) reduces to the form derived by Patil and Liao (1970).

4.5 Krishnan's Doubly Noncentral Bivariate t Distribution

If Y is a normal random variable with mean δ and unit variance, and S^2 is an independent noncentral chi-squared random variable with degrees of freedom ν and noncentrality parameter λ, then

$$X = \frac{Y\sqrt{\nu}}{S} \tag{4.9}$$

is said to have the doubly noncentral univariate t distribution with degrees of freedom ν and noncentrality parameters δ and λ. The properties of this distribution have been studied by several authors; see Robbins (1948), Patnaik (1955), Krishnan (1959), Krishnan (1967a), Bulgren and Amos (1968), and Krishnan (1968) – see also Chapter 31 in Johnson et al. (1995) for a summary. The pdf, the expectation, and the variance of X are given by

$$f(x) = \sum_{k=0}^{\infty}\sum_{l=0}^{\infty} \frac{\lambda^k \exp\left\{-\left(\lambda+\delta^2\right)/2\right\}}{\sqrt{\nu}2^k k!\Gamma\left(l/2+1\right) Be\left(l/2+1/2, \nu/2+k\right)} \times \left(\frac{\delta x}{\sqrt{2\nu}}\right)^l \left(1+\frac{x^2}{\nu}\right)^{-(\nu+1+2k+l)/2},$$

$$E(X) = \delta\sqrt{\frac{\nu}{2}}\Gamma\left(\frac{\nu-1}{2}\right) {}_1F_1\left(\frac{1}{2},\frac{\nu}{2};-\frac{\lambda}{2}\right) \Big/ \Gamma\left(\frac{\nu}{2}\right), \qquad \nu>1,$$

and

$$Var(X) = \left(1+\delta^2\right)\frac{\nu}{\nu-2}\, {}_1F_1\left(1,\frac{\nu}{2};-\frac{\lambda}{2}\right) - \left\{E(X)\right\}^2, \qquad \nu>2,$$

where ${}_1F_1$ denotes the confluent hypergeometric function.

A bivariate analog of (4.9) was defined by Krishnan (1970) as follows. Let (Y_1, Y_2) follow a bivariate normal distribution with zero means, unit variances, and correlation coefficient ρ. Let (S_1, S_2) follow independently a noncentral bivariate chi-squared distribution with degrees of freedom ν, noncentrality parameter λ, and correlation coefficient ρ (Krishnan, 1967b). Then the random vector

$$(X_1, X_2) = \left(\frac{Y_1\sqrt{\nu}}{S_1}, \frac{Y_2\sqrt{\nu}}{S_2}\right) \tag{4.10}$$

is said to have the doubly noncentral bivariate t distribution with degrees of freedom ν and noncentrality parameter λ. Krishnan (1970) derived the corresponding joint pdf of (X_1, X_2) and provided an application involving the sample means and variances from two correlated nonhomogeneous normal populations. The special case of (4.10) for $S_1 = S_2 = S$ was considered by Patil and Kovner (1969), who provided expressions for the joint cdf of (X_1, X_2) and showed that when the means of Y_j are zero the probabilities of (X_1, X_2) in rectangular regions are monotone functions of ρ. In the special case $S_1 = S_2 = S$ and $\lambda = 0$, the distribution of (X_1, X_2) reduces to that of the central bivariate t.

4.6 Bulgren et al.'s Bivariate t Distribution

Suppose $Y_1, \ldots, Y_m, Y_{m+1}, \ldots, Y_{m+n}$ denote iid normal random variables with common mean μ and common variance σ^2. Bulgren et al. (1974) considered the joint distribution of (X_1, X_2) defined by

$$(X_1, X_2) = \left(\frac{m\bar{Y}_1}{\sqrt{S_1^2}}, \frac{\sqrt{m+n}\bar{Y}_2}{\sqrt{\frac{(m-1)S_1^2+(n-1)S_*^2}{m+n-2}}} \right),$$

where

$$\bar{Y}_1 = \frac{1}{m} \sum_{i=1}^{m} Y_i,$$

$$\bar{Y}_* = \frac{1}{n} \sum_{i=m+1}^{m+n} Y_i,$$

$$\bar{Y}_2 = \frac{m\bar{Y}_1 + n\bar{Y}_*}{m+n},$$

$$S_1^2 = \frac{1}{m-1} \sum_{i=1}^{m} \left(Y_i - \bar{Y}_1\right)^2,$$

and

$$S_*^2 = \frac{1}{n-1} \sum_{i=m+1}^{m+n} \left(Y_i - \bar{Y}_*\right)^2.$$

The distribution of (X_1, X_2) is bivariate t with a different noncentrality parameter for each variable. Note also that X_1 and X_2 have, respectively, $m-1$ and $m+n-2$ degrees of freedom. Bulgren et al. (1974)

provided series representations for the joint pdf of (X_1, X_2). In the central case $\mu = 0$,

$$\begin{aligned} f(x_1, x_2) &= \frac{2^{(m+n)/2}}{A} \left(1 + \frac{x_2^2}{m+n-2}\right)^{-(m+n)/2} \\ &\quad \times \sum_{j=0}^{\infty} \frac{(-1)^j}{j!} \Gamma\left(\frac{m+n}{2} + j\right) \left(1 + \frac{x_2^2}{m+n-2}\right)^{-j} \\ &\quad \times \sum_{k=0}^{2j} \binom{2j}{k} B\left(\frac{m+k}{2}, \frac{n-1}{2}\right) \left(\frac{n+m}{n(m-1)}\right)^{k/2} \\ &\quad \times x_1^k \left(\frac{-\sqrt{m} x_2}{\sqrt{n(m+n-2)}}\right)^{2j-k}, \end{aligned} \tag{4.11}$$

where

$$A = 2^{(m+n)/2} \pi \sqrt{\frac{n(m-1)(m+n-2)}{m+n}} \Gamma\left(\frac{m-1}{2}\right) \Gamma\left(\frac{n-1}{2}\right).$$

In the noncentral case $\mu \neq 0$, the joint pdf is even more complicated. Letting $n = am$, $a > 0$, and taking $m \to \infty$ in (4.11), one observes that the limiting distribution in the central case is the bivariate normal distribution with zero means, unit variances, and correlation $\sqrt{1/(1+a)}$.

4.7 Siotani's Noncentral Bivariate t Distribution

Siotani (1976) considered the most general form of (4.6) introduced by Patil and Liao (1970). Let $\mathbf{Y}$ be a p-variate normal random vector with mean vector $\boldsymbol{\mu}$, unit variances, and correlation matrix $\mathbf{R}$. Let $S = \sqrt{(V_1^2 + V_2^2)/(2\nu)}$, where (V_1, V_2) has the central bivariate chi-squared distribution with degrees of freedom ν and correlation coefficient τ. Siotani derived the distribution of $\mathbf{X} = \mathbf{Y}/S$ for general p and $\mathbf{R}$. The derivation required the joint pdf of (V_1, V_2) that was given by Siotani (1959) in the form

$$\begin{aligned} f(v_1, v_2) &= \sum_{k=0}^{\infty} c_k(\tau) \frac{(v_1 v_2)^{(\nu+2k)/2-1}}{2^{\nu+2k} (1-\tau^2)^{\nu+2k}} \frac{1}{\Gamma^2((\nu+2k)/2)} \\ &\quad \times \exp\left\{-\frac{v_1 + v_2}{2(1-\tau^2)}\right\}, \end{aligned}$$

where

$$c_k(\tau) = \frac{\Gamma((\nu+2k)/2)}{k!\Gamma(\nu/2)} (1-\tau^2)^{\nu/2} \tau^{2k}. \tag{4.12}$$

From this, one can easily obtain the pdf of

$$W = \frac{S}{\sqrt{1-\tau^2}} = \sqrt{\frac{V_1 + V_2}{2\nu\left(1-\tau^2\right)}}$$

as

$$f(w) = \sum_{k=0}^{\infty} c_k(\tau) f_{2\nu+4k}(w), \tag{4.13}$$

where

$$f_{2\nu+4k}(w) = \frac{2(2\nu)^{\nu+4k}}{2^{\nu+4k}\Gamma(\nu+2k)} w^{2\nu+4k-1} \exp\left\{-\nu w^2\right\}. \tag{4.14}$$

Since $c_1(\tau) + \cdots + c_\infty(\tau) = 1$, (4.13) is a mixture of (4.14) with the weights given by (4.12). Thus the joint pdf of $\mathbf{X}$ is also obtained in the same form

$$f(\mathbf{x}) = \sum_{k=0}^{\infty} c_k(\tau) f_{2\nu+4k}^{**}(\mathbf{x}),$$

where $c_k(\tau)$ are given by (4.12) and

$$\begin{aligned} f_{2\nu+4k}^{**}(\mathbf{x}) &= \exp\left\{-\frac{1}{2}\boldsymbol{\mu}^T \mathbf{R} \boldsymbol{\mu}\right\} \frac{\Gamma(\nu+2k+p/2)(1-\tau)^{p/2}}{(2\nu\pi)^{p/2}\Gamma(\nu+2k)|\mathbf{R}|^{1/2}} \\ &\quad \times \left\{1 + \frac{1-\tau^2}{2\nu}\mathbf{x}^T\mathbf{R}^{-1}\mathbf{x}\right\}^{-(\nu+2k+p/2)} \\ &\quad \times \sum_{l=0}^{\infty} \frac{\Gamma(\nu+2k+(p+l)/2)}{l!\Gamma(\nu+2k+p/2)} \\ &\quad \times \left\{\frac{\sqrt{2\left(1-\tau^2\right)}\mathbf{x}^T\mathbf{R}^{-1}\boldsymbol{\mu}}{\sqrt{2\nu+\left(1-\tau^2\right)\mathbf{x}^T\mathbf{R}^{-1}\mathbf{x}}}\right\}^k. \end{aligned}$$

When $p = 2$, $\boldsymbol{\mu} = 0$, and $\rho = \tau$ (ρ is the correlation coefficient between Y_1 and Y_2) this coincides with the pdfs derived by Patil and Liao (1970).

4.8 Tiku and Kambo's Bivariate t Distribution

Suppose (X_1, X_2) has the bivariate normal distribution with means (μ_1, μ_2), variances (σ_1^2, σ_2^2), and correlation coefficient ρ. Its joint pdf can be factorized as

$$f(x_1, x_2) = f(x_1 \mid x_2) f(x_2),$$

where

$$f(x_1 \mid x_2) = \frac{1}{\sigma_1\sqrt{1-\rho^2}} \exp\left[-\frac{1}{2\sigma_1^2(1-\rho^2)}\left\{x_1-\mu_1 -\rho\frac{\sigma_1}{\sigma_2}(x_2-\mu_2)^2\right\}\right] \tag{4.15}$$

and

$$f(x_2) \propto \frac{1}{\sigma_2}\exp\left\{-\frac{1}{2\sigma_2^2}(x_2-\mu_2)^2\right\}. \tag{4.16}$$

Numerous nonnormal distributions can be generated by replacing either $f(x_1 \mid x_2)$ and/or $f(x_2)$ by nonnormal distributions. Tiku and Kambo (1992) studied the family of symmetric bivariate distributions obtained by replacing (4.16) by the Student's t density

$$f(x_2) \propto \frac{1}{\sqrt{k}\sigma_2}\left\{1+\frac{(x_2-\mu_2)^2}{k\sigma_2^2}\right\}^{-\nu}, \tag{4.17}$$

where $k = 2\nu - 3$ and $\nu \geq 2$. This is motivated by the fact that in many applications it is reasonable to assume that the difference $Y_1 - \mu_1 - \rho(\sigma_1/\sigma_2)(Y_2 - \mu_2)$ is normally distributed and the regression of Y_1 on Y_2 is linear. Moreover, in numerous applications Y_2 represents time-to-failure with a distribution (Tiku and Gill, 1989; Gill et al., 1990), which might be symmetric but is not normally distributed. Besides, most types of time-to-failure data are such that a transformation cannot be performed to impose normality on the underlying distribution (see, for example, Mann, 1982, page 262).

On replacing (4.16) by (4.17), the joint pdf of (X_1, X_2) becomes

$$f(x_1, x_2) = \frac{1}{\sigma_1\sigma_2\sqrt{k(1-\rho^2)}}\left\{1+\frac{(x_2-\mu_2)^2}{k\sigma_2^2}\right\}^{-\nu} \times \exp\left[-\frac{1}{2\sigma_1^2(1-\rho^2)}\left\{x_1-\mu_1-\rho\frac{\sigma_1}{\sigma_2}(x_2-\mu_2)^2\right\}\right]. \tag{4.18}$$

Limiting $\nu \to \infty$, (4.18) reduces to the bivariate normal pdf. Writing $\mu_{ij} = E((Y_1 - \mu_1)^i (Y_2 - \mu_2)^j)$ for the cross product moment of order $i + j$, one observes that all odd-order moments are zero and that the first few even-order moments are

$$\mu_{20} = \sigma_1^2,$$

$$\begin{aligned}
\mu_{11} &= \rho\sigma_1\sigma_2, \\
\mu_{02} &= \sigma_2^2, \\
\mu_{40} &= 3\sigma_1^4\left(1+\frac{2\rho^4}{2\nu-5}\right), \\
\mu_{31} &= 3\rho\sigma_1^3\sigma_2\left(1+\frac{2\rho^2}{2\nu-5}\right), \\
\mu_{22} &= \sigma_1^2\sigma_2^2\left(1+2\rho^2+\frac{6\rho^2}{2\nu-5}\right), \\
\mu_{13} &= 3\rho\sigma_1\sigma_2^3\left(1+\frac{2}{2\nu-5}\right),
\end{aligned}$$

and

$$\mu_{04} = \frac{3(2\nu-3)}{2\nu-5}\sigma_2^4.$$

In fact, the moment generating function (mgf) of (Y_1, Y_2) is given by

$$\begin{aligned}
E\left[\exp\left(\theta_1 X_1 + \theta_2 X_2\right)\right] &= \exp\left\{\left(\mu_1 - \frac{\rho\sigma_1}{\sigma_2}\mu_2\right)\theta_1 + \frac{\left(1-\rho^2\right)\sigma_1^2}{2}\theta_1^2\right\} \\
&\quad \times M_2\left(\theta_2 + \frac{\rho\sigma_1}{\sigma_2}\mu_2\theta_1\right),
\end{aligned}$$

where $M_2(\cdot)$ denotes the moment generating function of X_2. This moment generating function does not exist unless, of course, $\nu = \infty$. However, the characteristic function does exist and is given by Sutradhar (1986). Estimation issues of the distribution (4.18) are discussed in Section 10.1.

4.9 Conditionally Specified Bivariate t Distribution

Let (X, Y) be a continuous random vector with joint pdf $f_{X,Y}(x, y)$ over $\Re^2$. Let $f_X(x)$, $f_Y(y)$ and $f_{X|Y}(x \mid y)$, $f_{Y|X}(y \mid x)$ denote the associated marginal and conditional densities, respectively. Assume that $X \mid Y$ and $Y \mid X$ are Student's t-distributed with the pdfs

$$f_{X|Y}(x \mid y) = \frac{\Gamma((\nu+1)/2)}{\sqrt{\pi}\Gamma(\nu/2)}\sqrt{\sigma(y)}\left\{1+\sigma(y)x^2\right\}^{-(\nu+1)/2} \tag{4.19}$$

and

$$f_{Y|X}(y \mid x) = \frac{\Gamma((\nu+1)/2)}{\sqrt{\pi}\Gamma(\nu/2)} \sqrt{\tau(x)} \left\{1 + \tau(x) y^2\right\}^{-(\nu+1)/2}, \quad (4.20)$$

where $x \in \Re$, $y \in \Re$, $\nu > 0$, and $\sigma(y)$, $\tau(x)$ are some positive functions. Writing the joint pdf of (X, Y) as a product of marginal and conditional densities in both possible manners, one obtains

$$\begin{aligned} & f_Y(y) \sqrt{\sigma(y) \left\{1 + \sigma(y) x^2\right\}^{-(\nu+1)/2}} \\ & \quad = f_X(x) \sqrt{\tau(x) \left\{1 + \tau(x) y^2\right\}^{-(\nu+1)/2}}, \end{aligned} \quad (4.21)$$

where $x \in \Re$ and $y \in \Re$. Set

$$g(y) = \left\{f_Y(y)\sqrt{\sigma(y)}\right\}^{2/(\nu+1)}, \qquad h(x) = \left\{f_X(x)\sqrt{\tau(x)}\right\}^{2/(\nu+1)} \quad (4.22)$$

so that, after rearranging, (4.21) becomes

$$g(y) + y^2 g(y)\tau(x) - h(x) - x^2 h(x)\sigma(y) = 0, \quad (4.23)$$

which must be solved for σ, τ, g, and h. Kottas et al. (1999) recognized that (4.23) is a special case of the functional equation

$$\sum_{k=1}^{n} f_k(x) g_k(y) = 0,$$

whose most general solution is given in the classical book by Aczel (1966, page 161). Thus, with $h(x)$, $x^2 h(x)$ and $g(y)$, $y^2 g(y)$ being the systems of mutually independent functions, the solution of (4.23) is found to be

$$\tau(x) = \frac{\lambda_3 + \lambda_4 x^2}{\lambda_1 + \lambda_2 x^2}, \qquad \sigma(y) = \frac{\lambda_2 + \lambda_4 y^2}{\lambda_1 + \lambda_3 y^2} \quad (4.24)$$

and

$$h(x) = \frac{1}{\lambda_1 + \lambda_2 x^2}, \qquad g(y) = \frac{1}{\lambda_1 + \lambda_4 y^2} \quad (4.25)$$

for $\lambda_j \in \Re$, $j = 1, 2, 3, 4$. Finally, substituting (4.22), (4.24), and (4.25) into (4.21), the joint pdf is derived as

$$\begin{aligned} f_{X,Y}(x, y) &= N_\nu(\lambda_1, \lambda_2, \lambda_3, \lambda_4) \\ & \quad \times \left\{\lambda_1 + \lambda_2 x^2 + \lambda_3 y^2 + \lambda_4 x^2 y^2\right\}^{-(\nu+1)/2}, \end{aligned} \quad (4.26)$$

where $x \in \Re$, $y \in \Re$, and $N_\nu(\cdot)$ denotes the normalizing constant. Utilizing certain compatibility conditions given in Arnold and Press (1989), Kottas et al. found that (4.26) is a well-defined joint pdf if $\lambda_1 \in \Re_+ \cup \{0\}$

and $\lambda_j \in \Re_+$, $j = 2, 3, 4$. Moreover, if $\lambda_1 = 0$, then one must have $\nu \in (0, 1)$.

The normalizing constant is given by the integral

$$\frac{1}{N_\nu(\lambda_1, \lambda_2, \lambda_3, \lambda_4)} = \int_{-\infty}^{\infty}\int_{-\infty}^{\infty} \left(\lambda_1 + \lambda_2 x^2 + \lambda_3 y^2 + \lambda_4 x^2 y^2\right)^{-(\nu+1)/2} dx dy. \quad (4.27)$$

In the case $\lambda_1 \neq 0$, making the transformation $s = (\lambda_2/\lambda_1)x^2$, $t = (\lambda_3/\lambda_1)y^2$, letting $\phi = \lambda_1\lambda_4/(\lambda_2\lambda_3)$, and using the integral representation of the Beta function,

$$B(a, b) = \int_0^{\infty} x^{a-1}(1+x)^{-a-b} dx, \qquad a > 0, \quad b > 0,$$

one obtains

$$\frac{1}{N_\nu(\lambda_1, \lambda_2, \lambda_3, \lambda_4)} = \frac{B\left(\frac{1}{2}, \frac{\nu}{2}\right)}{\lambda_1^{(\nu-1)/2}\sqrt{\lambda_2\lambda_3}} \int_0^{\infty} \frac{dx}{(1+x)^{\nu/2}\sqrt{x(1+\phi x)}}.$$

Letting $w = x/(1+x)$ and manipulating, Kottas et al. obtained

$$N_\nu(\lambda_1, \lambda_2, \lambda_3, \lambda_4) = \frac{\lambda_1^{(\nu-1)/2}\sqrt{\lambda_2\lambda_3}}{B\left(\frac{1}{2}, \frac{\nu}{2}\right) I\left(\frac{1}{2}, \frac{1}{2}, \frac{\nu+1}{2}; 1-\phi\right)}, \quad (4.28)$$

where

$$I(a, b, c; x) = \int_0^1 w^{b-1}(1-w)^{c-b-1}(1-zw)^{-a} dw \quad (4.29)$$

for $c > b \geq 0$. In the case $\lambda_1 = 0$, similar arguments show that

$$N_\nu(0, \lambda_2, \lambda_3, \lambda_4) = \frac{(\lambda_2\lambda_3)^{\nu/2}\lambda_4^{(1-\nu)/2}}{B\left(\frac{1}{2}, \frac{\nu}{2}\right) B\left(\frac{1-\nu}{2}, \frac{\nu}{2}\right)},$$

where $0 < \nu < 1$. The integral (4.29) converges for $z < 1$. For $\mid z \mid > 1$, Kottas et al. provided an alternative representation of (4.28) in terms of the Gauss hypergeometric function (see, for example, Magnus et al., 1966, page 54). It is also possible to represent (4.28) in terms of elliptical integrals of the first and second kind (Carlson, 1977, Chapter 9). For example, if $\nu = 1$, then (4.28) can be easily rearranged to yield

$$N_\nu(\lambda_1, \lambda_2, \lambda_3, \lambda_4) = \frac{\sqrt{\lambda_1\lambda_4}}{\sqrt{2\pi} R_F(0, 1/\phi, 1)},$$

where R_F is the elliptical integral of the first kind defined by

$$R_F(a,b,c) = \frac{1}{2}\int_0^\infty \{(x+a)(x+b)(x+c)\}^{-1/2}\, dx$$

with a, b, c nonnegative, and at most one of them equal to zero.

If $0 < \nu < 1$, then (4.26) does not possess finite moments; thus, from here on we shall consider the case $\nu \geq 1$. If $\nu > \max(m, n)$, for non-negative integers m, n, then Kottas et al. showed that

$$E\left(X^m Y^n\right) = \frac{\Gamma\left(\frac{m+1}{2}\right)\Gamma\left(\frac{\nu-m}{2}\right) I\left(\frac{m+1}{2}, \frac{n+1}{2}, \frac{\nu+1}{2}; 1-\phi\right)}{\sqrt{\pi \mu_1^m \mu_2^n}\,\Gamma\left(\frac{\nu}{2}\right) I\left(\frac{1}{2}, \frac{1}{2}, \frac{\nu+1}{2}; 1-\phi\right)} \tag{4.30}$$

provided that both m and n are even or zero. The expectation is zero if at least one of m or n is odd. This suggests that the distribution may be an appropriate model for uncorrelated but nonindependent data.

From relations (4.21), (4.24), and (4.26) it is immediate that the marginal densities are

$$f_X(x) = \sqrt{\mu_1}\left\{I\left(\frac{1}{2}, \frac{1}{2}, \frac{\nu+1}{2}; 1-\phi\right)\sqrt{1+\phi\mu_1 x^2}(1+\mu_1 x^2)^{\nu/2}\right\}^{-1}$$

and

$$f_Y(y) = \sqrt{\mu_2}\left\{I\left(\frac{1}{2}, \frac{1}{2}, \frac{\nu+1}{2}; 1-\phi\right)\sqrt{1+\phi\mu_2 y^2}(1+\mu_2 y^2)^{\nu/2}\right\}^{-1},$$

where $x \in \Re$ and $y \in \Re$. Here, $\mu_j = \lambda_{j+1}/\lambda_1$, $j = 1, 2$ are the intensity parameters while ϕ and ν are the dependence and scale parameters, respectively. It is easily noted that X and Y are independent if and only if $\phi = 1$. The graph of the joint pdf is symmetric and bell-shaped and takes the standard form when $\mu_1 = \mu_2 = 1$.

From relations (4.19)–(4.20) and (4.24)–(4.25) it is immediate that $X \mid Y$ has the Student's t distribution with degrees of freedom ν and scale parameter $(1/\nu_1)(1+\mu_2 y^2)/(1+\phi\mu_2 y^2)$, and that $Y \mid X$ is also Student's t with degrees of freedom ν and scale parameter $(1/\nu_2)(1+\mu_1 x^2)/(1+\phi\mu_1 x^2)$, where $\mu_j = \lambda_{j+1}/\lambda_1$, $j = 1, 2$. Consequently, the conditional moments are

$$E(X^m \mid Y = y) = \frac{\Gamma\left(\frac{m+1}{2}\right)\Gamma\left(\frac{\nu-m}{2}\right)}{\sqrt{\pi}\,\Gamma\left(\frac{\nu}{2}\right)}\left(\frac{1+\mu_2 y^2}{\mu_1\left(1+\phi\mu_2 y^2\right)}\right)^{m/2}$$

and

$$E(Y^n \mid X = x) = \frac{\Gamma\left(\frac{m+1}{2}\right)\Gamma\left(\frac{\nu-m}{2}\right)}{\sqrt{\pi}\,\Gamma\left(\frac{\nu}{2}\right)}\left(\frac{1+\mu_1 x^2}{\mu_2\left(1+\phi\mu_1 x^2\right)}\right)^{m/2}$$

provided that m is an even number less than ν. If m is odd, then the corresponding conditional moments are zero.

In the special case $\nu = 1$, (4.26) reduces to the centered Cauchy conditionals model of Anderson and Arnold (1991). The limiting case $\phi \to 0$ gives the bivariate Pearson type VII distribution (Johnson, 1987, page 117) with location parameters equal to zero and uncorrelated components. If 2ν is a positive integer, then this limit distribution reduces to a special case of the general bivariate t distribution (see, for example, Johnson and Kotz, 1972, page 134, relation 1) with uncorrelated components and 2ν degrees of freedom. For $\mu_1 = \mu_2$ and $\nu = 2$, the limit distribution reduces to the bivariate Cauchy distribution (Mardia, 1970a, page 86) while for $\mu_1 = \mu_2$ and $\nu = v + 1$ it gives the bivariate t distribution (Johnson and Kotz, 1972, page 134, relation 2) with v degrees of freedom. In the latter case, the standard bivariate normal distribution with independent components arises as a further limiting case when $\nu \to \infty$. Other special cases of (4.26) are the centered normal conditionals model studied by Sarabia (1995) and the Beta conditionals model of the second kind (Castillo and Sarabia, 1990).

4.10 Jones' Bivariate t Distribution

Let Z_1, Z_2, W be mutually independent random variables with Z_i having the standard normal distribution and W having the chi-squared distribution with degrees of freedom n_1. Then the standard bivariate t distribution with degrees of freedom n_1 is the joint distribution of

$$\left(\frac{\sqrt{n_1} Z_1}{\sqrt{W}}, \frac{\sqrt{n_1} Z_2}{\sqrt{W}} \right). \tag{4.31}$$

One disadvantage of this model is that the two univariate marginals (which are Student's t) have the same degrees of freedom parameter and hence the same amount of tailweight. Jones (2002b) provided an alternative distribution with Student's t marginals, each with its own arbitrary degrees of freedom parameter. Precisely, if W_1, W_2 are independent chi-squared random variables (also independent of Z_1, Z_2) with degrees of freedom ν_1 and $\nu_2 - \nu_1$, respectively, then Jones (2002b) considered the joint distribution of

$$(X_1, X_2) = \left(\frac{\sqrt{\nu_1} Z_1}{\sqrt{W_1}}, \frac{\sqrt{\nu_2} Z_2}{\sqrt{W_1 + W_2}} \right). \tag{4.32}$$

Note that the ith marginal of this distribution is Student's t with degrees of freedom ν_i. It is easy to see that the correlation between X_1 and X_2

is zero, a property also shared by (4.31). If $r_1 < \nu_1$ and $r_1 + r_2 < \nu_2$, then the product moment is given by

$$E\left(X_1^{r_1} X_2^{r_2}\right) = \frac{\nu_1^{r_1/2} \nu_2^{r_2/2} \Gamma\left(\frac{1+r_1}{2}\right) \Gamma\left(\frac{1+r_2}{2}\right) \Gamma\left(\frac{\nu_1-r_1}{2}\right) \Gamma\left(\frac{\nu_2-r_1-r_2}{2}\right)}{\pi \Gamma\left(\frac{\nu_1}{2}\right) \Gamma\left(\frac{\nu_2-r_1}{2}\right)}$$

if r_1 and r_2 are even and is zero otherwise. The joint pdf of X_1 and X_2 is

$$\begin{aligned} f(x_1, x_2) = {} & C \left(1 + \frac{x_1^2}{\nu_1} + \frac{x_2^2}{\nu_2}\right)^{-(1+\nu_2/2)} {}_2F_1\left(1 + \frac{\nu_2}{2}, \frac{\nu_2 - \nu_1}{2}; \right. \\ & \left. \frac{1+\nu_2}{2}; \frac{x_1^2}{\nu_1}\left(1 + \frac{x_1^2}{\nu_1} + \frac{x_2^2}{\nu_2}\right)\right), \end{aligned} \tag{4.33}$$

where

$$C = \frac{1}{\pi} \frac{\Gamma\left(\frac{1+\nu_1}{2}\right) \Gamma\left(\frac{1+\nu_2}{2}\right)}{\sqrt{\nu_1 \nu_2} \Gamma\left(\frac{\nu_1}{2}\right) \Gamma\left(\frac{1+\nu_2}{2}\right)}$$

and ${}_2F_1$ denotes the Gauss hypergeometric function. The conditional pdf of X_2 given $X_1 = x_1$ is

$$\begin{aligned} f(x_2 \mid x_1) = {} & C \left(u_1 + \frac{x_2^2}{\nu_2}\right)^{-(1+\nu_2/2)} \\ & \times {}_2F_1\left(1 + \frac{\nu_2}{2}, \frac{\nu_2 - \nu_1}{2}; \frac{1+\nu_2}{2}; \frac{u_1 - 1}{u_1 + x_2^2/\nu_2}\right), \end{aligned} \tag{4.34}$$

where $u_1 = 1 + x_1^2/\nu_1$ and

$$C = \frac{u_1^{(1+\nu_1)/2} \Gamma\left(1 + \frac{\nu_2}{2}\right)}{\sqrt{\pi \nu_2} \Gamma\left(\frac{1+\nu_2}{2}\right)}.$$

If $\nu_2 + 1 > r$, then the conditional rth moment is given by

$$\begin{aligned} E\left(X_2^r \mid X_1 = x_1\right) = {} & \nu_2^{r/2} u_1^{(r-\nu_2+\nu_1)/2} \frac{\Gamma\left(\frac{1+r}{2}\right) \Gamma\left(\frac{1+\nu_2-r}{2}\right)}{\sqrt{\pi} \Gamma\left(\frac{1+\nu_2}{2}\right)} \\ & \times {}_2F_1\left(\frac{\nu_2 - r + 1}{2}, \frac{\nu_2 - \nu_1}{2}; \frac{1+\nu_2}{2}; \frac{u_1 - 1}{u_1}\right) \end{aligned} \tag{4.35}$$

if r is even and is zero otherwise. Setting $\nu_2 = \nu_1 = \nu$ in (4.34)–(4.35), one obtains the corresponding forms for the standard bivariate t distribution (see Section 1.11). Note that the conditional variance

$Var(X_2 \mid X_1 = x_1)$ increases with $\mid x_1 \mid$. In a parallel fashion, with $u_2 = 1 + x_2^2/\nu_2$, the conditional distribution of X_1 given $X_2 = x_2$ is

$$\begin{aligned} f(x_1 \mid x_2) &= C\left(u_2 + \frac{x_1^2}{\nu_1}\right)^{-(1+\nu_1/2)} \\ &\quad \times {}_2F_1\left(1 + \frac{\nu_2}{2}, \frac{\nu_2 - \nu_1}{2}; \frac{1+\nu_2}{2}; 1 - \frac{u_2}{u_2 + x_1^2/\nu_1}\right), \end{aligned}$$

where

$$C = \frac{u_2^{(1+\nu_2)/2}\Gamma\left(\frac{1+\nu_1}{2}\right)\Gamma\left(\frac{\nu_2}{2}\right)\Gamma\left(1 + \frac{\nu_2}{2}\right)}{\sqrt{\pi\nu_1}\Gamma\left(\frac{\nu_1}{2}\right)\Gamma\left(\frac{1+\nu_2}{2}\right)}.$$

This time, the conditional rth moments exist provided $\nu_1 > r$, unless $\nu_1 = \nu_2$, in which case one needs $1 + \nu_1 > r$. The odd conditional moments are again zero and the even conditional moments are given by the simpler form

$$E\left(X_1^r \mid X_2 = x_2\right) = \nu_1^{r/2} u_2^{r/2} \frac{\Gamma\left(\frac{1+r}{2}\right)\Gamma\left(\frac{\nu_1 - r}{2}\right)\Gamma\left(\frac{\nu_2}{2}\right)\Gamma\left(\frac{\nu_2 - r + 1}{2}\right)}{\sqrt{\pi}\Gamma\left(\frac{\nu_1}{2}\right)\Gamma\left(\frac{1+\nu_2}{2}\right)\Gamma\left(\frac{\nu_2 - r}{2}\right)}.$$

The construction (4.32) can be easily extended to the multivariate case. Two straightforward extensions are

- Let $Z_1, \ldots, Z_p$, $W_1, \ldots, W_p$ be mutually independent random variables with Z_i having the standard normal distribution and W_i having the chi-squared distribution with degrees of freedom $\nu_i - \nu_{i-1}$. Then,

$$\begin{aligned} &(X_1', X_2', \ldots, X_p') \\ &= \left(\frac{\sqrt{\nu_1} Z_1}{W_1}, \frac{\sqrt{\nu_2} Z_2}{\sqrt{W_1 + W_2}}, \ldots, \frac{\sqrt{\nu_p} Z_p}{\sqrt{W_1 + \cdots + W_p}}\right) \end{aligned} \quad (4.36)$$

  has a multivariate distribution with univariate marginals that are Student's t distributed with degrees of freedom ν_i, $i = 1, \ldots, p$. The bivariate marginals of (4.36) have the distribution of (4.32).
- With the notation as above,

$$\begin{aligned} &(X_1'', X_2'', \ldots, X_p'') \\ &= \left(\frac{\sqrt{\nu_1} Z_1}{W_1}, \frac{\sqrt{\nu_2} Z_2}{\sqrt{W_1 + U_2}}, \ldots, \frac{\sqrt{\nu_p} Z_p}{\sqrt{W_1 + U_p}}\right) \end{aligned} \quad (4.37)$$

  has a multivariate distribution with the same univariate and bivariate marginals. Here, U_i are independent chi-squared random variables (also independent of Z_i, W_i) with degrees of freedom $\nu_i - \nu_1$, $i = 1, \ldots, p$.

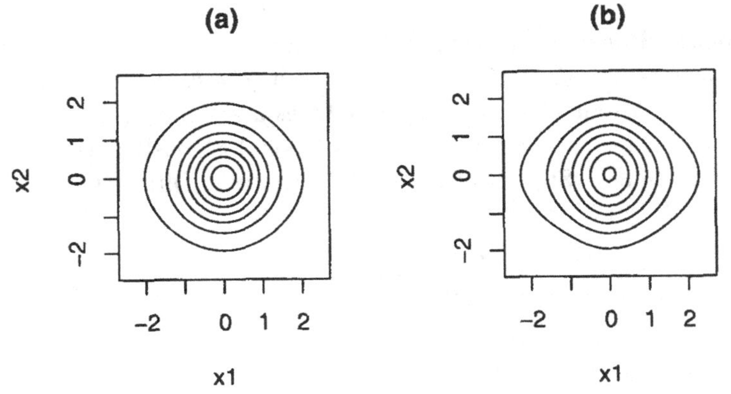

Fig. 4.1. Jones' bivariate skew t pdf (4.33) for (a) $\nu_1 = 2$ and $\nu_2 = 3$; and (b) $\nu_1 = 2$ and $\nu_2 = 20$

Further extensions of (4.32) arise by adding further independent chi-squared random variables inside the square roots in the denominators of all variables in (4.36) or by adding a single further independent chi-squared random variable inside the square root in the denominator of X_1'' in (4.37).

Jones (2002a) provided another bivariate generalization of (4.31). This generalization has the skew t distribution (Jones, 2001a) as its marginals. If U denotes a standard beta random variable with parameters a and c, then a skew t variate is defined by

$$X = \frac{\sqrt{a+c}(2U-1)}{2\sqrt{U(1-U)}}.$$

The corresponding pdf is

$$f(x;a,c) = \frac{2^{1-a-c}\Gamma(a+c)}{\sqrt{a+c}\Gamma(a)\Gamma(c)}\left\{1+\frac{x}{\sqrt{a+c+x^2}}\right\}^{a+1/2} \times\left\{1+\frac{x}{\sqrt{a+c+x^2}}\right\}^{c+1/2}. \quad (4.38)$$

The standard Student's t is the particular case for $\nu = 2a$ when $a = c$. If $a > c$, then (4.38) has positive skewness; also, $f(x;b,a) = f(-x;a,b)$. Further details about (4.38) are given in Jones and Faddy (2002). The bivariate generalization proposed in Jones (2002a) is constructed in the same way as (4.38): Specifically, if (U, V) denotes a Dirichlet random vector with the joint pdf

$$f(u,v) = \frac{\Gamma(a+b+c)}{\Gamma(a)\Gamma(b)\Gamma(c)}u^{a-1}v^{b-1}(1-u-v)^{c-1}$$

(where $u > 0$, $v > 0$, and $u + v < 1$), then define

$$(X_1, X_2) = \left(\frac{\sqrt{d}(2U-1)}{2\sqrt{U(1-U)}}, \frac{\sqrt{d}(1-2V)}{2\sqrt{U(1-U)}}\right), \quad (4.39)$$

where $d = a + b + c$. It can be verified that the joint pdf of (X_1, X_2) is

$$f(x_1, x_2) = \frac{d\Gamma(d+1)}{2^{d-1}\Gamma(a)\Gamma(b)\Gamma(c)}\left(1+\frac{x_1}{\sqrt{d+x_1^2}}\right)^{a-1} \times\left(1-\frac{x_2}{\sqrt{d+x_2^2}}\right)^{b-1}\frac{1}{(d+x_1^2)^{3/2}}\frac{1}{(d+x_2^2)^{3/2}} \times\left(\frac{x_2}{\sqrt{d+x_2^2}}-\frac{x_1}{\sqrt{d+x_1^2}}\right)^{c-1}. \quad (4.40)$$

Because of a direct analogy with the Dirichlet distribution, only one of the two marginals of (4.40) can be a symmetric Student's t distribution, the other necessarily being skewed. This Student's t distribution will have degrees of freedom d, and any skew t marginal will have a total parameter value of d, but divided up into unequal amounts. In this sense, marginals of (4.40) are most closely associated with Student's t distributions with degrees of freedom d.

Note that if instead of (X_1, X_2) in (4.39) the transformation was made to $(-X_1, X_2)$, then one would have obtained the equivalent distribution on $x_1 + x_2 > 0$. Also, $(-X_1, -X_2)$ would have given the equivalent distribution on $x_2 < x_1$ and $(X_1, -X_2)$ as the same on $x_2 + x_1 < 0$.

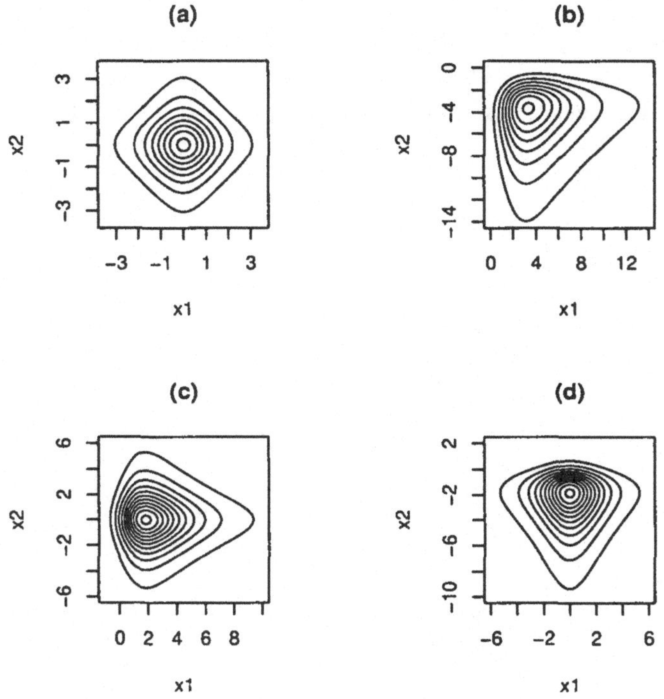

Fig. 4.2. Jones' bivariate skew t pdf (4.40) for (a) $a = 1$, $b = 1$, and $c = 1$; (b) $a = 3$, $b = 4$, and $c = 5$; (c) $a = 5$, $b = 1$, and $c = 1$; and (d) $a = 1$, $b = 5$, and $c = 1$

The corresponding changes to (4.40) would simply have been to make corresponding changes to the signs of x_1 and x_2.

The means and variances associated with (4.40) can be easily obtained from the results provided in Jones (2001a)

$$E(X_1) = \frac{\sqrt{d}}{2}(a-b-c)\frac{\Gamma(a-1/2)\Gamma(b+c-1/2)}{\Gamma(a)\Gamma(b+c)},$$

$$E(X_2) = \frac{\sqrt{d}}{2}(a-b-c)\frac{\Gamma(a+c-1/2)\Gamma(b-1/2)}{\Gamma(a+c)\Gamma(b)},$$

$$Var(X_1) = \frac{d}{4}\frac{(a-b-c)^2+d-2}{(a-1)(b+c-1)} - \{E(X_1)\}^2,$$

and

$$Var\,(X_2) \;=\; \frac{d}{4}\frac{(a-b+c)^2+d-2}{(a+c-1)(b-1)} - \{E\,(X_2)\}^2 .$$

The covariance between X_1 and X_2 appears not to be available in closed form.

5

Multivariate Generalizations and Related Distributions

This chapter contains a large number of modifications and extensions of the standard multivariate t distribution introduced in (1.1). Some of them are of somewhat complex nature. It thus requires a careful reading to see the forest behind the trees!

5.1 Kshirsagar's Noncentral Multivariate t Distribution

One of the earliest results in the area of noncentral multivariate t distributions is that due to Kshirsagar (1961). Let $\mathbf{Y}$ be a p-variate random vector having the normal distribution with mean vector $\boldsymbol{\mu}$, common variance σ^2, and correlation matrix $\mathbf{R}$. Let S^2 be distributed independently of $\mathbf{Y}$ according to a chi-squared distribution with degrees of freedom ν. Kshirsagar (1961) considered the distribution of $\mathbf{X} = \mathbf{Y}/S$ and showed that it has the joint pdf

$$\begin{aligned} f(\mathbf{x}) &= \exp\left\{-\frac{1}{2}\boldsymbol{\xi}^T\mathbf{R}\boldsymbol{\xi}\right\}\frac{\Gamma((\nu+p)/2)}{(\nu\pi)^{p/2}\Gamma(\nu/2)\,|\mathbf{R}|^{1/2}} \\ &\quad\times\left\{1+\frac{1}{\nu}\mathbf{x}^T\mathbf{R}^{-1}\mathbf{x}\right\}^{-(\nu+p)/2} \\ &\quad\times\sum_{k=0}^{\infty}\frac{\Gamma((\nu+p+k)/2)}{k!\Gamma((\nu+p)/2)}\left\{\frac{\sqrt{2}\mathbf{x}^T\mathbf{R}^{-1}\boldsymbol{\xi}}{\sqrt{\nu+\mathbf{x}^T\mathbf{R}^{-1}\mathbf{x}}}\right\}^k, \quad (5.1) \end{aligned}$$

where $\boldsymbol{\xi} = \boldsymbol{\mu}/\sigma$. This noncentral distribution reduces to the form of (1.1) when $\boldsymbol{\mu} = \mathbf{0}$.

We noted earlier in Section 1.11 that, if $\mathbf{X}$ has the central multivariate t distribution with degrees of freedom ν and correlation matrix $\mathbf{R}$, and if

$$\mathbf{X} = \begin{pmatrix} \mathbf{X}_1 \\ \mathbf{X}_2 \end{pmatrix}$$

and

$$\mathbf{R} = \begin{pmatrix} \mathbf{R}_{11} & \mathbf{R}_{12} \\ \mathbf{R}_{21} & \mathbf{R}_{22} \end{pmatrix},$$

where $\mathbf{X}_1$ is $p_1 \times 1$ and $\mathbf{R}_{11}$ is $p_1 \times p_1$, then

$$\mathbf{Z}_1 = \mathbf{X}_1$$

and

$$\mathbf{Z}_2 = \sqrt{\frac{\nu + p_1}{\nu}} \left(1 + \frac{1}{\nu}\mathbf{X}_1^T \mathbf{R}_{11}^{-1} \mathbf{X}_1\right)^{-1/2} \left(\mathbf{X}_2 - \mathbf{R}_{21}\mathbf{R}_{11}^{-1}\mathbf{X}_1\right)$$

are independently distributed, each according to a central multivariate t distribution. This result does not remain true to the noncentral distribution (5.1). Actually, Siotani (1976) showed that if

$$\boldsymbol{\xi} = \begin{pmatrix} \boldsymbol{\xi}_1 \\ \boldsymbol{\xi}_2 \end{pmatrix}$$

is the partition of $\boldsymbol{\xi}$ corresponding to that of $\mathbf{X}$,

$$\boldsymbol{\xi}_{2\cdot 1} = \boldsymbol{\xi}_2 - \mathbf{R}_{21}\mathbf{R}_{11}^{-1}\boldsymbol{\xi}_1$$

and

$$\mathbf{R}_{22\cdot 1} = \mathbf{R}_{22} - \mathbf{R}_{21}\mathbf{R}_{11}^{-1}\mathbf{R}_{12},$$

then the joint pdf of $\mathbf{Z}_1$ and $\mathbf{Z}_2$ is

$$\begin{aligned} f(\mathbf{z}_1, \mathbf{z}_2) &= K \exp\left(-\frac{1}{2}\boldsymbol{\xi}_1^T \mathbf{R}_{11}^{-1}\boldsymbol{\xi}_1 - \frac{1}{2}\boldsymbol{\xi}_{2\cdot 1}^T \mathbf{R}_{22\cdot 1}^{-1}\boldsymbol{\xi}_{2\cdot 1}\right) \\ &\quad \times t_\nu\left(\mathbf{z}_1; \mathbf{R}_{11}, p_1\right) t_{\nu+p_1}\left(\mathbf{z}_2; \mathbf{R}_{22\cdot 1}, p - p_1\right), \end{aligned}$$

where the last two terms denote the pdfs of central multivariate t distributions with appropriate parameters and K is given by the formidable expression

$$\begin{aligned} K &= \sum_{k=0}^{\infty}\sum_{l=0}^{\infty} \frac{\Gamma((\nu + p + k + l)/2)}{k!l!\Gamma((\nu + p)/2)} \left(\frac{2}{\nu}\right)^{k/2} \left(\frac{2}{\nu + p - p_1}\right)^{l/2} \\ &\quad \times \left\{\frac{\mathbf{z}_1^T \mathbf{R}_{11}^{-1}\boldsymbol{\xi}_1}{\sqrt{1 + \mathbf{z}_1^T \mathbf{R}_{11}^{-1}\mathbf{z}_1/\nu}}\right\}^k \end{aligned}$$

$$\times \frac{\left(\mathbf{z}_2^T \mathbf{R}_{22\cdot 1}^{-1} \boldsymbol{\xi}_{2\cdot 1}\right)^l}{\left\{1 + \mathbf{z}_2^T \mathbf{R}_{22\cdot 1}^{-1} \boldsymbol{\xi}_2 / (\nu + p_1)\right\}^{(k+l)/2}}.$$

Siotani (1976) also derived the corresponding noncentral distribution when $\mathbf{X}$ is partitioned into k sets of variates as in (1.20). Following the notations defined by (1.21), (1.22), (1.23), (1.24), and (1.25), the joint pdf of

$$\mathbf{Z}_1 = \mathbf{X}_1$$

and

$$\begin{aligned} \mathbf{Z}_{l+1} &= \sqrt{\frac{\nu + q_l}{\nu}} \left(1 + \frac{1}{\nu} \mathbf{X}_{(l)}^T \mathbf{R}_l^{-1} \mathbf{X}_{(l)}\right)^{-1/2} \\ &\quad \times \left(\mathbf{X}_{(l+1)} - \mathbf{R}_{(l)}^{(l+1)^T} \mathbf{R}_{(l)}^{-1} \mathbf{X}_{(l)}\right), \\ &\quad l = 1, \ldots, k-1, \end{aligned}$$

is given by the lengthy expression

$$\begin{aligned} f(\mathbf{z}_1, \ldots, \mathbf{z}_k) &= \exp\left(-\frac{1}{2} \sum_{l=1}^k \delta_l^2\right) \prod_{l=1}^k \frac{\Gamma((\nu + q_l)/2) \left|\mathbf{R}_{ll\cdot(l-1)}\right|^{-1/2}}{\sqrt{(\nu + q_{l-1})\pi} \Gamma((\nu + q_{l-1})/2)} \\ &\quad \times \prod_{l=1}^k \left(1 + \frac{\mathbf{z}_l^T \mathbf{R}_{ll\cdot(l-1)}^{-1} \mathbf{z}_l}{\nu + q_{l-1}}\right)^{-(\nu + q_l)/2} \\ &\quad \times \sum_{l_1=0}^\infty \cdots \sum_{l_k=0}^\infty \frac{\Gamma((\nu + p + l_1 + \cdots + l_k)/2)}{\Gamma((\nu + p)/2) l_1! \cdots l_k!} \\ &\quad \times \prod_{m=1}^k \left(\frac{2}{\nu + q_{m-1}}\right)^{l_m/2} \\ &\quad \times \prod_{m=1}^k \frac{\left(\mathbf{z}_m^T \mathbf{R}_{mm\cdot(m-1)}^{-1} \boldsymbol{\xi}_{m\cdot(m-1)}\right)^{l_m}}{\left(1 + \mathbf{z}_m^T \mathbf{R}_{mm\cdot(m-1)}^{-1} \mathbf{z}_m\right)^{l_1 + \cdots + l_m}}, \end{aligned}$$

where

$$\boldsymbol{\xi}_{m\cdot(m-1)} = \boldsymbol{\xi}_m - \mathbf{R}_{(m-1)}^{(m)^T} \mathbf{R}_{(m-1)}^{-1} \boldsymbol{\xi}_{(m-1)}$$

and

$$\delta_m^2 = \boldsymbol{\xi}_{m\cdot(m-1)}^T \mathbf{R}_{mm\cdot(m-1)}^{-1} \boldsymbol{\xi}_{m\cdot(m-1)}.$$

5.2 Miller's Noncentral Multivariate t Distribution

Let $\mathbf{Y}$ have the p-variate normal distribution with mean vector $\boldsymbol{\mu}$ and correlation matrix $\mathbf{R} > 0$. Let $\mathbf{S}$ be distributed independently of $\mathbf{Y}$ according to a ν-variate normal distribution with mean vector $\boldsymbol{\lambda}$ and correlation matrix $m\mathbf{I}_p$. Miller (1968) considered the joint distribution of

$$\mathbf{X}^T = (X_1, X_2, \ldots, X_p) = \left(\frac{Y_1}{\mid \mathbf{S} \mid}, \frac{Y_2}{\mid \mathbf{S} \mid}, \ldots, \frac{Y_p}{\mid \mathbf{S} \mid} \right),$$

which he referred to as the generalized p-dimensional t random vector. Assuming $\mid \mathbf{S} \mid^2$ has the chi-squared distribution, Miller showed that the joint pdf of $\mathbf{X}$ is given by

$$\begin{aligned} f(\mathbf{x}) &= \frac{2^{1-(\nu+p)/2} m^{-\nu/2} \Gamma(\nu+p)}{\Gamma(\nu/2) \pi^{p/2} |\mathbf{R}|^{1/2}} \left(\frac{1}{m} + \mathbf{x}^T \mathbf{R}^{-1} \mathbf{x} \right)^{-(\nu+p)/2} \\ & \quad \times \exp\left\{ -\frac{1}{2} \boldsymbol{\mu}^T \mathbf{R}^{-1} \boldsymbol{\mu} \right\} \exp\left\{ \frac{m \left(\mathbf{x}^T \mathbf{R}^{-1} \boldsymbol{\mu} \right)^2}{4 \left(m \mathbf{x}^T \mathbf{R}^{-1} \mathbf{x} + 1 \right)} \right\} \\ & \quad \times D_{-(\nu+p)} \left[- \frac{\sqrt{m} \mathbf{x}^T \mathbf{R}^{-1} \boldsymbol{\mu}}{\sqrt{m \mathbf{x}^T \mathbf{R}^{-1} \mathbf{x} + 1}} \right], \end{aligned} \tag{5.2}$$

where $D_{-(\nu+p)}(\cdot)$ is the parabolic cylinder function (see, for example, Erdélyi et al., 1953). If $\boldsymbol{\mu} = \mathbf{0}$, then (5.2) reduces to

$$\begin{aligned} f(\mathbf{x}) &= \frac{m^{-\nu/2} \Gamma((\nu+p)/2)}{\Gamma(\nu/2) \pi^{p/2} |\mathbf{R}|^{1/2}} \left(\frac{1}{m} + \mathbf{x}^T \mathbf{R}^{-1} \mathbf{x} \right)^{-(\nu+p)/2} \\ & \quad \times \exp\left\{ -\frac{|\boldsymbol{\lambda}|^2}{2m} \right\} {}_1F_1 \left[\frac{\nu+p}{2}, \frac{\nu}{2}; \frac{|\boldsymbol{\lambda}|^2}{2m \left(m \mathbf{x}^T \mathbf{R}^{-1} \mathbf{x} + 1 \right)} \right], \end{aligned}$$

where ${}_1F_1$ is the confluent hypergeometric function (see, for example, Erdélyi et al., 1953). If both $\boldsymbol{\mu} = \mathbf{0}$ and $\boldsymbol{\lambda} = \mathbf{0}$, then (5.2) reduces to the usual central multivariate t distribution (1.1) with degrees of freedom ν and correlation matrix $\mathbf{R}$. To the best of our knowledge, this interesting distribution given by (5.2) has not been pursued further since its introduction some 35 years ago.

5.3 Stepwise Multivariate t Distribution

Let $\mathbf{Y}$ be a p-variate normal random vector with mean vector $\boldsymbol{\mu}$, common variance σ^2, and correlation matrix $\mathbf{R} > 0$. Let $\nu S^2/\sigma^2$ be a chi-squared random variable with degrees of freedom ν, distributed independently of $\mathbf{Y}$. Then the joint distribution of

$$X_1 = \frac{Y_1}{S}$$

and

$$X_k = \sqrt{\frac{\nu + k - 1}{\nu (1 - \bar{r}_k^2)}} \frac{Y_k - \mathbf{r}_{(k-1)}^T \mathbf{R}_{k-1}^{-1} \mathbf{Y}_{(k-1)}}{S} \Big/ \sqrt{1 + \frac{1}{\nu S^2} \mathbf{Y}_{(k-1)}^T \mathbf{R}_{k-1}^{-1} \mathbf{Y}_{(k-1)}},$$
$$k = 2, \ldots, p, \tag{5.3}$$

where $\bar{r}_k$ denotes the multiple correlation coefficient between Y_k and $(Y_1, \ldots, Y_{k-1})$,

$$\mathbf{Y}_{(k)}^T = (Y_1, Y_2, \ldots, Y_k),$$

$$\mathbf{R}_{(k)} = \begin{pmatrix} 1 & r_{12} & \cdots & r_{kk} \\ r_{21} & 1 & \cdots & r_{2k} \\ \vdots & \vdots & \ddots & \vdots \\ r_{k1} & r_{k2} & \cdots & 1 \end{pmatrix},$$

and

$$\mathbf{r}_{(k)}^T = (r_{1,k+1}, r_{2,k+1}, \ldots, r_{k,k+1}),$$

is known as the stepwise multivariate t distribution. This distribution has applications in linear multiple regression analysis; for instance, suppose that a random sample $Y_1, \ldots, Y_n$, corresponding to some non-random values (z_{1i}, z_{2i}), $i = 1, \ldots, n$, is available. The null hypothesis to be tested is that the slopes of the two simple regression lines, Y on z_1 and Y on z_2, are both zero. Then, the X_1 and X_2 above could correspond to the usual t statistics for testing the two regression coefficients (Steffens, 1974).

Steffens (1969a) studied the distribution of (5.3) for the special case $\mathbf{R} = \mathbf{I}_p$, the $p \times p$ identity matrix. In this case, since $\bar{r}_k = 0$, $\mathbf{r}_{(k-1)} = \mathbf{0}$ and $\mathbf{R}_{(k-1)} = \mathbf{I}_{(k-1)}$, (5.3) reduces to

$$X_k = \sqrt{\frac{\nu + k - 1}{\nu}} \frac{Y_k}{S} \Big/ \sqrt{1 + \frac{1}{\nu S^2} \mathbf{Y}_{(k-1)}^T \mathbf{Y}_{(k-1)}}$$

$$
\begin{aligned}
&= \frac{Y_k}{\sqrt{\nu+k-1}S}\sqrt{\nu+\sum_{l=1}^{k-1}\left(\frac{Y_l}{S}\right)^2} \\
&= \frac{\sqrt{\nu+k-1}Y_k}{\sqrt{\nu S^2+X_1^2+\cdots+X_{k-1}^2}}.
\end{aligned}
$$

If $\boldsymbol{\mu} = \mathbf{0}$, Steffens (1969a) showed that $X_1, X_2, \ldots, X_p$ are independent Student's t random variables with degrees of freedom $\nu, \nu+1, \ldots, \nu+p-1$, respectively. This result also holds for general $\mathbf{R}$ (Siotani, 1976, Corollary 3.1). In the noncentral case $\boldsymbol{\mu} \neq \mathbf{0}$, the X_j's are still independent, but X_1 has the noncentral distribution with degrees of freedom ν and noncentrality parameter μ_1/σ while the X_j's $(j = 2, 3, \ldots, p)$ have the doubly noncentral t distributions with degrees of freedom $\nu + j - 1$ and noncentrality parameter μ_j/σ in the numerator and $(\mu_1/\sigma)^2 + (\mu_2/\sigma)^2 + \cdots + (\mu_{j-1}/\sigma)^2$ in the denominator. Steffens derived the joint pdf of the X_j's in the bivariate case as the double infinite series

$$
\begin{aligned}
f(x_1, x_2) &= \frac{\exp\left\{-\left(\delta_1^2+\delta_2^2\right)/2\right\}}{\pi\sqrt{\nu(\nu+1)}\Gamma(\nu/2)}\sum_{k=0}^{\infty}\sum_{l=0}^{\infty}\frac{\left(\sqrt{2}\delta_1\right)^l}{\nu^{1/(2l)}}\frac{\left(\sqrt{2}\delta_2\right)^k}{(\nu+1)^{1/(2k)}} \\
&\quad\times\Gamma\left(\frac{\nu}{2}+\frac{1}{2k}+\frac{1}{2l}+1\right)\frac{x_1^k}{k!}\left(1+\frac{x_1^2}{\nu}\right)^{-(\nu+k+1)/2} \\
&\quad\times\frac{x_2^l}{l!}\left(1+\frac{x_2^2}{\nu+1}\right)^{-(\nu+k+l+2)/2},
\end{aligned}
$$

where $\delta_j = \mu_j/\sigma$, $j = 1, 2$. For general p and $\mathbf{R}$, Siotani (1976) showed that if $\boldsymbol{\mu} = \mathbf{0}$, then $X_1, X_2, \ldots, X_p$ are still independent Student's t random variables with degrees of freedom $\nu, \nu+1, \ldots, \nu+p-1$, respectively. In the noncentral case, the joint pdf of the X_j in (5.3) generalizes to

$$
\begin{aligned}
f(x_1, \ldots, x_p) &= \exp\left(-\frac{1}{2}\sum_{k=1}^{p}\tau_k^2\right)\prod_{k=1}^{p}\frac{\Gamma((\nu+k)/2)}{\sqrt{\nu+k-1}\Gamma((\nu+k-1)/2)} \\
&\quad\times\prod_{l=1}^{p}\left\{1+\frac{x_k^2}{\nu+k-1}\right\}^{-(\nu+k)/2} \\
&\quad\times\sum_{k_1=0}^{\infty}\cdots\sum_{k_p=0}^{\infty}\frac{\Gamma((\nu+k_1+\cdots+k_p)/2)}{\Gamma((\nu+p)/2)\,k_1!\cdots k_p!} \\
&\quad\times\prod_{l=1}^{p}\left(\frac{2}{\nu+l-1}\right)^{k_l/2}
\end{aligned}
$$

$$\times \prod_{l=1}^{p} \frac{(x_l \tau_l)^{k_l}}{\{1 + x_l^2/(\nu + l - 1)\}^{k_1 + \cdots + k_l}},$$

where the τ_k's are the noncentrality parameters given by

$$\tau_k = \frac{\xi_k^2 - \mathbf{r}_{(k-1)}^T \mathbf{R}_{(k-1)}^{-1} \boldsymbol{\xi}_{(k-1)}}{\sqrt{1 - \bar{r}_k^2}},$$

$$\boldsymbol{\xi}_{(k)} = (\xi_1, \xi_2, \ldots, \xi_k),$$

and $\xi_j = \mu_j / \sigma$.

5.4 Siotani's Noncentral Multivariate t Distribution

In Section 4.5, we discussed a bivariate generalization of the doubly noncentral univariate t distribution given by (4.9). Siotani (1976) provided a multivariate generalization of (4.9) by observing that the pdf of $S^* = S/\sqrt{\nu}$ (where S is a noncentral chi-squared random variable with degrees of freedom ν and noncentrality parameter λ) can be expressed as the Poisson mixture

$$f(s^*) = \sum_{k=0}^{\infty} P_k(\lambda) f_{\nu+2k}(s^*),$$

where

$$P_k(\lambda) = \frac{\exp(-\lambda)\lambda^k}{k!}$$

is the kth probability of the Poisson distribution with parameter λ and

$$f_{\nu+2k}(s^*) = \frac{2\nu^{(\nu+2k)/2}}{2^{(\nu+2k)/2}\Gamma((\nu+2k)/2)} (s^*)^{\nu+2k-1} \exp\left(-\frac{1}{2}\nu s^{*2}\right).$$

He defined $\mathbf{X} = \mathbf{Y}/S^*$ to have the doubly noncentral multivariate t distribution, where $\mathbf{Y}$ is a multivariate normal random vector with mean vector $\boldsymbol{\mu}$, unit variances, and correlation matrix $\mathbf{R}$. The joint pdf of $\mathbf{X}$ is easily obtained as a Poisson mixture of the noncentral pdf (5.1) with $\nu + 2k$ in place of ν in the arguments of gamma functions and in the power of $1 + \mathbf{x}^T\mathbf{R}^{-1}\mathbf{x}/\nu$, that is,

$$f(\mathbf{x}) = \sum_{k=0}^{\infty} P_k(\lambda) f_{\nu+2k}^*(\mathbf{x}), \tag{5.4}$$

where

$$f^*_{\nu+2k}(\mathbf{x}) = \exp\left\{-\frac{1}{2}\boldsymbol{\xi}^T\mathbf{R}\boldsymbol{\xi}\right\} \frac{\Gamma((\nu+2k+p)/2)}{(\nu\pi)^{p/2}\Gamma((\nu+2k)/2)|\mathbf{R}|^{1/2}}$$
$$\times\left\{1+\frac{1}{\nu}\mathbf{x}^T\mathbf{R}^{-1}\mathbf{x}\right\}^{-(\nu+2k+p)/2}$$
$$\times\sum_{l=0}^{\infty}\frac{\Gamma((\nu+2k+p+l)/2)}{l!\Gamma((\nu+2k+p)/2)}\left\{\frac{\sqrt{2}\mathbf{x}^T\mathbf{R}^{-1}\boldsymbol{\xi}}{\sqrt{\nu+\mathbf{x}^T\mathbf{R}^{-1}\mathbf{x}}}\right\}^k.$$

5.5 Arellano-Valle and Bolfarine's Generalized t Distribution

Arellano-Valle and Bolfarine (1995) considered what is being referred to as a generalized t distribution within the class of elliptical distributions. The distribution is defined by

$$\mathbf{X} = \boldsymbol{\mu} + V^{1/2}\mathbf{Y}, \tag{5.5}$$

where V has the inverse gamma distribution given by the pdf

$$f(v) = \frac{(\lambda/2)^{\nu/2}}{\Gamma(\nu/2)}\left(\frac{1}{v}\right)^{(\nu/2)+1}\exp\left(-\frac{\lambda}{2v}\right), \qquad v>0$$

and $\mathbf{Y}$ is distributed independently of V according to a p-dimensional normal distribution with mean vector $\mathbf{0}$ and covariance matrix $\mathbf{R}$. We shall write $\mathbf{X} \sim t_p(\boldsymbol{\mu}, \mathbf{R}; \lambda, \nu)$. When $\lambda = \nu$, this distribution reduces to the usual multivariate t distribution (1.1) with mean vector $\boldsymbol{\mu}$, correlation matrix $\mathbf{R}$, and degrees of freedom ν. For $\mathbf{R} > \mathbf{0}$, the joint pdf of $\mathbf{X} \sim t_p(\boldsymbol{\mu}, \mathbf{R}; \lambda, \nu)$ is

$$f(\mathbf{x}) = \frac{\Gamma((\nu+p)/2)}{(\pi\lambda)^{p/2}\Gamma(\nu/2)|\mathbf{R}|^{1/2}}\left[1+\frac{1}{\lambda}(\mathbf{x}-\boldsymbol{\mu})^T\mathbf{R}^{-1}(\mathbf{x}-\boldsymbol{\mu})\right]^{-(\nu+p)/2}. \tag{5.6}$$

It is easy to observe from (5.5) that

$$E(\mathbf{X}) = \boldsymbol{\mu}, \qquad \nu>1 \tag{5.7}$$

and

$$Var(\mathbf{X}) = \frac{\lambda}{\nu-2}\mathbf{R}, \qquad \nu>2. \tag{5.8}$$

Furthermore, for an $m\times 1$ vector $\boldsymbol{\eta}$ and an $m \times p$ matrix $\mathbf{B}$,

$$\mathbf{Z} = \boldsymbol{\eta} + \mathbf{B}\mathbf{X}$$

$$\begin{aligned} &= (\eta + \mathbf{B}\mu) + V^{1/2}\mathbf{B}\mathbf{Y} \\ &\sim t_m\left(\eta + \mathbf{B}\mu, \mathbf{B}\mathbf{R}\mathbf{B}^T; \lambda, \nu\right) \end{aligned} \tag{5.9}$$

since $\mathbf{BY}$ has the m-dimensional normal distribution with mean vector $\mathbf{0}$ and correlation matrix $\mathbf{BRB}^T$. Now let

$$\mathbf{X} = \begin{pmatrix} \mathbf{X}_1 \\ \mathbf{X}_2 \end{pmatrix}, \tag{5.10}$$

$$\mu = \begin{pmatrix} \mu_1 \\ \mu_2 \end{pmatrix}, \tag{5.11}$$

and

$$\mathbf{R} = \begin{pmatrix} \mathbf{R}_{11} & \mathbf{R}_{12} \\ \mathbf{R}_{21} & \mathbf{R}_{22} \end{pmatrix}, \tag{5.12}$$

where $\mathbf{X}_1$ is $m \times 1$, $\mathbf{R}_{11}$ is $m \times m$ and so on. Taking $\mathbf{B} = [\mathbf{I}_m, \mathbf{0}]$ in (5.9), note that $\mathbf{X}_1 \sim t_m(\mu_1, \mathbf{R}_{11}; \lambda, \nu)$. By symmetry, $\mathbf{X}_2 \sim t_{p-m}(\mu_2, \mathbf{R}_{22}; \lambda, \nu)$. Assuming $\mathbf{R} > \mathbf{0}$, let

$$\begin{aligned} \mu_1(\mathbf{x}_2) &= \mu_1 + \mathbf{R}_{12}\mathbf{R}_{22}^{-1}(\mathbf{x}_2 - \mu_2), \\ \mathbf{R}_{11\cdot 2} &= \mathbf{R}_{11} - \mathbf{R}_{12}\mathbf{R}_{22}^{-1}\mathbf{R}_{21}, \end{aligned}$$

and

$$q(\mathbf{x}_2) = (\mathbf{x}_2 - \mu_2)^T \mathbf{R}_{22}^{-1}(\mathbf{x}_2 - \mu_2).$$

Using the fact that

$$|\mathbf{R}| = |\mathbf{R}_{11\cdot 2}|\,|\mathbf{R}_{22}|$$

and

$$(\mathbf{x} - \mu)^T \mathbf{R}^{-1}(\mathbf{x} - \mu) = (\mathbf{x}_1 - \mu_1)^T \mathbf{R}_{11\cdot 2}^{-1}(\mathbf{x}_2 - \mu_1(\mathbf{x}_2)) + q(\mathbf{x}_2),$$

note that the conditional pdf of $\mathbf{X}_1$ given $\mathbf{X}_2 = \mathbf{x}_2$ is given by

$$\begin{aligned} f(\mathbf{x}_1 \mid \mathbf{x}_2) = &\frac{\Gamma((\nu + p)/2)\{\lambda + q(\mathbf{x}_2)\}^{(\nu+p-m)/2}}{\pi^{m/2}\Gamma((\nu + p - m)/2)|\mathbf{R}|^{1/2}} \Big[\lambda + q(\mathbf{x}_2) \\ &+ (\mathbf{x}_1 - \mu_1(\mathbf{x}_2))^T \mathbf{R}_{11\cdot 2}^{-1}(\mathbf{x}_1 - \mu_1(\mathbf{x}_2))\Big]^{-(\nu+p)/2}. \end{aligned}$$

This means that

$$\mathbf{X}_1 \mid \mathbf{X}_2 = \mathbf{x}_2 \sim t_m\left(\mu_1(\mathbf{x}_2), \mathbf{R}_{11\cdot 2}; \lambda + q(\mathbf{x}_2), \nu + p - m\right). \tag{5.13}$$

Note that when $\lambda = \nu$, $\mathbf{X}_1 \mid \mathbf{X}_2 = \mathbf{x}_2 \sim t_m(\boldsymbol{\mu}_1(\mathbf{x}_2), \mathbf{R}_{11\cdot 2}; \nu + q(\mathbf{x}_2), \nu + p - m)$. Since $q(\mathbf{x}_2) \neq p - m$, this shows that the usual t distribution does not retain its conditional distributions (see Section 1.11). Finally, it follows from (5.13) and (5.7)–(5.8) that

$$E(\mathbf{X}_1 \mid \mathbf{X}_2) = \boldsymbol{\mu}_1 + \mathbf{R}_{11}\mathbf{R}_{22}^{-1}(\mathbf{X}_2 - \boldsymbol{\mu}_2)$$

and

$$Cov(\mathbf{X}_1 \mid \mathbf{X}_2) = \frac{\lambda + (\mathbf{X}_2 - \boldsymbol{\mu}_2)^T \mathbf{R}_{22}^{-1}(\mathbf{X}_2 - \boldsymbol{\mu}_2)}{\nu + p - m - 2} \times \left(\mathbf{R}_{11} - \mathbf{R}_{12}\mathbf{R}_{22}^{-1}\mathbf{R}_{21}\right).$$

Arellano-Valle and Bolfarine (1995) also presented characterizations of the generalized t distribution (5.5) in terms of marginal distributions, conditional distributions, quadratic forms, and within the class of compound normal distributions. Briefly, these characterizations are

- Let $\mathbf{X}$ have the p-variate elliptically symmetric distribution with mean vector $\boldsymbol{\mu}$ and covariance matrix $\mathbf{R}$ (for a definition of an elliptically symmetric distribution see, for example, Fang et al., 1990). Then, any marginal distribution is a generalized t distribution if and only if $\mathbf{X}$ has a generalized t distribution.
- Let $\mathbf{X} = (\mathbf{X}_1^T, \mathbf{X}_2^T)^T$ have the p-variate elliptically symmetric distribution with mean vector $\boldsymbol{\mu}$ and covariance matrix $\mathbf{R}$, where $\mathbf{X}_1$ is $m \times 1$. Then, the conditional distribution of $\mathbf{X}_1$ given $\mathbf{X}_2$ is the generalized m-variate t distribution if and only if the distribution of $\mathbf{X}$ is the generalized p-variate t distribution. The proof of this result, which assumes the existence of a density, is similar to the proof considered in the pioneering paper by Kelker (1970) for the characterization of the multivariate normal distribution.
- Let $\mathbf{X}$ have the p-variate elliptically symmetric distribution with mean vector $\mathbf{0}$ and covariance matrix $\mathbf{I}_p$, and let $\mathbf{A}$ be a symmetric $p \times p$ matrix. Then, $\mathbf{X}^T\mathbf{A}\mathbf{X} \sim (m\lambda/\nu)F_{m,\nu}$ if and only if $\mathbf{X} \sim t_p(\mathbf{0}, \mathbf{I}_p; \lambda, \nu)$, $\mathbf{A}^2 = \mathbf{A}$, and $rank(\mathbf{A}) = m$. This result is proved by utilizing Anderson and Fang's (1987) assertion on the spherical distributions that put zero mass in the origin.

The fourth characterization within the class of compound normal distributions is a consequence of a well known theorem due to Diaconis and Ylvisaker (1979), which asserts that, in the regular exponential family with the natural parameterization, if the posterior expectation is linear, then the prior distribution must be conjugated. It states that if

$X_1, X_2, \ldots$ is an infinite sequence of orthogonally invariant random variables (which means that for each p, $\mathbf{X} = (X_1, \ldots, X_p)^T$ and $\mathbf{\Gamma X}$ are identically distributed, for all $p \times p$ orthogonal matrices $\mathbf{\Gamma}$) such that $X_1 = 0$ with probability zero and

$$Var(X_2 \mid X_1) = b + aX_1^2, \qquad 0 < a < 1, \quad b > 0, \qquad (5.14)$$

then $\mathbf{X}$ is distributed as $t_p(\mathbf{0}, \mathbf{I}_p; b/a, (a+1)/a)$. The converse also holds. Arellano-Valle et al. (1994) pointed out that (5.14) could be extended to yield a location mixture of generalized t distributions as follows. Let $X_1, X_2, \ldots$ be an infinite sequence random variables such that for each p, $\mathbf{X} = (X_1, \ldots, X_p)^T$ and $\mathbf{\Gamma X}$ are identically distributed, for all $p \times p$ orthogonal matrices $\mathbf{\Gamma}$ satisfying $\mathbf{\Gamma 1}_p = \mathbf{1}_p$ (where $\mathbf{1}_p$ is a p-dimensional vector of 1's). Under this assumption there exists random variables M and V such that, conditional on M and V, $X_1, X_2, \ldots$ are independent and normally distributed with mean M and variance V. Actually, M and V can be interpreted as the limits

$$\bar{X}_n = \frac{1}{n} \sum_{i=1}^{n} X_i \quad \rightarrow \quad M$$

and

$$S_n^2 = \frac{1}{n} \sum_{i=1}^{n} \left(X_i - \bar{X}_n\right)^2 \quad \rightarrow \quad V$$

as $n \rightarrow \infty$, where the convergence is with probability 1. Furthermore, if

$$Var\left\{(X_2 - M)^2 \mid X_1, M\right\} = a\left(X_1 - M\right)^2 + b, \qquad 0 < a < 1, \quad b > 0, \quad (5.15)$$

then $\mathbf{X}$ is a location mixture of $t_p(M\mathbf{1}_p, \mathbf{I}_p; b/a, (a+1)/a)$ and, in addition, M and V are independent. Because of the form of the conditions (5.14)–(5.15), these two results are known as the predictivistic characterizations of the generalized t distribution. These results could be extended further to the matrix-variate t distributions (see Section 5.11).

5.6 Fang et al.'s Asymmetric Multivariate t Distribution

Fang et al. (2002) introduced an asymmetric p-variate t distribution with degrees of freedom $(m, m_1, \ldots, m_p)$. Its joint pdf is given by

$$f(\mathbf{x}) = \phi\left(T_m^{-1}\left(T_{m_1}(x_1)\right), \ldots, T_m^{-1}\left(T_{m_p}(x_1)\right); \mathbf{R}\right) \prod_{i=1}^{p} t_{m_i}(x_i),$$

(5.16)

where

$$\phi(y_1, \ldots, y_p; \mathbf{R}) = \frac{\Gamma((m+p)/2)\,\Gamma^{p-1}(m/2)}{\Gamma^p((m+1)/2)\,|\mathbf{R}|^{1/2}} \left(1 + \frac{\mathbf{y}^T \mathbf{R}^{-1} \mathbf{y}}{m}\right)^{-\frac{m+p}{2}} \times \prod_{i=1}^{p} \left(1 + \frac{y_i^2}{m}\right)^{(m+1)/2}.$$

Here, $\mathbf{R}$ denotes the correlation matrix, and t_m and T_m, respectively, denote the pdf and the cdf of the Student's t distribution with degrees of freedom m. Note that the marginals of (5.16) have different degrees of freedom. In the particular case $m_i = m$, $i = 1, \ldots, p$, (5.16) reduces to the usual p-variate t distribution with degrees of freedom m.

5.7 Gupta's Skewed Multivariate t Distribution

In the next four sections (starting with this section), we shall discuss skewed multivariate t distributions – a topic that has received special attention in the last few years, following the introduction of the skewed multivariate normal distribution in the classical paper by Azzalini and Dalla Valle (1996). A careful reader will observe that the possibilities of constructing skewed multivariate t distributions are practically limitless.

A p-variate random vector $\mathbf{Y} = (Y_1, Y_2, \ldots, Y_p)^T$ is said to have the skewed normal distribution if its joint pdf is given by

$$f_{\mathbf{Y}}(\mathbf{y}) = 2\phi_p(\mathbf{y}; \mathbf{\Sigma})\,\Phi\left(\boldsymbol{\alpha}^T \mathbf{y}\right), \qquad \mathbf{y} \in \Re^p, \tag{5.17}$$

where $\mathbf{\Sigma} > \mathbf{0}$ (with $\mathbf{R}$ denoting the corresponding correlation matrix), $\boldsymbol{\alpha} \in \Re^p$, $\phi_p(\mathbf{y}; \mathbf{\Sigma})$ is the p-dimensional normal density with zero means and covariance matrix $\mathbf{\Sigma}$, and $\Phi(\cdot)$ is the cdf of the standard normal distribution. Let W be a chi-squared random variable with degrees of freedom ν, distributed independently of $\mathbf{Y}$. Gupta (2000) defined the joint distribution of

$$X_j = \frac{Y_j}{\sqrt{W/\nu}}, \qquad j = 1, 2, \ldots, p \tag{5.18}$$

as the skewed multivariate t distribution with degrees of freedom ν. The joint pdf of (5.18) is given by

$$f_{\mathbf{X}}(\mathbf{x}; \boldsymbol{\alpha}) = 2 f_\nu(\mathbf{x}) F_{\nu+p}\left(\sqrt{\nu+p}\,\frac{\boldsymbol{\alpha}^T \mathbf{x}}{\sqrt{\nu + \mathbf{x}^T \mathbf{\Sigma}^{-1} \mathbf{x}}}\right), \tag{5.19}$$

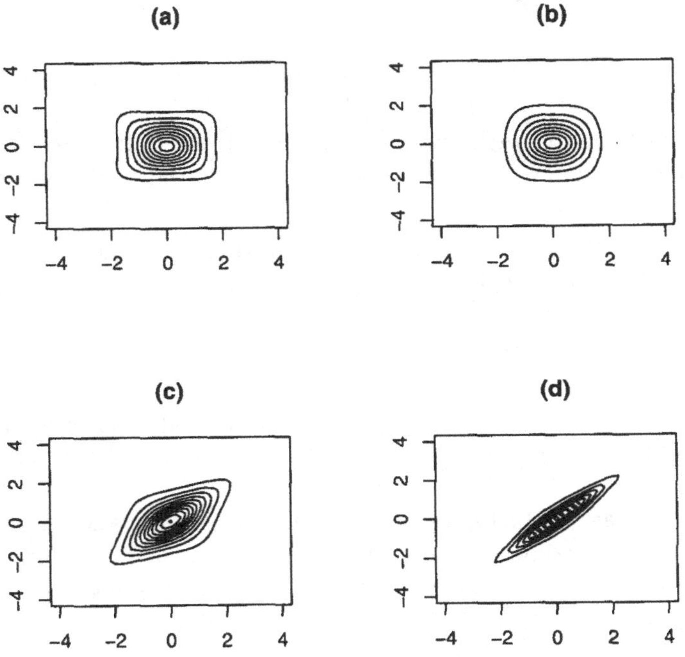

Fig. 5.1. Fang et al.'s asymmetric t pdf (5.16) in the bivariate case (a) $m = 2$, $m_1 = 10$, $m_2 = 10$, and $r_{12} = 0$; (b) $m = 2$, $m_1 = 10$, $m_2 = 2$, and $r_{12} = 0$; (c) $m = 2$, $m_1 = 10$, $m_2 = 10$, and $r_{12} = 0.5$; and (d) $m = 2$, $m_1 = 10$, $m_2 = 10$, and $r_{12} = 0.9$

where $\mathbf{x} \in \Re^p$. Here, f_k and F_k, respectively, denote the joint pdf of the central p-variate t distribution with correlation matrix $\mathbf{R}$ and degrees of freedom k and the cdf of the Student's t distribution with degrees of freedom k. From the definition (5.18) and the joint pdf (5.19), Gupta noted the following properties

- If $\boldsymbol{\alpha} = \mathbf{0}$, then (5.19) reduces to the central p-variate t distribution with correlation matrix $\mathbf{R}$ and degrees of freedom ν.
- The skewed multivariate t distribution approaches the skewed multivariate normal distribution as $\nu \to \infty$, that is,

$$\lim_{\nu \to \infty} f_{\mathbf{X}}(\mathbf{x}; \boldsymbol{\alpha}) = 2\phi_p(\mathbf{x}; \boldsymbol{\Sigma}) \Phi\left(\boldsymbol{\alpha}^T \mathbf{x}\right).$$

- Since Y_j^2 is a chi-squared random variable with degree of freedom 1,

$$X_j^2 = \frac{Y_j^2}{W/\nu} \sim F(1,\nu),$$

 the F distribution with degrees of freedom 1 and ν; furthermore, the joint distribution of $(X_1^2, X_2^2, \ldots, X_p^2)$ is multivariate F with parameters $1, 1, \ldots, 1, \nu + p$.

- Since $\mathbf{y}^T \boldsymbol{\Sigma}^{-1} \mathbf{y}$ is a chi-squared random variable with degrees of freedom p, the quadratic form

$$\mathbf{x}\boldsymbol{\Sigma}^{-1}\mathbf{x} = \frac{\mathbf{y}\boldsymbol{\Sigma}^{-1}\mathbf{y}}{W/\nu} \sim pF_{p,\nu};$$

 note that – as in the case for multivariate normal – the distribution of this quadratic form does not depend on $\boldsymbol{\alpha}$.

The special case of (5.19) for $\boldsymbol{\Sigma} = \mathbf{I}_p$ is called the standard skewed multivariate t distribution. If, in addition, $\nu = 1$, then it is defined as the skewed multivariate Cauchy distribution with the joint pdf

$$\begin{aligned} f_{\mathbf{X}}(\mathbf{x};\boldsymbol{\alpha}) &= 2\pi^{-(p+1)/2}\Gamma\left(\frac{p+1}{2}\right)\left(1+\sum_{j=1}^{p} x_j^2\right)^{-(p+1)/2} \\ &\quad \times F_{p+1}\left(\frac{\sqrt{p+1}\boldsymbol{\alpha}^T\mathbf{x}}{\sqrt{1+\sum_{j=1}^{p} x_j^2}}\right), \qquad \mathbf{x} \in \Re^p. \end{aligned}$$

It should be noted that the above does not belong to the class of elliptically symmetric distributions, whereas the multivariate Cauchy does.

Using results in Azzalini and Dalla Valle (1996) and Gupta and Kollo (2000), the mean vector and the covariance matrix associated with (5.19) are calculated as

$$E(\mathbf{X}) = \sqrt{\frac{2\nu}{\pi}} \frac{\boldsymbol{\Sigma}\boldsymbol{\alpha}}{(\nu-2)\sqrt{1+\boldsymbol{\alpha}^T\boldsymbol{\Sigma}\boldsymbol{\alpha}}}, \quad \nu > 2,$$

and

$$Cov(\mathbf{X}) = \frac{\nu}{(\nu-2)(\nu-4)}\boldsymbol{\Sigma}\left[\mathbf{I}_p - \frac{2(\nu+4)\boldsymbol{\alpha}\boldsymbol{\alpha}^T\boldsymbol{\Sigma}}{\pi(\nu-2)(1+\boldsymbol{\alpha}^T\boldsymbol{\Sigma}\boldsymbol{\alpha})}\right], \quad \nu > 4,$$

respectively. Furthermore, using the definition (5.18), the product moments are easily obtained as

$$\begin{aligned} \mu'_{r_1,r_2,\ldots,r_p} &= E\left[\prod_{j=1}^{p} X_j^{r_j}\right] \\ &= \nu^{r/2} E\left(W^{-r}\right) E\left(\prod_{j=1}^{p} Y_j^{r_j}\right) \\ &= \frac{\sqrt{\nu}}{2} \frac{\Gamma\left((\nu/2) - r\right)}{\Gamma(\nu/2)} E\left(\prod_{j=1}^{p} Y_j^{r_j}\right) \end{aligned}$$

for $r < \nu/2$, where $r = r_1 + r_2 + \cdots + r_p$. If $Y_1, Y_2, \ldots, Y_p$ are mutually independent, then the right-hand side can be easily calculated.

Branco and Dey (2001) noted that the joint pdf (5.19) is a particular case of a general class of skewed multivariate elliptical distributions. Actually, the joint pdf of the general class takes the form

$$f_{\mathbf{X}}(\mathbf{x}) = f_{\nu,\tau}(\mathbf{x}) F_{\nu^*,\tau^*}\left(\boldsymbol{\lambda}^T (\mathbf{x} - \boldsymbol{\mu})\right), \tag{5.20}$$

where $\nu^* = \nu + p$,

$$\tau^* = \tau + (\mathbf{x} - \boldsymbol{\mu})^T \mathbf{R}^{-1} (\mathbf{x} - \boldsymbol{\mu}),$$

$$\boldsymbol{\lambda} = \frac{\boldsymbol{\alpha}^T \mathbf{R}^{-1}}{\sqrt{1 - \boldsymbol{\alpha}^T \mathbf{R}^{-1} \boldsymbol{\alpha}}},$$

$$f_{\nu,\tau}(\mathbf{x}) = \frac{\Gamma\left((\nu + p)/2\right)}{(\pi\lambda)^{p/2} \Gamma(\nu/2) |\mathbf{R}|^{1/2}} \left[1 + \frac{(\mathbf{x} - \boldsymbol{\mu})^T \mathbf{R}^{-1} (\mathbf{x} - \boldsymbol{\mu})}{\nu}\right]^{-\frac{\nu+p}{2}},$$

and

$$F_{\nu^*,\tau^*}(x) = \frac{(\tau^*)^{\nu^*/2} \Gamma\left((\nu^* + 1)/2\right)}{\sqrt{\pi} \Gamma(\nu^*/2)} \int_{-\infty}^{x} \left(\tau^* + y^2\right)^{-(\nu^*+1)/2} dy.$$

Note that $f_{\nu,\tau}(\mathbf{x})$ is the generalized t pdf described in equation (5.6), and that $F_{\nu^*,\tau^*}(x)$ is the cdf of a generalized version of the Student's t distribution. The mean and the variance of the univariate marginals of (5.20) are

$$E(X) = \frac{\alpha \Gamma\left((\nu - 1)/2\right)}{\Gamma(\nu/2)} \sqrt{\frac{\nu}{\pi}}$$

(provided $\nu > 1$) and

$$Var(X) = \frac{\nu}{\nu-2} - \frac{\alpha^2 \nu}{\pi} \left[\frac{((\nu-1)/2)}{\Gamma(\nu/2)} \right]^2$$

(provided $\nu > 2$), respectively.

5.8 Sahu et al.'s Skewed Multivariate t Distribution

Using transformation and conditioning, Sahu et al. (2000) obtained a skewed multivariate t distribution given by the joint pdf

$$\begin{aligned} f(\mathbf{x}) &= 2^m t_{m,\nu}\left(\mathbf{x}; \boldsymbol{\mu}, \mathbf{R} + \mathbf{D}^2\right) T_{m,\nu+m}\left[\sqrt{\frac{\nu+p}{\nu+q(\mathbf{y})}}\right. \\ &\quad \times \left.\left(\mathbf{I} - \mathbf{D}\left(\mathbf{R}+\mathbf{D}^2\right)^{-1}\mathbf{D}\right)^{-1/2} \mathbf{D}\left(\mathbf{R}+\mathbf{D}^2\right)^{-1}\mathbf{y}\right], \end{aligned} \tag{5.21}$$

where $\mathbf{y} = \mathbf{x} - \boldsymbol{\mu}$, $q(\mathbf{y}) = \mathbf{y}^T(\mathbf{R}+\mathbf{D}^2)^{-1}\mathbf{y}$, and $\mathbf{D}$ is a diagonal matrix with the skewness parameters $\delta_1, \ldots, \delta_m$. In (5.21), $t_{m,\nu}(\boldsymbol{\mu}, \boldsymbol{\Omega})$ denotes the usual m-variate t density with mean vector $\boldsymbol{\mu}$, correlation matrix $\boldsymbol{\Omega}$, and degrees of freedom ν. Furthermore, $T_{m,\nu+m}(\cdot)$ denotes the joint cdf of $t_{m,\nu}(\mathbf{0}, \mathbf{I})$. The mean and the variance of this skewed t distribution are given by

$$E(\mathbf{X}) = \boldsymbol{\mu} + \sqrt{\frac{\nu}{\pi}} \frac{\Gamma((\nu-1)/2)}{\Gamma(\nu/2)} \delta$$

and

$$Cov(\mathbf{X}) = \frac{\nu}{\nu-2}\left(\mathbf{R}+\mathbf{D}^2\right) - \frac{\nu}{\pi}\left(\frac{\Gamma((\nu-1)/2)}{\Gamma(\nu/2)}\right)^2 \mathbf{D}^2$$

(provided $\nu > 2$), respectively, where $\boldsymbol{\delta} = (\delta_1, \ldots, \delta_m)^T$. The multivariate skewness measure $\beta_{1,m}$ (Mardia, 1970b) can be calculated in analytic form. The expression does not simplify and involves nonlinear interactions between the degrees of freedom (ν) and the skewness parameter δ when $\mathbf{D} = \delta\mathbf{I}$. However, $\beta_{1,m}$ approaches ± 1 as $\delta \to \pm\infty$. Sahu et al. (2000) discussed an application of this model in Bayesian regression models.

5.9 Azzalini and Capitanio's Skewed Multivariate t Distribution

A slight extension of the skewed normal distribution given in (5.17) is

$$f_{\mathbf{Y}}(\mathbf{y}) = \phi_p(\mathbf{y}-\boldsymbol{\xi};\boldsymbol{\Sigma})\,\Phi\left(\boldsymbol{\alpha}^T\mathbf{W}^{-1}(\mathbf{y}-\boldsymbol{\xi})\right), \qquad (5.22)$$

where $\mathbf{y} \in \Re^p$, $\boldsymbol{\xi} \in \Re^p$, $\mathbf{W} = \mathrm{diag}(\sqrt{\sigma_{11}}, \ldots, \sqrt{\sigma_{pp}})$, and the rest is as defined in (5.17). In the particular case $\boldsymbol{\xi} = \mathbf{0}$, (5.22) reduces to (5.17).

Starting with a random vector $\mathbf{Y}$ having the pdf (5.22) with $\boldsymbol{\xi} = \mathbf{0}$, Azzalini and Capitanio (2002) defined a skewed t variate as the scale mixture

$$\mathbf{X} = \boldsymbol{\xi} + \mathbf{Y}/\sqrt{V}, \qquad (5.23)$$

where νV is distributed independently of $\mathbf{Y}$ according to a chi-squared distribution with degrees of freedom ν. Simple calculations using a preliminary result on Gamma variates show that the joint pdf of $\mathbf{X}$ is

$$f_{\mathbf{X}}(\mathbf{x}) = 2 f_\nu(\mathbf{x})\, F_{\nu+p}\left(\boldsymbol{\alpha}^T\mathbf{W}^{-1}(\mathbf{x}-\boldsymbol{\xi})\sqrt{\frac{\nu+p}{Q+\nu}}\right),$$

where $Q = (\mathbf{x}-\boldsymbol{\xi})^T\mathbf{R}^{-1}(\mathbf{x}-\boldsymbol{\xi})$ and f_k, F_k are as defined in (5.19). Note that this pdf coincides with that of Branco and Dey (2001) given in (5.20). In the standard case $\boldsymbol{\xi} = \mathbf{0}$ and $\boldsymbol{\Sigma} = \mathbf{R}$, the joint cdf of $\mathbf{X}$ can be represented as

$$F_{\mathbf{X}}(\mathbf{x}) = 2\Pr\left(-U_0/\sqrt{V} \leq 0, \mathbf{U}/\sqrt{V} \leq \mathbf{x}\right), \qquad (5.24)$$

where $(U_0, U^T)^T$ has the $(p+1)$-dimensional normal distribution with zero means and covariance matrix given by

$$\boldsymbol{\Sigma}^* = \begin{pmatrix} 1 & \boldsymbol{\delta}^T \\ \boldsymbol{\delta} & \mathbf{R} \end{pmatrix},$$

where

$$\boldsymbol{\delta} = \frac{\mathbf{R}\boldsymbol{\alpha}}{\sqrt{1+\boldsymbol{\alpha}^T\mathbf{R}\boldsymbol{\alpha}}}.$$

The representation (5.24) can also be written in terms of a $(p+1)$-dimensional t distribution.

In the case $\boldsymbol{\xi} = \mathbf{0}$, simple expressions for the moments of $\mathbf{X}$ can be obtained. Defining

$$\boldsymbol{\mu} = \sqrt{\frac{\nu}{\pi}}\frac{\Gamma((\nu-1)/2)}{\Gamma(\nu/2)}\boldsymbol{\delta},$$

and provided that $\nu > 1$, one obtains

$$
\begin{aligned}
E(X_k) &= w_{kk}\mu_k, \\
E(X_k^2) &= \frac{\nu}{\nu-2} w_{kk}^2, \\
E(X_k^3) &= \frac{\nu\left(3-\delta^T\delta\right)}{\nu-3} w_{kk}^3 \mu_k, \\
E(X_k^4) &= \frac{3\nu^2}{(\nu-2)(\nu-4)} w_{kk}^4, \\
E(\mathbf{X}) &= \mathbf{W}\boldsymbol{\mu}, \\
E(\mathbf{X}\mathbf{X}^T) &= \frac{\nu}{\nu-2}\boldsymbol{\Sigma}, \quad \nu > 2, \\
\text{Skewness}(X_k) &= \mu_k \left\{ \frac{\nu\left(3-\delta^T\delta\right)}{\nu-3} - \frac{3\nu}{\nu-2} + 2\boldsymbol{\mu}^T\boldsymbol{\mu} \right\} \\
&\quad \Big/ \left\{ \frac{\nu}{\nu-2} - \boldsymbol{\mu}^T\boldsymbol{\mu} \right\}^{3/2}, \quad \nu > 3,
\end{aligned}
$$

and

$$
\begin{aligned}
\text{Kurtosis}(X_k) &= \left\{ \frac{3\nu^2}{(\nu-2)(\nu-4)} - \frac{4\nu\boldsymbol{\mu}^T\boldsymbol{\mu}\left(3-\delta^T\delta\right)}{\nu-3} \right. \\
&\quad \left. + \frac{6\nu\boldsymbol{\mu}^T\boldsymbol{\mu}}{\nu-2} - 3\left(\delta^T\delta\right)^2 \right\} \Big/ \left\{ \frac{\nu}{\nu-2} - \boldsymbol{\mu}^T\boldsymbol{\mu} \right\}^2, \\
&\quad \nu > 4.
\end{aligned}
$$

Properties concerning linear and quadratic forms of $\mathbf{X}$ can also be derived. For example, if $\mathbf{a} \in \Re^m$ and $\mathbf{A}$ is a $m \times p$ constant matrix of rank m, then the affine transformation $\mathbf{a} + \mathbf{A}\mathbf{X}$ will also follow the skewed t distribution given by (5.23) with the parameters $\boldsymbol{\xi}$, $\boldsymbol{\Sigma}$, and $\boldsymbol{\alpha}$ replaced by $\mathbf{a} + \mathbf{A}\boldsymbol{\xi}$, $\boldsymbol{\Sigma}'$, and $\boldsymbol{\alpha}'$, respectively (the degrees of freedom ν remains unchanged), where

$$
\boldsymbol{\Sigma}' = \mathbf{A}\boldsymbol{\Sigma}\mathbf{A}^T,
$$

$$\boldsymbol{\alpha}' = \frac{\mathbf{W}'\left(\boldsymbol{\Sigma}'\right)^{-1}\mathbf{B}^T\boldsymbol{\alpha}}{\sqrt{1+\boldsymbol{\alpha}^T\left(\boldsymbol{\Sigma}''-\mathbf{B}\left(\boldsymbol{\Sigma}'\right)^{-1}\mathbf{B}^T\right)\boldsymbol{\alpha}}},$$

$\mathbf{W}' = \sqrt{\boldsymbol{\Sigma}'}$, $\mathbf{B} = \mathbf{W}^{-1}\boldsymbol{\Sigma}\mathbf{A}^T$, and $\boldsymbol{\Sigma}''$ is given by $\boldsymbol{\Sigma} = \mathbf{W}\boldsymbol{\Sigma}''\mathbf{W}$. Also, for appropriate choices of $\mathbf{B}$ (a symmetric $p \times p$ matrix), the quadratic form $Q = (\mathbf{X}-\boldsymbol{\xi})^T\mathbf{B}(\mathbf{X}-\boldsymbol{\xi})$ can be shown to have the $fF_{f,\nu}$ distribution for some degrees of freedom f. For details see Azzalini and Capitanio (1999) and and Capitanio et al. (2002).

A further extension of (5.17) examined independently by Arnold and Beaver (2000) and Capitanio et al. (2002) is of the form

$$f_{\mathbf{Y}}(\mathbf{y}) = \phi_p(\mathbf{y}-\boldsymbol{\xi};\boldsymbol{\Sigma})\,\Phi\left(\alpha_0+\boldsymbol{\alpha}^T\mathbf{W}^{-1}(\mathbf{y}-\boldsymbol{\xi})\right)\Big/\Phi(\tau), \quad (5.25)$$

where $\mathbf{y} \in \Re^p$, $\tau \in \Re$, $\alpha_0 = \tau/\sqrt{1-\boldsymbol{\delta}^T\mathbf{R}^{-1}\boldsymbol{\delta}}$, and the rest are as defined in (5.22). In the particular case $\tau = 0$, (5.25) reduces to (5.22). Taking $\mathbf{Y}$ in (5.23) to have the pdf (5.25) with $\boldsymbol{\xi} = \mathbf{0}$, one obtains an extended skewed t distribution for $\mathbf{X}$. The corresponding joint pdf for $\mathbf{X}$ is quite complicated, but the joint cdf can be represented as

$$F(\mathbf{x}) = \Pr\left(-(U_0+\tau)/\sqrt{V} \le 0, \mathbf{U}/\sqrt{V} \le \mathbf{x}\right)\Big/\Phi(\tau)$$

(compare with (5.24)). Moreover, for the particular case $\boldsymbol{\xi} = \mathbf{0}$, the first- and second-order moments are

$$E(\mathbf{X}) = E\left(1/\sqrt{V}\right)\eta_1(\tau)\mathbf{W}\boldsymbol{\delta}$$

(provided $\nu > 1$) and

$$E\left(\mathbf{X}\mathbf{X}^T\right) = \frac{\nu}{\nu-2}\left[\boldsymbol{\Sigma}+\left\{\eta_2(\tau)+\eta_1^2(\tau)\right\}\mathbf{W}\boldsymbol{\delta}\left(\mathbf{W}\boldsymbol{\delta}^T\right)^T\right]$$

(provided $\nu > 2$), respectively, where

$$\eta_k(x) = \frac{d^k}{dx^k}\log\{2\Phi(x)\}$$

for $k = 0, 1, 2, \ldots$.

5.10 Jones' Skewed Multivariate t Distribution

The univariate Student's t distribution has the pdf

$$\frac{\Gamma((\nu+1)/2)}{\sqrt{\nu\pi}\Gamma(\nu/2)}\left\{1+\frac{x^2}{\nu}\right\}^{-(\nu+1)/2}. \quad (5.26)$$

By replacing (5.26) with the skewed univariate t pdf (4.38) in a multivariate distribution, Jones (2002c) introduced a new skewed multivariate t distribution that we shall describe in this section. Let $\mathbf{X}$ be a p-variate random vector having the standard multivariate t distribution with the joint pdf given by

$$\frac{\Gamma((\nu+p)/2)}{\sqrt{\nu\pi}\Gamma(\nu/2)}\left\{1+\frac{\mathbf{x}^T\mathbf{x}}{\nu}\right\}^{-(\nu+p)/2} \tag{5.27}$$

The univariate marginals of this are (5.26). Multiplying (5.27) by (4.38) and dividing by (5.26) yields Jones' (2002c) skewed multivariate t distribution. The corresponding joint pdf is

$$\begin{aligned}&\frac{2^{1-a-c}\Gamma((\nu+p)/2)\Gamma(a+c)}{(\nu\pi)^{(p-1)/2}\sqrt{a+c}\Gamma((\nu+1)/2)\Gamma(a)\Gamma(c)}\\&\times\left\{1+\frac{x_1^2}{\nu}\right\}^{(\nu+1)/2}\left\{1+\frac{x_1}{\sqrt{a+c+x_1^2}}\right\}^{a+1/2}\\&\times\left\{1-\frac{x_1}{\sqrt{a+c+x_1^2}}\right\}^{c+1/2}\left\{1+\frac{\mathbf{x}^T\mathbf{x}}{\nu}\right\}^{-(\nu+p)/2}.\end{aligned} \tag{5.28}$$

This reduces to (5.27) for $a = c = \nu/2$. In the bivariate case, (5.28) is a distribution with (i) a skewed t marginal with parameters a and c in the x_1 direction; (ii) conditional distributions of $X_2 \mid X_1$ that match those of the bivariate t distribution being t distributions on $\nu+1$ degrees of freedom scaled by a factor of $\sqrt{(\nu+x_1^2)/(\nu+1)}$; and (iii) a diagonal correlation matrix. Another new multivariate distribution can be obtained by replacing (5.26) by the pdf of the Gumbel distribution: $\exp(-x_1-\exp(-x_1))$. This results in the joint pdf

$$\begin{aligned}&\frac{\Gamma((\nu+p)/2)}{(\nu\pi)^{(p-1)/2}\Gamma((\nu+1)/2)}\exp\{-x_1-\exp(-x_1)\}\\&\times\left(1+\frac{x_1^2}{\nu}\right)^{(\nu+1)/2}\left\{1+\frac{\mathbf{x}^T\mathbf{x}}{\nu}\right\}^{-(\nu+p)/2}.\end{aligned} \tag{5.29}$$

With respect to the correlation structure, this pdf has much in common with (5.28). But the conditional distribution of X_1 given $X_2, \ldots, X_p$ and the marginals are different.

Jones (2001a) noted that, if Y has the beta distribution with parameters a and c, then $X = \sqrt{a+c}Y/\sqrt{1-Y^2}$ has the skewed univariate t distribution given in (4.38). Jones (2002c) observed a similar relation-

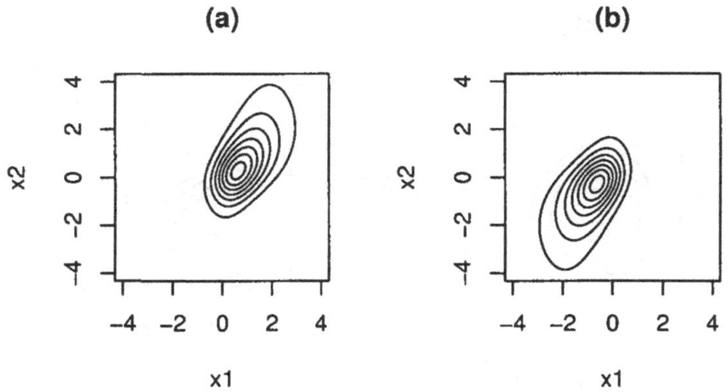

Fig. 5.2. Jones' skewed multivariate t pdf (5.28) for $p = 2$ and (a) $a = 6$, $\nu = 3$, and $c = 2$; and (b) $a = 2$, $\nu = 3$, and $c = 6$

ship between the joint beta pdf

$$\frac{2^{1-a-c}\Gamma(a+c)\Gamma(b)}{\pi\Gamma(a)\Gamma(b-1/2)\Gamma(c)}(1+y_1)^{a-b}(1-y_1)^{c-b}\left(1-y_1^2-y_2^2\right)^{b-1},$$
$$a>0, \quad b>1/2, \quad c>0$$

and the skewed multivariate t distribution given in (5.28) when $p = 2$ and $b = \nu/2 + 1$; namely, if (Y_1, Y_2) have the former distribution, then

$$(X_1, X_2) = \left(\frac{\sqrt{Y_1\sqrt{a+c}}}{\sqrt{1-Y_1^2}}, \frac{\sqrt{Y_2\sqrt{\nu+Y_1^2(a+c-\nu)}}}{\sqrt{1-Y_1^2-Y_2^2}\sqrt{1-Y_1^2}}\right)$$

has the distribution (5.28) for $p = 2$.

In the univariate case, F and skewed t (equation (4.38)) distributions are linked in two ways that produce identical results: (i) A random

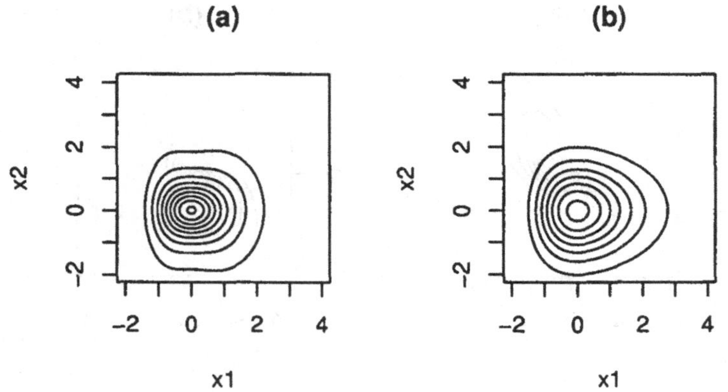

Fig. 5.3. Jones' skewed multivariate t pdf (5.29) for $p = 2$ and (a) $\nu = 1$; and (b) $\nu = 20$

variable with any one distribution can be obtained by transforming a random variable from the other; (ii) a random variable with each distribution can be written as a function of two independent chi-squared random variables. If $W_i \sim \chi^2_{2\nu_i}$, $F_i \sim F_{2\nu_i,\nu_0}$, and T_i is a random variable with the pdf (4.38), then

$$T_i = \frac{\sqrt{w_i}}{2}\left(\sqrt{\frac{\nu_i}{\nu_0}}\sqrt{F_i} - \sqrt{\frac{\nu_0}{\nu_i}}\frac{1}{\sqrt{F_i}}\right), \tag{5.30}$$

$$F_i = \frac{\nu_0}{w_i\nu_i}\left(T_i + \sqrt{w_i + T_i^2}\right)^2,$$

$$T_i = \frac{\sqrt{w_i}}{2}\left(\sqrt{\frac{W_i}{W_0}} - \sqrt{\frac{W_0}{W_i}}\right),$$

and

$$F_i = \frac{\nu_0}{\nu_i} \frac{W_i}{W_0},$$

where $w_i = \nu_0 + \nu_i$. By extending this relationship between the univariate F and the skewed univariate t, Jones (2001b) introduced another skewed multivariate t distribution. It is known (see, for example, Johnson and Kotz, 1972, Chapter 40, and Hutchinson and Lai, 1990, Section 6.3) that the joint pdf of the random variables F_i, $i = 1, \ldots, p$ is

$$f(f_1, \ldots, f_p) = \frac{\Gamma(n)}{\Gamma(\nu_0) \cdots \Gamma(\nu_p)} \prod_{k=0}^{p} \frac{\nu_k^{\nu_k} \prod_{j=1}^{p} f_j^{\nu_j - 1}}{\left(\nu_0 + \sum_{j=1}^{p} \nu_j f_j\right)^n}, \quad f_1 > 0, \ldots, f_p > 0, \tag{5.31}$$

where $n = \nu_0 + \cdots + \nu_p$. Applying the transformation (5.30) to (5.31), Jones (2001b) obtained the joint pdf of T_i as

$$\begin{aligned} f(t_1, \ldots, t_p) = {} & \frac{\Gamma(n)}{\Gamma(\nu_0) \cdots \Gamma(\nu_p)} \prod_{k=0}^{p} \left\{ \frac{2\left(t_k + \sqrt{w_k + t_k^2}\right)^{2\nu_k}}{w_k^{\nu_k} \sqrt{w_k + t_k^2}} \right\} \\ & \Bigg/ \left\{ 1 + \sum_{j=1}^{p} \frac{\left(t_j + \sqrt{w_j + t_j^2}\right)^2}{w_j} \right\}^n, \\ & t_1 \in \Re, \ldots, t_p \in \Re. \end{aligned} \tag{5.32}$$

The univariate marginals of this pdf take the form of (4.38). The conditional pdf of T_i given any subset $T_{i_1}, \ldots, T_{i_{p_2}}$ of the other variables, $p_2 < p$, is proportional to

$$\frac{\left(t_i + \sqrt{w_i + t_i^2}\right)^{2\nu_i}}{\sqrt{w_i + t_i^2} \left\{ 1 + K^{-1} \left(t_i + \sqrt{w_i + t_i^2}\right)^2 \right\}^{w_1 + \nu_{i_1} + \cdots + \nu_{i_{p_2}}}},$$

where

$$K = w_i \left\{ 1 + \sum_{l=1}^{p_2} \frac{1}{w_l} \left(t_{i_l} + \sqrt{w_l + t_{i_l}^2}\right)^2 \right\}.$$

The regression of T_i given $T_{i_1}, \ldots, T_{i_{p_2}}$ takes the nonlinear form

$$E\left(T_i \mid T_{i_1}, \ldots, T_{i_{p_2}}\right) = \frac{\Gamma(\psi - 1/2)\, \Gamma(\nu_i - 1/2)}{2\Gamma(\psi)\, \Gamma(\nu_i)} \left\{ \frac{(\nu_i - 1/2)\sqrt{\psi K}}{\sqrt{\nu_0}} \right.$$

$$-\frac{\sqrt{\nu_0} w_i (\psi - 1/2)}{\sqrt{\psi K}}\Bigg\},$$

where $\psi = \nu_0 + \nu_{i_1} + \cdots + \nu_{i_{p_2}}$. Note that the corresponding relation for the multivariate F distribution in (5.31) is linear. If $T_1, \ldots, T_m$ denote any m of the p T_i's (with their degrees of freedom correspondingly renumbered as $\nu_1, \ldots, \nu_m$ along with ν_0), then the product moment of $T_1, \ldots, T_m$ is

$$\begin{aligned} E\left(\prod_{i=1}^{m} T_i^{\lambda_i}\right) &= \frac{w_1^{\lambda_1/2} \cdots w_m^{\lambda_m/2}}{2^{\lambda_1+\cdots+\lambda_m}\Gamma(\nu_1)\cdots\Gamma(\nu_m)} \prod_{i=1}^{p} \Gamma\left(\frac{\lambda_i}{2} - j_i + \nu_i\right) \\ &\quad \times \sum_{j_1=0}^{\lambda_1} \cdots \sum_{j_m=0}^{\lambda_m} (-1)^{j_1+\cdots+j_m} \binom{\lambda_1}{j_1} \cdots \binom{\lambda_m}{j_m} \\ &\quad \times \Gamma\left(\sum_{i=1}^{m} j_i - \frac{\sum_{i=1}^{m} \lambda_i}{2} + \nu_0\right), \end{aligned}$$

provided that $\nu_i > \lambda_i/2$, $i = 1, \ldots, m$ and $\nu_0 > (\lambda_1 + \cdots + \lambda_m)/2$. In particular, the variances and the covariances are given by

$$\begin{aligned} Var(T_i) &= \frac{w_i}{4}\Bigg[(\nu_i - \nu_0)^2 \Bigg\{\frac{1}{(\nu_0 - 1)(\nu_i - 1)} \\ &\quad - \left(\frac{\Gamma(\nu_0 - 1/2)\Gamma(\nu_i - 1/2)}{\Gamma(\nu_0)\Gamma(\nu_i)}\right)^2\Bigg\} \\ &\quad + \frac{w_i - 2}{(\nu_0 - 1)(\nu_i - 1)}\Bigg] \end{aligned}$$

(provided $\nu_0 > 1$ and $\nu_i > 1$) and

$$\begin{aligned} Cov(T_i, T_j) &= \frac{\sqrt{w_i w_j}\Gamma(\nu_i - 1/2)\Gamma(\nu_j - 1/2)}{4\Gamma(\nu_i)\Gamma(\nu_j)} \Bigg\{1 + \nu_0 - \nu_i - \nu_j \\ &\quad + \frac{(\nu_i - 1/2)(\nu_j - 1/2)}{(\nu_0 - 1/2)} \\ &\quad - (\nu_i - \nu_0)(\nu_j - \nu_0)\left(\frac{\Gamma(\nu_0 - 1/2)}{\Gamma(\nu_0)}\right)^2\Bigg\} \end{aligned}$$

(provided $\nu_i > 1/2$, $\nu_j > 1/2$ and $\nu_0 > 1$), respectively. In the particular case $\nu_0 = \cdots = \nu_p = \nu/2$, (5.32) reduces to

$$f(t_1, \ldots, t_p) = \frac{\Gamma((p+1)\nu/2)}{\Gamma^{p+1}(\nu/2)} \prod_{k=1}^{p} \left\{ \frac{\left(t_k + \sqrt{\nu + t_k^2}\right)^{\nu}}{\sqrt{\nu + t_k^2}} \right\} \Bigg/ \left\{ \nu + \sum_{j=1}^{p} \left(t_j + \sqrt{\nu + t_j^2}\right)^2 \right\}^{(p+1)\nu/2},$$
$$t_1 \in \Re, \ldots, t_p \in \Re. \quad (5.33)$$

Jones (2001b) referred to this distribution as the symmetric multivariate t distribution. Note that all of the marginals of (5.33) have the Student's t distribution with degrees of freedom ν. The correlation between any two T_i's in (5.33) takes the simple form

$$\rho = \frac{2\nu - 3}{8} \left(\frac{\Gamma((\nu - 1)/2)}{\Gamma(\nu/2)} \right)^2$$

provided that $\nu > 2$.

The limiting form of (5.32) as $\nu_0 \to \infty$ and $\nu_i > 1$ remains fixed, $i = 1, \ldots, p$ can be shown to be

$$\left(\prod_{i=1}^{p} \sigma_i \right) g(\mu_1 + \sigma_1 t_1, \ldots, \mu_p + \sigma_p t_p),$$

where

$$g(t_1, \ldots, t_p) = 2^p \prod_{i=1}^{p} \frac{t_i^{-(2\nu_i+1)}}{\Gamma(\nu_i)} \exp\left(-\frac{1}{t_i^2} \right),$$

$$\mu_i = \frac{\Gamma(\nu_i - 1/2)}{\Gamma(\nu_i)},$$

and

$$\sigma_i = \sqrt{\frac{1}{\nu_i - 1} - \frac{\Gamma^2(\nu_i - 1/2)}{\Gamma^2(\nu_i)}}.$$

Note that μ_i and σ_i are the mean and the standard deviation of $\sqrt{2/\chi^2_{2\nu_i}}$ distribution. When ν_0 remains fixed but $\nu_1, \ldots, \nu_p \to \infty$, the marginals of (5.32) tend to $\sqrt{2/\chi^2_{2\nu_0}}$ distribution, but the correlations between the T_i's tend to 1 and the joint distribution becomes degenerate. When all $\nu_0, \nu_1, \ldots, \nu_p \to \infty$, all of the marginals tend to the normal distribution – but the form of the limiting joint distribution will depend on the

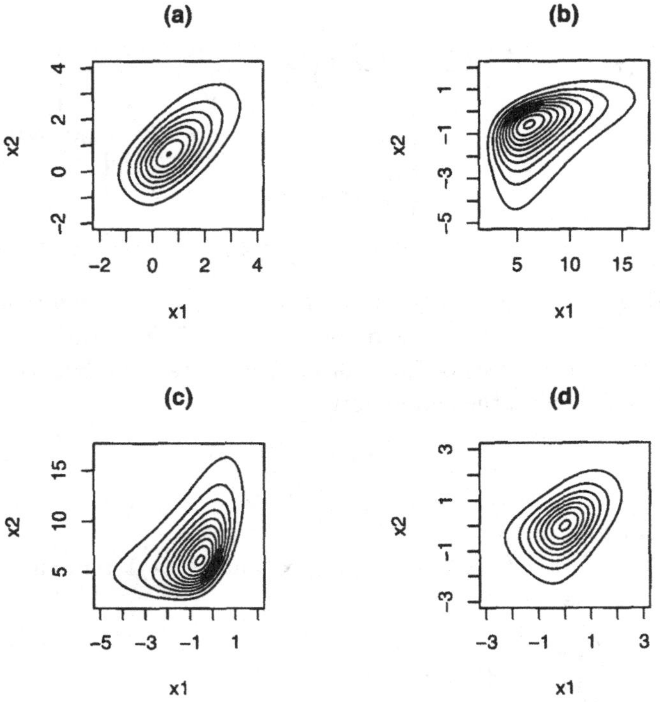

Fig. 5.4. Jones' skewed multivariate t pdf (5.32) for $p = 2$ and (a) $\nu_0 = 2$, $\nu_1 = 4$, and $\nu_2 = 4$; (b) $\nu_0 = 2$, $\nu_1 = 20$, and $\nu_2 = 1$; (c) $\nu_0 = 2$, $\nu_1 = 1$, and $\nu_2 = 20$; and (d) $\nu_0 = \nu_1 = \nu_2 = 2$

specific relationships between the ν's. The limit of (5.33) as $\nu \to \infty$ is the multivariate normal distribution with zero means, unit variances, and an intraclass correlation structure with correlation 1/2.

5.11 Matrix-Variate t Distribution

The matrix-variate t distribution, motivated by applications in Bayesian inference, is the product of James Dickey's research in the mid-1960s. We need the following terminology to discuss its mathematical properties. Let $\boldsymbol{\mu}$ be a $p \times q$ constant matrix, let $\mathbf{R} > 0$ be a $p \times p$ matrix, and let

$\mathbf{Q} > 0$ be a $q \times q$ matrix. For $m > p + q - 1$ define

$$k(m,p,q) = \pi^{pq/2} \frac{\Gamma_q((m-p)/2)}{\Gamma_q(m/2)}, \tag{5.34}$$

where

$$\Gamma_r(x) = \pi^{r(r-1)/4} \Gamma(x) \Gamma\left(x - \frac{1}{2}\right) \cdots \Gamma\left(x - \frac{r}{2} + \frac{1}{2}\right)$$

is the generalized gamma function. Furthermore, for real or complex constants $a_1, \ldots, a_p$ and $b_1, \ldots, b_q$ and for random matrices $\mathbf{S}$ and $\mathbf{T}$, define the general hypergeometric functions (see Constantine, 1963)

$$\begin{aligned} & {}_pF_q\left(a_1, \ldots, a_p; b_1, \ldots, b_q; \mathbf{S}\right) \\ = & \sum_{k=0}^{\infty} \sum_{\kappa} \frac{\left(a_1\right)_\kappa \cdots \left(a_p\right)_\kappa}{\left(b_1\right)_\kappa \cdots \left(b_q\right)_\kappa} \frac{C_\kappa\left(\mathbf{S}\right)}{k!} \end{aligned} \tag{5.35}$$

and

$$\begin{aligned} & {}_pF_q\left(a_1, \ldots, a_p; b_1, \ldots, b_q; \mathbf{S}, \mathbf{T}\right) \\ = & \sum_{k=0}^{\infty} \sum_{\kappa} \frac{\left(a_1\right)_\kappa \cdots \left(a_p\right)_\kappa}{\left(b_1\right)_\kappa \cdots \left(b_q\right)_\kappa} \frac{C_\kappa\left(\mathbf{S}\right) C_\kappa\left(\mathbf{T}\right)}{k!}, \end{aligned} \tag{5.36}$$

where $\kappa = \{k_1, \ldots, k_m\}$, $k_1 \geq k_2 \geq \cdots \geq k_m \geq 0$, $k_1 + k_2 + \cdots + k_m = k$,

$$(x)_\kappa = \frac{\Gamma_m(x, \kappa)}{\Gamma_m(x)},$$

$$\Gamma_r(x, \kappa) = \pi^{\frac{r(r-1)}{4}} \Gamma\left(x + k_1\right) \Gamma\left(x + k_2 - \frac{1}{2}\right) \cdots \Gamma\left(x + k_r - \frac{r-1}{2}\right),$$

and $C_\kappa(\mathbf{S})$ and $C_\kappa(\mathbf{T})$ are symmetric homogeneous polynomials of degree k in the latent roots of $\mathbf{S}$ and $\mathbf{T}$, respectively.

A $p \times q$ random matrix $\mathbf{X}$ is said to have the matrix-variate t distribution with parameters $\boldsymbol{\mu}$, $\mathbf{R}$, $\mathbf{Q}$, and m if its joint pdf is

$$\begin{aligned} f(\mathbf{X}) = & \frac{1}{k(m,p,q)} |\mathbf{Q}|^{(m-p)/2} |\mathbf{R}|^{-q/2} \\ & \times \left|\mathbf{Q} + (\mathbf{X} - \boldsymbol{\mu})^T \mathbf{R}^{-1} (\mathbf{X} - \boldsymbol{\mu})\right|^{-m/2} \end{aligned} \tag{5.37}$$

(Dickey, 1966a, 1967b). If $\boldsymbol{\mu} = \mathbf{0}$, then we say that $\mathbf{X}$ has the central matrix-variate t distribution with parameters $\mathbf{R}$, $\mathbf{Q}$, and m. Otherwise, we refer to the distribution as a noncentral matrix-variate t. The usual multivariate t distribution (1.1) is the special case of (5.37) for $p = 1$

(single row) or $q = 1$ (single column). It is also known that the particular case of (5.37) for $\boldsymbol{\mu} = \mathbf{0}$ and $\mathbf{R} = \mathbf{I}_p$ is a mixture of the normal density with zero means and covariance matrix $\mathbf{I}_p \otimes \mathbf{V}$ – in the $q \times q$ positive definite scale matrix $\mathbf{V}$. Densities of the form (5.37) appear in the frequentist approach to normal regression as the distribution of the Studentized error, both the error in the least squares estimate of the coefficients matrix and the error in the corresponding predictor of a future data array (Cornish, 1954; Kshirsagar, 1961; Kiefer and Schwartz, 1965). In the Bayesian conjugate-prior and diffuse-prior analyses for the same sampling models, it arises as the marginal prior or posterior distribution of the unknown coefficients matrix, and also as the predictive distribution of a future data array (Geisser and Cornfield, 1963; Ando and Kaufman, 1965; Geisser, 1965; Dickey, 1967b, Section 4; Zellner, 1971, Chapter 8; Press, 1972, Section 8.6). More recently, Van Dijk (1985, 1986) discussed applications of (5.37) in the linear simultaneous equation (SEM) model, which is one of the best-known models in econometrics. The SEM model is used in several areas, for instance, in microeconomic modeling for the description of the operation of a market for a particular economic commodity and in macroeconomic modeling for the description of the interrelations between a large number of macroeconomic variables.

If $\mathbf{X}$ has the central matrix-variate distribution with parameters $\mathbf{R}$, $\mathbf{Q}$ and m, then it can be represented in numerous ways, as described by Dickey (1967b) and Dawid (1981). The following results (due to Dickey, 1967b, and Dickey et al., 1986) concern the conditional and the marginal distributions of $\mathbf{X}$

- If $\mathbf{X} = (\mathbf{X}_1, \mathbf{X}_2)^T$, then the conditional distribution of $\mathbf{X}_1$, given $\mathbf{X}_2$, is the matrix-variate t with parameters $-\mathbf{R}_{11}\mathbf{R}_{12}^{-1}\mathbf{X}_2$, $\mathbf{R}_{11}^{-1}$, $\mathbf{Q} + \mathbf{X}_2^T\mathbf{R}_{22}^{-1}\mathbf{X}_2$, and m.
- If $\mathbf{X} = (\mathbf{X}_1, \mathbf{X}_2)$, then the conditional distribution of $\mathbf{X}_1$, given $\mathbf{X}_2$, is a matrix-variate t with parameters $\mathbf{X}_2\mathbf{Q}_{22}^{-1}\mathbf{Q}_{21}$, $(\mathbf{R} + \mathbf{X}_2\mathbf{Q}_{22}^{-1}\mathbf{X}_2^T)^{-1}$, $\mathbf{Q}_{11} - \mathbf{Q}_{12}\mathbf{Q}_{22}^{-1}\mathbf{Q}_{21}$, and m.
- If $\mathbf{X} = (\mathbf{X}_1, \mathbf{X}_2)^T$, where $\mathbf{X}_i$ is $p_i \times q$, then the marginal distribution of $\mathbf{X}_2$ is a central matrix-variate t with parameters $\mathbf{R}_{22}^{-1}$, $\mathbf{Q}$ and $m - p_1$. In the particular case $\mathbf{X}^T = (\mathbf{x}_1, \ldots, \mathbf{x}_p)$, each row $\mathbf{x}_i^T$ has the central multivariate t distribution with degrees of freedom $m - p - q + 1$ and correlation matrix $r_{ii}\mathbf{Q}/(m - p - q + 1)$. A consequence of this is that the density (5.37) of $\mathbf{X}$ can be written as the product of conditional

multivariate t distributions of the rows of $\mathbf{X}$, that is,

$$f(\mathbf{X}) = f(\mathbf{x}_1) f(\mathbf{x}_2 \mid \mathbf{x}_1) \cdots f(\mathbf{x}_p \mid \mathbf{x}_1, \ldots, \mathbf{x}_{p-1}).$$

- If $\mathbf{X} = (\mathbf{X}_1, \mathbf{X}_2)$, where $\mathbf{X}_j$ is $p \times q_j$, then the marginal distribution of $\mathbf{X}_2$ is a central matrix-variate t with parameters $\mathbf{R}^{-1}$, $\mathbf{Q}_{22}$, and $m - q_1$. In the particular case $\mathbf{X} = (\mathbf{x}_1, \ldots, \mathbf{x}_q)$, each column $\mathbf{x}_j$ has the central multivariate t distribution with degrees of freedom $m - p - q + 1$ and correlation matrix $q_{ii}\mathbf{R}/(m - p - q + 1)$. A consequence of this is that the density (5.37) of $\mathbf{X}$ can be written as the product of conditional multivariate t distributions of the columns of $\mathbf{X}$, that is,

$$f(\mathbf{X}) = f(\mathbf{x}_1) f(\mathbf{x}_2 \mid \mathbf{x}_1) \cdots f(\mathbf{x}_q \mid \mathbf{x}_1, \ldots, \mathbf{x}_{q-1}).$$

- If $\mathbf{X}$ is doubly partitioned as

$$\mathbf{X} = \begin{pmatrix} \mathbf{X}_{11} & \mathbf{X}_{12} \\ \mathbf{X}_{21} & \mathbf{X}_{22} \end{pmatrix},$$

  where $\mathbf{X}_{ij}$ is $p_i \times q_j$ with $p_1 + p_2 = p$ and $q_1 + q_2 = q$, then the conditional distribution of $\mathbf{X}_{12}^T$ given $\mathbf{X}_{11}$ and $\mathbf{X}_{21}$ is a matrix-variate t with parameters $\mathbf{R}_{12}^T(\mathbf{R}_{11}^T)^{-1}\mathbf{X}_{11}^T$, $\mathbf{Q}_{11} + \mathbf{X}_{11}^T\mathbf{R}_{11}^{-1}\mathbf{X}_{11}$, $\mathbf{R}_{22} - \mathbf{R}_{21}\mathbf{R}_{11}^{-1}\mathbf{R}_{12}$, and $m + q_1 - p - q + 1$. (Here, the partitions of $\mathbf{R}$ and $\mathbf{Q}$ correspond to the partition of $\mathbf{X}$.). Since this depends only on $\mathbf{X}_{11}$, it follows that $\mathbf{X}_{12}$ and $\mathbf{X}_{21}$ given $\mathbf{X}_{11}$ are conditionally independent.

The following results (due to Javier and Gupta, 1985, and Dickey et al., 1986) concern the distributions of the quadratic forms $\mathbf{XAX}^T$ and $\mathbf{AXB}$ when $\mathbf{X}$ has the central matrix-variate t distribution with parameters $\mathbf{R}$, $\mathbf{Q}$, and m.

- If $\mathbf{A} > \mathbf{0}$ is $q \times q$, then the pdf of $\mathbf{W} = \mathbf{XAX}^T$ is given by

$$\begin{aligned} f(\mathbf{W}) = {} & \frac{1}{k(m,p,q)} |\mathbf{A}|^{-p/2} |\mathbf{R}|^{(m-q)/2} |\mathbf{Q}|^{-p/2} |\mathbf{W}|^{(q-p-1)/2} \\ & \times |\mathbf{R} + \mathbf{W}|^{-m/2} \\ & \times {}_1F_0\left(\frac{m}{2}; (\mathbf{R} + \mathbf{W})^{-1}\mathbf{W}, \mathbf{I}_q - (\mathbf{QA})^{-1}\right), \end{aligned}$$

  where $\mathbf{W} > \mathbf{0}$, $k(m,p,q)$ is given by (5.34), and ${}_1F_0$ is as defined in (5.35). An immediate consequence of this result is that

$$\begin{aligned} \int_{\mathbf{W}>\mathbf{0}} & |\mathbf{W}|^{(q-p-1)/2} |\mathbf{R} + \mathbf{W}|^{-m/2} \\ & \times {}_1F_0\left(\frac{m}{2}; (\mathbf{R} + \mathbf{W})^{-1}\mathbf{W}, \mathbf{I}_q - (\mathbf{QA})^{-1}\right) d\mathbf{W} \end{aligned}$$

$$= \frac{\Gamma(q/2)\Gamma((m-q)/2)}{\Gamma(m/2)} |\mathbf{A}|^{p/2} |\mathbf{R}|^{-(m-q)/2} |\mathbf{Q}|^{p/2}.$$

Hence the hth moment of $|\mathbf{W}|$ is

$$E\left[|\mathbf{W}|^h\right] = \frac{\Gamma((q+2h)/2)\Gamma((m-q-2h)/2)}{\Gamma(m/2)} \times |\mathbf{A}|^{p/2} |\mathbf{R}|^{-(m-q-2h)/2} |\mathbf{Q}|^{p/2}. \quad (5.38)$$

Further using the fact that an F-distribution is uniquely determined by its moments, it follows that $|\mathbf{W}|$ can be written as a product of q independent univariate F's, that is,

$$|\mathbf{W}| \sim \prod_{j=1}^{q} F(q-(j-q), m-q-(j-1)).$$

For the special case $\mathbf{A} = \mathbf{I}_p$ and $p = q$, (5.38) gives the hth moment of $\mathbf{X}\mathbf{X}^T$.

- If $\mathbf{A} > \mathbf{0}$ is $p \times p$ and $\mathbf{B} > \mathbf{0}$ is $q \times q$, then $\mathbf{AXB}$ has the central matrix-variate t distribution with parameters $\mathbf{B}^T\mathbf{QB}$, $\mathbf{AR}^{-1}\mathbf{A}^T$, and m, $m > p+q-1$.
- If $\mathbf{A} > \mathbf{0}$ is $p \times p$ and $\mathbf{B}$ is a $q \times r$ rectangular matrix, then $\mathbf{AXB}$ has the central matrix-variate t distribution with parameters $\mathbf{B}^T\mathbf{QB}$, $\mathbf{AR}^{-1}\mathbf{A}^T$, and m, $m > p+r-1$.
- If $\mathbf{a}$ is a $q \times 1$ vector, then $\mathbf{a}^T\mathbf{X}^T$ has the central multivariate t distribution with degrees of freedom $m-p-q+1$ and correlation matrix $\mathbf{a}^T\mathbf{Q}\mathbf{a}\mathbf{R}/(m-p-q+1)$.
- If $\mathbf{a}$ is a $q \times 1$ vector such that $\mathbf{a}^T\mathbf{a} = 1$ and $\mathbf{b}$ is a $p \times 1$ vector, then $\mathbf{a}^T\mathbf{X}^T\mathbf{b}$ is a linear combination of Student's t random variables.
- If $\mathbf{b}$ is a $p \times 1$ vector, then $\mathbf{X}^T\mathbf{b}$ has the central multivariate t distribution with degrees of freedom $m-p-q+1$ and correlation matrix $\mathbf{b}^T\mathbf{R}\mathbf{b}\mathbf{Q}/(m-p-q+1)$.
- If $\mathbf{a}$ is a $q \times 1$ vector and $\mathbf{b}$ is a $p \times 1$ vector, then

$$\frac{(m-p-q+1)\mathbf{a}^T\mathbf{X}^T\mathbf{b}}{(\mathbf{a}^T\mathbf{Q}\mathbf{a})(\mathbf{b}^T\mathbf{R}\mathbf{b})}$$

  has the Student's t distribution with degrees of freedom $m-p-q+1$.
- In the special case $\mathbf{R} = \mathbf{I}_p$ and $\mathbf{Q} = 1$, if a is a real number and $\mathbf{b}$ is a $q \times 1$ vector such that $a^2\mathbf{b}^T\mathbf{b} = 1$, then $a\mathbf{X}\mathbf{b}$ has the Student's t distribution with degrees of freedom $m-q$.

Javier and Gupta (1985) also derived a useful factorization of the central matrix-variate t density in terms of the product of $q-1$ independent univariate F densities and q independent multivariate t densities – paralleling the result of Tan (1969a) for matrix-variate beta distributions. Let $\mathbf{X}$ be a $p \times q$ random matrix having the density (5.37) with $\boldsymbol{\mu} = \mathbf{0}$. Set

$$\mathbf{U} = \left(\mathbf{R}^{-1/2}\mathbf{X}\mathbf{Q}^{-1/2}\right)\left(\mathbf{R}^{-1/2}\mathbf{X}\mathbf{Q}^{-1/2}\right)^T,$$

so that $\mathbf{U}$ is $p \times p$, symmetric, and $\mathbf{U} > \mathbf{0}$. Partition $\mathbf{U}$ as

$$\mathbf{U} = \begin{pmatrix} \mathbf{U}_{11} & \mathbf{U}_{12} \\ \mathbf{U}_{21} & \mathbf{U}_{22} \end{pmatrix}$$

so that $\mathbf{U}_{11}$ is 1×1 and $\mathbf{U}_{22}$ is $(p-1) \times (p-1)$. Abbreviating $\mathbf{D}_{22} - \mathbf{D}_{21}\mathbf{D}_{11}^{-1}\mathbf{D}_{12}$ by $\mathbf{D}_{22\cdot 1}$, define the following submatrices

$$\mathbf{U}_{22\cdot 1}^{(j)} = \left(\mathbf{U}_{22\cdot 1}^{(j-1)}\right)_{22\cdot 1}, \quad j = 1, 2, \ldots, p-1,$$

$$\mathbf{U}_{22\cdot 1}^{(0)} = \mathbf{U}_{22\cdot 1},$$

$$\mathbf{U}_{11}^{(j)} = \left(\mathbf{U}_{22\cdot 1}^{(j-1)}\right)_{11}, \quad j = 1, 2, \ldots, p-1,$$

and

$$\mathbf{U}_{11}^{(0)} = \mathbf{U}_{11},$$

so that $\mathbf{U}_{22\cdot 1}^{(j)}$ is $(p-j) \times (p-j)$ and $\mathbf{U}_{11}^{(j)}$ is 1×1. With all of this notation the factorization of the density of $\mathbf{X}$ (due to Javier and Gupta) can be stated as

$$f(\mathbf{X}) = \prod_{j=0}^{q-2} F\left(\frac{p-j+2}{2}, \frac{m-jq-2(p-j)-2}{2}\right)$$
$$\times \prod_{j=1}^{p} t_{\mathbf{U}_{11}^{(j)}\left\{1+\mathbf{U}_{11}^{(j)}\right\}}\left(\mathbf{I}_{p-j} + \mathbf{U}_{22\cdot 1}^{(j)}; m-(j-1)q\right),$$

where $t_\nu(\mathbf{T}; r)$ is the joint pdf of a central multivariate t distribution with degrees of freedom ν and correlation matrix $\mathbf{T}$ and $F(\alpha, \beta)$ is the pdf of a univariate F distribution with degrees of freedom α and β.

The two predictivistic characterizations of the multivariate t distribution based on (5.14) and (5.15) have the following matrix-variate generalizations

- Let $\mathbf{X}_1, \mathbf{X}_2, \ldots$ be an infinite sequence of q-dimensional random column vectors that are orthogonally invariant (which means that, for each k, $\mathbf{X}^{(k)} = (\mathbf{X}_1, \ldots, \mathbf{X}_k)^T$ and $\mathbf{\Gamma X}^{(k)}$ are identically distributed, for all $k \times k$ orthogonal matrices $\mathbf{\Gamma}$) and, for k fixed, let $\mathbf{X}_i^{(k)} = (\mathbf{X}_{(i-1)k+1}, \ldots, \mathbf{X}_{ik})^T$, $i = 1, 2, \ldots$. If $\mathbf{X}_1, \ldots, \mathbf{X}_q$ are linearly independent with probability 1 and

$$E\left[\mathbf{X}_2^{(q)^T} \mathbf{X}_2^{(q)} \middle| \mathbf{X}_1^{(q)}\right] = a\mathbf{X}_1^{(q)^T} \mathbf{X}_1^{(q)} + \mathbf{B},$$

where $0 < a < 1$ and $\mathbf{B}$ is a $q \times q$ positive definite matrix, then the distribution of $\mathbf{X}^{(p)}$ is the matrix-variate t with $\boldsymbol{\mu} = \mathbf{0}$, $\mathbf{R} = \mathbf{I}_p$, $\mathbf{Q} = (1/a)\mathbf{B}$, and $m = 1 + (p/a) - p$.

- Let $\mathbf{X}_1, \mathbf{X}_2, \ldots$ be an infinite sequence of q-dimensional random column vectors such that, for each p, $\mathbf{X}^{(p)}$ and $\mathbf{\Gamma X}^{(p)}$ are identically distributed, for all $p \times p$ orthogonal matrices $\mathbf{\Gamma}$ satisfying $\mathbf{\Gamma 1}_p = \mathbf{1}_p$ (where $\mathbf{1}_p$ is a p-dimensional vector of 1's). Under this assumption there exists a σ-algebra $\mathcal{T}$ of events such that

$$\begin{aligned} \bar{\mathbf{X}}_n &= \frac{1}{n}\sum_{i=1}^{n} \mathbf{X}_i \\ &\to E(\mathbf{X}_1 \mid \mathcal{T}) = \mathbf{M} \end{aligned}$$

and

$$\begin{aligned} \mathbf{S}_n &= \frac{1}{n}\sum_{i=1}^{n} (\mathbf{X}_i - \bar{\mathbf{X}}_n)(\mathbf{X}_i - \bar{\mathbf{X}}_n)^T \\ &\to E(\mathbf{X}_1\mathbf{X}_1^T \mid \mathcal{T}) - E(\mathbf{X}_1 \mid \mathcal{T})\{E(\mathbf{X}_1^T \mid \mathcal{T})\}^T = \mathbf{V} \end{aligned}$$

as $n \to \infty$ (Chow and Teicher, 1978), where the convergence is almost everywhere. Moreover, if

$$\begin{aligned} & E\left[\left(\mathbf{X}_2^{(q)} - \mathbf{1}_q\mathbf{M}^T\right)^T \left(\mathbf{X}_2^{(q)} - \mathbf{1}_q\mathbf{M}^T\right) \mid \mathbf{X}_i^{(q)}, \mathbf{M}\right] \\ &= a\left(\mathbf{X}_1^{(q)} - \mathbf{1}_q\mathbf{M}^T\right)^T \left(\mathbf{X}_1^{(q)} - \mathbf{1}_q\mathbf{M}^T\right) + \mathbf{B}, \end{aligned}$$

where $0 < a < 1$ and $\mathbf{B}$ is a $q \times q$ symmetric positive definite matrix, then $\mathbf{X}^{(p)}$ is a location mixture of the matrix-variate t distribution with $\boldsymbol{\mu} = \mathbf{1}_p^T\mathbf{M}$, $\mathbf{R} = \mathbf{I}_p$, $\mathbf{Q} = (1/a)\mathbf{B}$, and $m = 1 + (p/a) - p$. In addition, $\mathbf{M}$ and $\mathbf{V}$ are independent.

Dawid (1981) provided a different but more convenient parameterization of (5.37). If $\mathbf{Y}$ $(p \times p)$ has the standard matrix inverse Wishart distribution with parameter δ and if, given $\mathbf{Y}$, $\mathbf{X}$ $(n \times p)$ has the matrix normal distribution with parameters $\mathbf{I}_n$ and $\mathbf{Y}$, then $\mathbf{X}$ is termed as having the standard matrix t distribution. In the notation of (5.37), this would correspond to $\boldsymbol{\mu} = \mathbf{0}$, $\mathbf{R} = \mathbf{I}_p$, $\mathbf{Q} = \mathbf{I}_n$, and $m = \delta + n + p - 1$. Under Dawid's parameterization, if $\mathbf{X}^*$ is a $n^* \times p^*$ submatrix of $\mathbf{X}$, then $\mathbf{X}^*$ has the matrix t distribution with parameters $\mathbf{I}_{n^*}$, $\mathbf{I}_{p^*}$, and δ: Note that δ is unchanged. This kind of consistency enabled Dawid (1981) to construct what is termed as the standard infinite matrix t distribution. Namely, $\mathbf{X} = \{x_{ij}, i \geq 1, j \geq 1\}$ is said to have the above-named distribution if it has the property that for all (n, p) the leading $n \times p$ submatrix of $\mathbf{X}$ has the standard matrix t distribution with parameter δ. The standard matrix t distribution also has the attractive property of being spherical, that is, if $\mathbf{P}$ $(n \times n)$ and $\mathbf{Q}$ $(p \times p)$ are two orthogonal matrices, then both $\mathbf{PX}$ and $\mathbf{XQ}$ have the same distribution as $\mathbf{X}$.

5.12 Complex Multivariate t Distribution

A complex normal random vector $\mathbf{Y} = V + \sqrt{-1}W$ is a complex random variable whose real and imaginary parts possess the bivariate normal distribution. A complex p-variate normal random vector

$$\begin{aligned} \mathbf{Y} &= \mathbf{V} + \sqrt{-1}\mathbf{W} \\ &= \left(V_1 + \sqrt{-1}W_1,\ V_2 + \sqrt{-1}W_2, \ldots, V_p + \sqrt{-1}W_p\right)^T \end{aligned} \tag{5.39}$$

is a p-tuple of complex normal random vectors such that the vector of real and imaginary parts $(V_1, W_1, \ldots, V_p, W_p)$ has the $2p$-variate normal distribution (Goodman, 1963). Section 45.13 of Kotz et al. (2000) provides an account of this distribution. It is usually assumed that the $2p$-variate normal distribution of $(V_1, W_1, \ldots, V_p, W_p)$ has zero means and covariance matrix given by

$$\frac{1}{2}\begin{pmatrix} \boldsymbol{\Sigma}_1 & -\boldsymbol{\Sigma}_2 \\ \boldsymbol{\Sigma}_2 & \boldsymbol{\Sigma}_1 \end{pmatrix},$$

where $\boldsymbol{\Sigma}_1$ is symmetric (matrix A is symmetric if $A^T = A$) and $\boldsymbol{\Sigma}_2$ is skew-symmetric (matrix A is skew-symmetric if $A = -A^T$). From the given structure it is easily seen that the covariances of the p-variate vectors $\mathbf{V}$ and $\mathbf{W}$ are each equal to $\boldsymbol{\Sigma}_1/2$ and the covariance between $\mathbf{V}$ and $\mathbf{W}$ is equal to $\boldsymbol{\Sigma}_2/2$. Hence the covariance of the complex p-variate normal random vector $\mathbf{Y}$ in (5.39) is $\boldsymbol{\Sigma}_1 + \sqrt{-1}\boldsymbol{\Sigma}_2 = \boldsymbol{\Sigma}$, say. The properties of the distribution of $\mathbf{Y}$ have been studied by many authors. The joint pdf of $\mathbf{Y}$ is given by

$$f(\mathbf{y}) = \frac{1}{\pi^p |\boldsymbol{\Sigma}|} \exp\left\{-\bar{\mathbf{y}}^T \boldsymbol{\Sigma}^{-1} \mathbf{y}\right\}, \tag{5.40}$$

where $\bar{\mathbf{y}}$ denotes the complex conjugate of $\mathbf{y}$ (Goodman, 1963). For example, for the complex univariate normal distribution, $\mathbf{y} = v_1 + \sqrt{-1}w_1$ and the covariance matrix $\mathbf{\Sigma} = \sigma^2$, and thus the joint pdf of $\mathbf{Y}$ becomes

$$f(\mathbf{y}) = \frac{1}{\pi\sigma^2} \exp\left(-\frac{v_1^2 + w_1^2}{\sigma^2}\right).$$

The characteristic function of $\mathbf{Y}$ can be shown to be

$$E\left[\exp\left\{i\left(\mathbf{s}^T\mathbf{V} + \mathbf{t}^T\mathbf{W}\right)\right\}\right] = \exp\left\{-\frac{1}{4}\bar{\mathbf{u}}^T\mathbf{\Sigma}\mathbf{u}\right\},$$

where $\mathbf{u} = \mathbf{s} + \sqrt{-1}\mathbf{t}$ (Wooding, 1956). Explicit expressions for the moments of $\mathbf{Y}$ have been derived by Sultan and Tracy (1996). The complex multivariate normal distribution has applications in describing the statistical variability of estimators for the spectral density matrix of a multiple stationary normal time series and in describing the statistical variability of estimators for functions of the elements of a spectral density matrix of a multiple stationary normal time series.

Relatively few results are available that deal with complex multivariate t distributions. Originally, the complex multivariate t distribution was introduced by Gupta (1964). Let $\mathbf{Y}$ have the complex p-variate normal distribution with zero means, common variance σ^2, and covariance matrix $\sigma^2\mathbf{R}$. Let $2\nu S^2/\sigma^2$ have the chi-squared distribution with degrees of freedom 2ν, distributed independently of $\mathbf{Y}$. Then $\mathbf{X} = \mathbf{Y}/S$ is said to have the complex p-variate t distribution with degrees of freedom ν and correlation matrix $\mathbf{R}$. By writing down the joint distribution of S and $\mathbf{X}$ and then integrating out S, the pdf of $\mathbf{X}$ can be obtained as

$$f(\mathbf{x}) = \frac{\Gamma(\nu + p)}{(\pi\nu)^p\Gamma(\nu)\,|\mathbf{R}|^p}\left\{1 + \frac{1}{\nu}\bar{\mathbf{x}}^T\mathbf{R}^{-1}\mathbf{x}\right\}^{-(\nu+p)}.$$

Tan (1973) discussed some properties of this distribution. Tan (1969b) provided a brief discussion of a complex analog of the matrix-variate t distribution given by (5.37).

5.13 Steyn's Nonnormal Distributions

Strictly speaking, this section does not deal with multivariate t distributions per se. This section is about nonnormal distributions arising from the class of multivariate elliptical distributions that contains the multivariate t as a particular case.

One weakness of the class of multivariate elliptical distributions is that all fourth-order cumulants are expressed in terms of a single kurtosis parameter (moreover, the univariate marginals have zero skewness and the same kurtosis). In fact, the cumulant generating function (cgf) and the moment generating function (mgf) of a p-variate elliptical distribution with zero means and correlation matrix $\mathbf{R}$ are

$$\begin{aligned} & K(t_1, \ldots, t_p) \\ = {} & \frac{1}{2}\left(\mathbf{t}^T\mathbf{R}\mathbf{t}\right) + \frac{1}{2}\kappa\left\{\frac{1}{2}\left(\mathbf{t}^T\mathbf{R}\mathbf{t}\right)\right\}^2 + \sum_{k\geq 3} A_k\left(\mathbf{t}^T\mathbf{R}\mathbf{t}\right)^k \end{aligned} \tag{5.41}$$

and

$$\begin{aligned} & M(t_1, \ldots, t_p) \\ = {} & \exp\left(\frac{\mathbf{t}^T\mathbf{R}\mathbf{t}}{2}\right)\left[1 + \frac{\kappa}{2}\frac{\left(\mathbf{t}^T\mathbf{R}\mathbf{t}\right)^2}{1!} + \sum_{k\geq 3} B_k\left(\mathbf{t}^T\mathbf{R}\mathbf{t}\right)^k\right], \end{aligned} \tag{5.42}$$

respectively, where A_k, B_k are constants and κ is the kurtosis parameter. Steyn (1993) attempted to introduce meaningful multivariate distributions that are related to the elliptical distributions and that contain more than one kurtosis parameter.

As an example, consider a random vector (X_1, X_2, X_3) possessing the three-dimensional normal distribution with the mgf

$$\begin{aligned} & M(t_1, t_2, t_3) \\ = {} & \exp\left\{\frac{1}{2}\left(t_1^2 + t_2^2 + t_3^2 + 2r_{12}t_1t_2 + 2r_{13}t_1t_3 + 2r_{23}t_2t_3\right)\right\}. \end{aligned} \tag{5.43}$$

Suppose this model is placed in a changing environment that favors a change in one of the random variables, say X_1, in such a way that the kurtosis should be taken into consideration. Specifically, assume that the marginal distribution of X_1 is elliptical with the kurtosis parameter κ_1, while the conditional distribution of (X_2, X_3) given $X_1 = x_1$ remains unchanged. Note that (5.43) can be written as

$$\begin{aligned} & M(t_1, t_2, t_3) \\ = {} & \frac{1}{\sqrt{2\pi}}\exp\left[\frac{1}{2}\left\{\left(1 - r_{12}^2\right)t_2^2 + \left(1 - r_{13}^2\right)t_3^2 + 2\left(r_{23} - r_{12}r_{13}\right)t_1t_2\right\}\right] \\ & \times \int_{-\infty}^{\infty}\exp\left\{-\frac{x_1^2}{2} + \left(t_1 + r_{12}t_2 + r_{13}t_3\right)x_1\right\}dx_1. \end{aligned} \tag{5.44}$$

Changing the probability element in the integrand in (5.44) to that of the elliptical distribution in (5.42), one can show that the mgf changes to

$$M_1(t_1,t_2,t_3) = M(t_1,t_2,t_3)\exp\left\{\frac{\kappa_1}{8}(t_1+r_{12}t_2+r_{13}t_3)^4+\cdots\right\}.$$

The corresponding cgf becomes

$$\begin{aligned} K_1(t_1,t_2,t_3) &= \frac{1}{2}\left(t_1^2+t_2^2+t_3^2+2r_{12}t_1t_2+2r_{13}t_1t_3+2r_{23}t_2t_3\right) \\ &\quad +\frac{1}{8}\kappa_1(t_1+r_{12}t_2+r_{13}t_3)^4+\cdots. \end{aligned} \tag{5.45}$$

Setting $t_2 = t_3 = 0$, the cgf of the marginal distribution of X_1 is given by

$$K_1(t_1,0,0) = \frac{1}{2}t_1^2+\frac{1}{8}\kappa_1t_1^4+\cdots,$$

which shows (as it should) that the marginal distribution of X_1 is elliptical with kurtosis parameter κ_1 (compare with equation (5.41)). However, for $t_1 = t_3 = 0$ and $t_1 = t_2 = 0$, one obtains

$$K_1(0,t_2,0) = \frac{1}{2}t_2^2+\frac{1}{8}\kappa_1(r_{12}t_2)^4+\cdots$$

and

$$K_1(0,0,t_3) = \frac{1}{2}t_3^2+\frac{1}{8}\kappa_1(r_{13}t_3)^4+\cdots;$$

thus, the marginal distributions of X_2 and X_3 are also elliptical but with kurtosis parameters $\kappa_1r_{12}^4$ and $\kappa_1r_{13}^4$, respectively. Furthermore, for $t_1 = 0$,

$$K_1(0,t_2,t_3) = \frac{1}{2}\left(t_2^2+2r_{23}t_2t_3+t_3^2\right)+\frac{1}{8}\kappa_1(r_{12}t_2+r_{13}t_3)^4+\cdots,$$

which shows that the joint marginal distribution of (X_2, X_3) is not elliptical. The fourth-order cumulants are easily obtained from (5.45) as $\kappa_{ijk} = 3\kappa_1r_{12}^jr_{13}^k$, where $i+j+k=4$.

Suppose now that the model given by (5.43) is placed in an environment that favors a change in not only X_1 but also influences (X_2, X_3). Assume – in particular – that the conditional distributions of X_2 given $X_1 = x_1$ and X_3 given $(X_1, X_2) = (x_1, x_2)$ are elliptical with kurtosis parameters κ_2 and κ_3, respectively. Then calculations similar to those above show that the mgf (5.43) changes to

$$M_2(t_1,t_2,t_3) = M(t_1,t_2,t_3)\exp\left\{\frac{1}{2}\kappa_1\left(\frac{u_1^2}{2}\right)^2+\frac{1}{2}\kappa_2\left(\frac{u_2^2\sigma_{2\cdot1}^2}{2}\right)^2\right.$$

$$+\frac{1}{2}\kappa_3\left(\frac{t_3^2\sigma_{3\cdot 12}^2}{2}\right)^2+\cdots\Bigg\},\qquad(5.46)$$

where

$$u_1 = t_1 + r_{12}t_2 + r_{13}t_3,$$

$$u_2 = t_2 + \frac{r_{23} - r_{12}r_{13}}{1 - r_{12}^2}t_3,$$

$$\sigma_{2\cdot 1}^2 = 1 - r_{12}^2,$$

and

$$\sigma_{3\cdot 12}^2 = 1 - r_{13}^2 - \frac{(r_{23} - r_{12}r_{13})^2}{1 - r_{12}^2}.$$

It is easily seen that the marginal distributions of X_1, X_2, and X_3 are elliptical with kurtosis parameters given by κ_1, $r_{12}^4\kappa_1 + \sigma_{2\cdot 1}^4\kappa_2$, and $r_{13}^4\kappa_1 + (\sigma_{3\cdot 1}^2 - \sigma_{3\cdot 12}^4)\kappa_3$, where $\sigma_{3\cdot 1}^2 = 1 - r_{13}^2$. This time, the fourth-order cumulants are given by

$$\kappa_{ijk} = 3\left\{\kappa_1 r_{12}^j r_{13}^k + \kappa_2\sigma_{2\cdot 1}^{2(j-2)}(r_{23} - r_{12}r_{13})^k\right\},$$

where $i + j + k = 4$. In the case of κ_{004}, $\kappa_3\sigma_{3\cdot 12}^4$ should be added.

Similar constructions can be performed when $\mathbf{X} = (X_1, \ldots, X_p)^T$ has a p-variate normal distribution with zero means, covariance matrix $\mathbf{R}$, and the corresponding mgf

$$M_2(t_1, \ldots, t_p) = \exp\left(\frac{1}{2}\mathbf{t}^T\mathbf{R}\mathbf{t}\right).\qquad(5.47)$$

Consider two environments similar to those considered above for the trivariate normal model. First, divide $\mathbf{X}$ into two random vectors $\mathbf{X}^{(1)} = (X_1, \ldots, X_h)^T$ and $\mathbf{X}^{(2)} = (X_{h+1}, \ldots, X_p)^T$, and let

$$\mathbf{R} = \begin{pmatrix} \mathbf{R}_{11} & \mathbf{R}_{12} \\ \mathbf{R}_{12}^T & \mathbf{R}_{22} \end{pmatrix}$$

be the corresponding partition of the correlation matrix. Also let $\mathbf{t}^{(1)} = (t_1, \ldots, t_h)^T$ and $\mathbf{t}^{(2)} = (t_{h+1}, \ldots, t_p)^T$ be the corresponding partition of $\mathbf{t}$. Now assume that the marginal distribution of $\mathbf{X}^{(1)}$ is changed to an h-dimensional elliptical distribution with kurtosis parameter κ_1

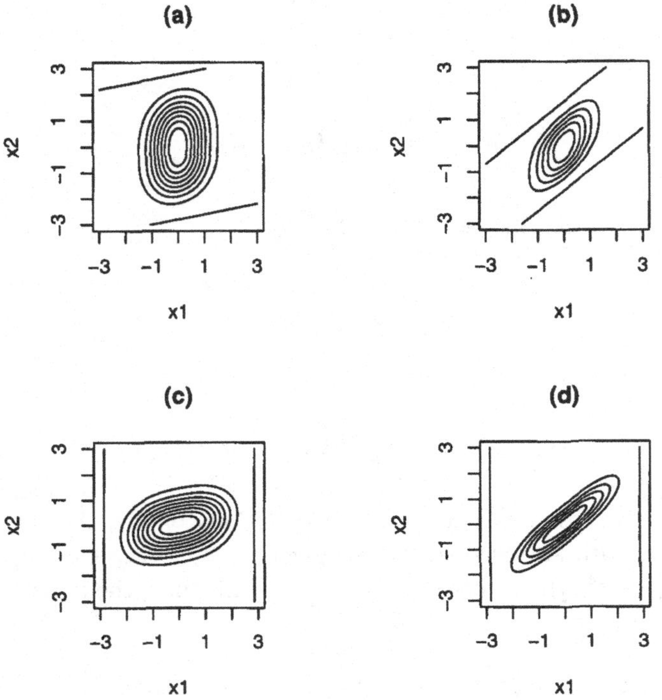

Fig. 5.5. Steyn's bivariate pdf corresponding to (5.46) for $t_3 = 0$ and (a) $\kappa_1 = 0.8$, $\kappa_2 = -0.4$, and $r_{12} = 0.2$; (b) $\kappa_1 = 0.8$, $\kappa_2 = -0.4$, and $r_{12} = 0.8$; (c) $\kappa_1 = -0.4$, $\kappa_2 = 0.8$, and $r_{12} = 0.2$; and (d) $\kappa_1 = -0.4$, $\kappa_2 = 0.8$, and $r_{12} = 0.8$

and that the conditional distribution of $\mathbf{X}^{(2)}$ given $\mathbf{X}^{(1)} = \mathbf{x}^{(1)}$ remains unchanged. Then calculations show that (5.47) changes to

$$\begin{aligned} M_1(\mathbf{t}) &= M(\mathbf{t}) \exp\left[\frac{\kappa_1}{8}\left\{\left(\mathbf{t}^{(1)} + \mathbf{R}_{11}^{-1}\mathbf{R}_{12}\mathbf{t}^{(2)}\right)^T\right.\right. \\ &\qquad \left.\left. \times \mathbf{R}_{11}\left(\mathbf{t}^{(1)} + \mathbf{R}_{11}^{-1}\mathbf{R}_{12}\mathbf{t}^{(2)}\right)\right\}^2 + \cdots\right]. \quad (5.48) \end{aligned}$$

Clearly,

$$M_1\left(\mathbf{t}^{(1)}, \mathbf{0}\right) = \exp\left\{\frac{1}{2}\mathbf{t}^{(1)^T}\mathbf{R}_{11}\mathbf{t}^{(1)} + \frac{1}{8}\kappa_1\left(\mathbf{t}^{(1)^T}\mathbf{R}_{11}\mathbf{t}^{(1)}\right)^2 + \cdots\right\}$$

and

$$M_1\left(\mathbf{0},\mathbf{t}^{(2)}\right) = \exp\left\{\frac{1}{2}\mathbf{t}^{(2)^T}\mathbf{R}_{22}\mathbf{t}^{(2)} + \frac{1}{8}\kappa_1\left(\mathbf{t}^{(2)^T}\mathbf{R}_{22}\mathbf{R}_{11}^{-1}\mathbf{R}_{12}\mathbf{t}^{(2)}\right)^2 + \cdots\right\},$$

which shows that the marginal distribution of $\mathbf{X}^{(1)}$ is an h-dimensional elliptical distribution (as it should) while that of $\mathbf{X}^{(1)}$ is not elliptical. The second-order cumulants of (5.48) are the same as those for (5.47).

For the second construction, partition $\mathbf{X}_2$ into $\mathbf{X}^{(3)} = (X_h, \ldots, X_{h+s})^T$ and $\mathbf{X}^{(4)} = (X_{h+s+1}, \ldots, X_p)^T$, and let $\mathbf{t}^{(2)}$ be partitioned correspondingly into $\mathbf{t}^{(3)}$ and $\mathbf{t}^{(4)}$. Let $\mathbf{C}$ denote the conditional covariance matrix of $\mathbf{X}_2$ given $\mathbf{X}_1 = \mathbf{x}_1$, that is,

$$\mathbf{C} = \mathbf{R}_{22} - \mathbf{R}_{21}\mathbf{R}_{11}^{-1}\mathbf{R}_{12},$$

and let $\mathbf{C}$ be partitioned as

$$\mathbf{C} = \begin{pmatrix} \mathbf{C}_{11} & \mathbf{C}_{12} \\ \mathbf{C}_{12}^T & \mathbf{C}_{22} \end{pmatrix},$$

so that $\mathbf{C}_{11}$ is $s\times s$, $\mathbf{C}_{12}$ is $s\times(p-h-s)$, and $\mathbf{C}_{22}$ is $(p-h-s)\times(p-h-s)$. Now, assuming that the distribution of $\mathbf{X}_1$ is elliptical with kurtosis parameter κ_1 and that of $\mathbf{X}^{(3)} - E(\mathbf{X}^{(3)} \mid \mathbf{X}^{(1)} = \mathbf{x}^{(1)})$ is elliptical with kurtosis parameter κ_2, one can show that the mgf (5.47) changes to

$$\begin{aligned} M_2(\mathbf{t}) = M(\mathbf{t})\exp\Bigg[&\left(\frac{1}{8}\kappa_1\left\{\left(\mathbf{t}^{(1)} + \mathbf{R}_{11}^{-1}\mathbf{R}_{12}\mathbf{t}^{(2)}\right)^T \times\mathbf{R}_{11}\left(\mathbf{t}^{(1)} + \mathbf{R}_{11}^{-1}\mathbf{R}_{12}\mathbf{t}^{(2)}\right)\right\}^2 + \cdots\right) \\ &+\left(\frac{1}{8}\kappa_2\left\{\left(\mathbf{t}^{(3)} + \mathbf{C}_{11}^{-1}\mathbf{C}_{12}\mathbf{t}^{(4)}\right)^T \times\mathbf{C}_{11}\left(\mathbf{t}^{(3)} + \mathbf{C}_{11}^{-1}\mathbf{C}_{12}\mathbf{t}^{(4)}\right)\right\}^2 + \cdots\right)\Bigg]. \end{aligned} \tag{5.49}$$

This defines the mgf of a multivariate distribution that is equal to the product of the mgf of the multivariate normal and a function of two quadratic forms in $\mathbf{t}$ depending on the two kurtosis parameters κ_i, $i =$

1, 2, and on the elements of the normal covariance matrix. Setting $\mathbf{t}^{(2)} = \mathbf{0}$ into (5.49), we see that $\mathbf{X}^{(1)}$ has an h-dimensional elliptical distribution with zero means, covariance matrix $\mathbf{R}_{11}$, and kurtosis parameter κ_1 (as it should). If either $\mathbf{t}^{(1)} = \mathbf{0}$ or $\mathbf{t}^{(1)} = \mathbf{0}$ and $\mathbf{t}^{(3)} = \mathbf{0}$, then $M_2(\mathbf{t})$ becomes a function of three different forms.

5.14 Inverted Dirichlet Distribution

There is a close connection between the multivariate t distribution defined by (1.1) and the inverted Dirichlet distribution (Cornish, 1954; Dunnett and Sobel, 1954). To see this, consider the central p-variate t distribution with the pdf

$$f(\mathbf{x}) = \frac{\Gamma((\nu+p)/2)}{(\pi\nu)^{p/2}\Gamma(\nu/2)|\mathbf{R}|^{1/2}}\left[1+\frac{\mathbf{x}^T\mathbf{R}^{-1}\mathbf{x}}{\nu}\right]^{-(\nu+p)/2}.$$

Upon transforming to the canonical variables $\mathbf{Z} = (Z_1, \ldots, Z_p)$, $\mathbf{Z} = \mathbf{PX}$, where $\mathbf{P}$ is a $p \times p$ matrix such that $\mathbf{P}^T\mathbf{P} = \mathbf{R}^{-1}$, it is easily seen that

$$f(\mathbf{z}) = \frac{\Gamma((\nu+p)/2)}{(\pi\nu)^{p/2}\Gamma(\nu/2)}\left[1+\frac{\mathbf{z}^T\mathbf{z}}{\nu}\right]^{-(\nu+p)/2}. \tag{5.50}$$

In (5.50) now perform a further transformation $T_i = Z_i^2/\nu$, which is one-to-one in each of 2^p regions with the Jacobian

$$|J| = 2^{-p}\nu^{p/2}/\sqrt{t_1\cdots t_p}.$$

Consequently, the joint pdf of $\mathbf{T}^T = (T_1, \ldots, T_p)$ becomes

$$f(\mathbf{t}) = \frac{\Gamma((\nu+p)/2)}{\pi^{p/2}\Gamma(\nu/2)}t_1^{-1/2}\cdots t_p^{-1/2}\left(1+\sum_{i=1}^{p}t_i\right)^{-(\nu+p)/2},$$

which is the inverted p-dimensional Dirichlet distribution $D'(1/2, \ldots, 1/2; \nu/2)$; see, for example, Kotz et al. (2000, Chapter 49).

6
Probability Integrals

There has been a very substantial amount of research carried out on probability integrals of multivariate t distributions. Most of the work was done during the pre-computer era, but recently several computer programs have been written to evaluate probability integrals.

Sections 6.1 to 6.7 by now may have lost some of their usefulness but are still of substantial historical interest in addition to their mathematical value. We have decided to record these results in some detail in this book in spite of the fact that some of the expressions are quite lengthy and cumbersome. Sections 6.8 to 6.13 contain more practically relevant and modern results.

6.1 Dunnett and Sobel's Probability Integrals

One of the earliest results on probability integrals is that due to Dunnett and Sobel (1954). Let (X_1, X_2) have the central bivariate t distribution with degrees of freedom ν and the equicorrelation structure $r_{ij} = \rho$, $i \neq j$. The corresponding bivariate pdf is

$$f(x_1, x_2; \nu, \rho) = \frac{1}{2\pi\sqrt{1-\rho^2}} \left\{1 + \frac{x_1^2 + x_2^2 - 2\rho x_1 x_2}{\nu(1-\rho^2)}\right\}^{-(\nu+2)/2} \quad (6.1)$$

with the probability integral

$$P(y_1, y_2; \nu, \rho) = \int_{-\infty}^{y_2} \int_{-\infty}^{y_1} f(x_1, x_2; \nu, \rho) dx_1 dx_2. \quad (6.2)$$

Let

$$x(m, y_1, y_2) = \frac{(y_1 - \rho y_2)^2}{(y_1 - \rho y_2)^2 + (1-\rho^2)(m + y_2^2)}$$

and let

$$I_{x(m,y_1,y_2)}(a,b) = \int_0^{x(m,y_1,y_2)} \frac{\Gamma(a+b)}{\Gamma(a)\Gamma(b)} y^{a-1}(1-y)^{b-1} dy$$

denote the incomplete beta function. Dunnett and Sobel (1954) evaluated exact expressions for (6.2) when ν takes on positive integer values. For even ν and odd ν, they obtained

$$\begin{aligned} P(y_1, y_2; \nu, \rho) &= \frac{1}{2\pi} \arctan \frac{\sqrt{1-\rho^2}}{-\rho} \\ &\quad + \frac{y_2}{4\sqrt{\nu\pi}} \sum_{j=1}^{\nu/2} \frac{\Gamma(j-1/2)}{\Gamma(j)} \left(1 + \frac{y_2^2}{\nu}\right)^{1/2-j} \\ &\quad \times \left[1 + \operatorname{sgn}(y_1 - \rho y_2) I_{x(\nu,y_1,y_2)}\left(\frac{1}{2}, j - \frac{1}{2}\right)\right] \\ &\quad + \frac{y_1}{4\sqrt{\nu\pi}} \sum_{j=1}^{\nu/2} \frac{\Gamma(j-1/2)}{\Gamma(j)} \left(1 + \frac{y_1^2}{\nu}\right)^{1/2-j} \\ &\quad \times \left[1 + \operatorname{sgn}(y_2 - \rho y_1) I_{x(\nu,y_2,y_1)}\left(\frac{1}{2}, j - \frac{1}{2}\right)\right] \end{aligned} \tag{6.3}$$

and

$$\begin{aligned} P(y_1, y_2; \nu, \rho) &= \frac{1}{2\pi} \arctan \left\{ -\sqrt{\nu} \left[\frac{\alpha\beta + \gamma\delta}{\gamma\beta - \nu\alpha\delta} \right] \right\} \\ &\quad + \frac{y_2}{4\sqrt{\nu\pi}} \sum_{j=1}^{(\nu-1)/2} \frac{\Gamma(j)}{\Gamma(j+1/2)} \left(1 + \frac{y_2^2}{\nu}\right)^{-j} \\ &\quad \times \left[1 + \operatorname{sgn}(y_1 - \rho y_2) I_{x(\nu,y_1,y_2)}\left(\frac{1}{2}, j\right)\right] \\ &\quad + \frac{y_1}{4\sqrt{\nu\pi}} \sum_{j=1}^{(\nu-1)/2} \frac{\Gamma(j)}{\Gamma(j+1/2)} \left(1 + \frac{y_1^2}{\nu}\right)^{-j} \\ &\quad \times \left[1 + \operatorname{sgn}(y_2 - \rho y_1) I_{x(\nu,y_2,y_1)}\left(\frac{1}{2}, j\right)\right], \end{aligned} \tag{6.4}$$

respectively. Here,

$$\alpha = y_1 + y_2,$$

$$\beta = y_1 y_2 + \rho\nu,$$

$$\gamma = y_1 y_2 - \nu,$$

and

$$\delta = \sqrt{y_1^2 - 2\rho y_1 y_2 + y_2^2 + \nu(1-\rho^2)}.$$

In the special case $y_1 = y_2 = 0$, both (6.3) and (6.4) reduce to the neat expression

$$P(0,0;\nu,\rho) = \arctan \frac{\sqrt{1-\rho^2}}{-\rho}, \tag{6.5}$$

which is independent of ν and is therefore identical with the corresponding result for the bivariate normal integral. Since the number of terms in (6.3) and (6.4) increases with ν, the usefulness of these expressions is confined to small values of ν. Dunnett and Sobel (1954) also derived an asymptotic expansion in powers of $1/\nu$, the first few terms of which yield a good approximation to the probability integral even for moderately small values of ν. The method of derivation is essentially the same as that used by Fisher (1925) to approximate the probability integral of the univariate Student's t distribution: Express the difference $f(x_1, x_2; \nu, \rho) - f(x_1, x_2; \infty, \rho)$ as a power series in $1/\nu$ and then integrate this series term by term over the desired region of integration. Setting

$$r^2 = \frac{y_1^2 - 2\rho y_1 y_2 + y_2^2}{1-\rho^2},$$

Dunnett and Sobel obtained

$$\begin{aligned}
\frac{f(y_1, y_2; \nu, \rho)}{f(y_1, y_2; \infty, \rho)} &= 1 + \left(\frac{r^2}{4} - r^2\right)\frac{1}{\nu} + \left(\frac{r^8}{32} - \frac{5r^6}{12} + r^4\right)\frac{1}{\nu^2} \\
&\quad + \left(\frac{r^{12}}{384} - \frac{7r^{10}}{96} + \frac{13r^8}{24} - r^8\right)\frac{1}{\nu^3} \\
&\quad + \left(\frac{r^{16}}{6144} - \frac{r^{14}}{128} + \frac{17r^{12}}{144} - \frac{77r^{10}}{120} + r^8\right)\frac{1}{\nu^4} \\
&= 1 + D(r),
\end{aligned}$$

say. Thus, the desired probability integral is

$$\begin{aligned}
P(y_1, y_2; \nu, \rho) &= \int_{-\infty}^{y_2}\int_{-\infty}^{y_1} f(x_1, x_2; \infty, \rho)\, dx_1\, dx_2 \\
&\quad + \int_{-\infty}^{y_2}\int_{-\infty}^{y_1} D(r) f(x_1, x_2; \infty, \rho)\, dx_1\, dx_2. \qquad (6.6)
\end{aligned}$$

The first term on the right-hand side of (6.6) is the integral of the bivariate normal pdf, and it has been tabulated by Pearson (1931) with a series of correction terms. The second term can be integrated term by term to obtain an asymptotic expansion in powers of $1/\nu$. Dunnett and Sobel gave expressions for the coefficients A_k of the terms $1/\nu^k$ for $k = 1, 2, 3, 4$. The first of these coefficients takes the form

$$\begin{aligned} A_1 &= \frac{ay_2}{4}\phi(a)\phi(y_2) + \frac{by_1}{4}\phi(b)\phi(y_1) - \frac{y_2(y_2^2+1)}{4}\phi(y_2)\,\Phi(a) \\ &\quad - \frac{y_1(y_1^2+1)}{4}\phi(y_1)\,\Phi(b), \end{aligned}$$

where ϕ and Φ are, respectively, the pdf and the cdf of the standard normal distribution, and

$$\begin{aligned} a &= \frac{y_1 - \rho y_2}{\sqrt{1-\rho^2}}, \\ b &= \frac{y_2 - \rho y_1}{\sqrt{1-\rho^2}}. \end{aligned}$$

In the special case $y_1 = y_2 = y$, (6.6) reduces to

$$\begin{aligned} P(y, y; \nu, \rho) &= \int_{-\infty}^{y}\int_{-\infty}^{y} f(x_1, x_2; \infty, \rho)dx_1 dx_2 \\ &\quad + \frac{A_1}{\nu} + \frac{A_2}{\nu^2} + \frac{A_3}{\nu^3} + \frac{A_4}{\nu^4} + \cdots, \end{aligned} \tag{6.7}$$

with the first two coefficients A_1 and A_2 now taking the forms

$$A_1 = -\frac{y\phi(y)}{2}\left\{\left(y^2+1\right)\Phi(cy) - y\Phi'(cy)\right\}$$

and

$$\begin{aligned} A_2 &= -\frac{y\phi(y)}{48}\Big\{\left(3y^6 - 7y^4 - 5y^2 - 3\right)\Phi(cy) \\ &\quad - y\Phi'(cy)\left[3y^4\left(c^4+3c^2+3\right) - y^2\left(c^2+5\right) - 3\right]\Big\}, \end{aligned}$$

where $c = \sqrt{(1-\rho)/(1+\rho)}$. In this special case, Dunnett and Sobel (1954) tabulated numerical values of the coefficients A_k for selected values of ρ, y, and ν. The following table gives the values for $\rho = 0.5$

Coefficients of the asymptotic expansion (6.7) for $\rho = 0.5$

y	ν	A_1	A_2	A_3	A_4
0.25	4	-0.025870	0.003371	0.003816	-0.001050
0.50	4	-0.057784	0.008999	0.006868	-0.002155
0.75	6	-0.100016	0.021983	0.006891	-0.001879
1.00	5	-0.150182	0.047374	-0.006835	0.007991
1.25	6	-0.198378	0.079687	-0.033130	0.036817
1.50	6	-0.231628	0.096254	-0.038696	0.032808
1.75	9	-0.240531	0.067469	0.052274	-0.191482
2.00	12	-0.223682	-0.020268	0.293449	-0.819219
2.25	13	-0.187525	-0.149011	0.623867	-1.618705
2.50	12	-0.142571	-0.276255	0.858993	-1.765249
3.00	18	-0.062685	-0.376815	0.432592	2.236773

These values can be used to construct tables for the probability integral in (6.7).

6.2 Gupta and Sobel's Probability Integrals

Gupta and Sobel (1957) investigated the special case when $\mathbf{X}$ follows the central p-variate t distribution with degrees of freedom ν and the correlation structure $r_{ij} = \rho = 1/2$, $i \neq j$. If $Y_1, Y_2, \ldots, Y_n, Y$ are independent normal random variables with common mean and common variance σ^2, and if $\nu S^2/\sigma^2$ is a chi-squared random variable with degrees of freedom ν, independent of $Y_1, Y_2, \ldots, Y_n, Y$, then one can rewrite the probability integral as

$$\begin{aligned} P(d) &= \int_{-\infty}^{d} \cdots \int_{-\infty}^{d} f(x_1, \ldots, x_p; \nu, \rho) dx_p \cdots dx_1 \\ &= \Pr\left\{ \frac{\max(Y_1, Y_2, \ldots, Y_p) - Y}{S} \leq \sqrt{2} d \right\} \\ &= \Pr\left(\frac{M_p - Y}{S} < \sqrt{2} d \right) \\ &= \Pr\left(Z < \sqrt{2} d \right), \end{aligned} \tag{6.8}$$

where $M_p = \max(Y_1, Y_2, \ldots, Y_p)$ and $Z = (M_p - Y)/S$. Gupta and Sobel (1957) provided four useful expressions for $P(d)$. These are by now classical results applicable in statistical inference. The first expression is derived by fixing Y and S in (6.8) and integrating with respect to M_p

$$P(d) = \int_0^\infty h(s)\left[\int_{-\infty}^\infty \Phi^p(y)\phi\left(y - \sqrt{\frac{2}{\nu}}ds\right)dy\right]ds, \qquad (6.9)$$

where ϕ and Φ are, respectively, the pdf and cdf of the standard normal distribution and h is the pdf of the chi-squared distribution with ν degrees of freedom. Based on the fact that the pdf ϕ admits an expansion about $d = 0$, it easy to justify a term-by-term integration of (6.9) to obtain the second expression

$$P(d) = \frac{1}{p+1}\sum_{k=0}^\infty \frac{2^{k/2}d^k}{k!}A_k E\left\{H_k\left(\frac{\max(X_1, X_2, \ldots, X_{p+1})}{\sigma}\right)\right\},$$

where

$$A_k = E\left\{\left(\frac{S}{\sigma}\right)^k\right\} = \frac{\Gamma\left(\frac{\nu+k}{2}\right)}{\left(\frac{\nu}{2}\right)^{k/2}\Gamma\left(\frac{\nu}{2}\right)} \qquad (6.10)$$

is the kth moment of $\chi_\nu/\sqrt{\nu}$ (provided that $k > -\nu$) and H_k is the kth Hermite polynomial defined by

$$\left(-\frac{d}{dx}\right)^k \exp\left(-\frac{x^2}{2}\right) = H_k(x)\exp\left(-\frac{x^2}{2}\right). \qquad (6.11)$$

A third expression for $P(d)$ is derived by first expanding ϕ about $S = \sigma$ and then integrating term by term, obtaining

$$\begin{aligned} P(d) &= \int_{-\infty}^\infty \Phi^p(y)\Big[\phi\left(y - \sqrt{2}d\right) - \sqrt{2}d\phi^{(1)}\left(y - \sqrt{2}d\right)E\left(\frac{S}{\sigma} - 1\right) \\ &\quad + d^2\phi^{(2)}\left(y - \sqrt{2}d\right)E\left(\frac{S}{\sigma} - 1\right)^2 - \cdots\Big]dy \\ &= \int_{-\infty}^\infty \Phi^p(y)\phi\left(y - \sqrt{2}d\right)dy \\ &\quad - \sqrt{2}d\left(1 - A_1\right)\int_{-\infty}^\infty \left(y - \sqrt{2}d\right)\Phi^p(y)\phi\left(y - \sqrt{2}d\right)dy \\ &\quad + 2d^2\left(1 - A_1\right)\int_{-\infty}^\infty \left\{y^2 - 2\sqrt{2}dy + 2d^2 - 1\right\} \\ &\quad \times \Phi^p(y)\phi\left(y - \sqrt{2}d\right)dy + \cdots, \end{aligned}$$

where A_1 is given by (6.10). Each of the integrals above can be evaluated by expanding the pdf ϕ about $d = 0$, as was done in (6.9). The fourth and final expression for $P(d)$ given by Gupta and Sobel (1957) uses the result of Seal (1954) that the distribution of $D = (M_p - Y)/\sigma$ is asymptotically normal as p tends to infinity. It follows directly from Seal's result that the third and higher central moments of D tend to the corresponding moments of the standard normal distribution. Since the coefficients involving ν in A_{-k} in (6.10) tend to unity as $\nu \to \infty$, it follows that the third and higher central moments of $Z = (M_p - Y)/S$ tend to the corresponding moments of the standard normal distribution as both ν and p tend to infinity. It is therefore reasonable to approximate the distribution of $W = (Z - E(Z))/\sqrt{Var(Z)}$ by a Gram-Charlier expansion in the Edgeworth form, where

$$E(Z) = A_{-1} a_{p,1}$$

and

$$Var(Z) = A_{-2}\left(a_{p,2} + 1\right) - \left(A_{-1} a_{p,1}\right)^2.$$

Here, $a_{p,i}$ denotes the ith moment of the largest of p independent standard normal random variables. Using equation (17.7.3) of Cramér (1951) and letting $d_s = (\sqrt{2}d - E(Z))/\sqrt{Var(Z)}$, Gupta and Sobel obtained

$$\begin{aligned} P(d) &= \Pr\left(Z < \sqrt{2}d\right) \\ &= \Pr\left(W < d_s\right) \\ &= \Phi\left(d_s\right) - \frac{\alpha_3}{3!}\phi^{(2)}\left(d_s\right) \\ &\quad + \frac{\alpha_4}{4!}\phi^{(3)}\left(d_s\right) + \frac{10\alpha_3^2}{6!}\phi^{(5)}\left(d_s\right) \\ &\quad - \frac{\alpha_5}{5!}\phi^{(4)}\left(d_s\right) - \frac{35\alpha_3\alpha_4}{7!}\phi^{(6)}\left(d_s\right) - \frac{280\alpha_3^3}{9!}\phi^{(8)}\left(d_s\right) + \cdots, \end{aligned}$$

where

$$\alpha_k = \frac{\kappa_k}{\sqrt{\kappa_2}} \qquad (6.12)$$

is the kth standardized cumulant of Z obtained from the moments around the origin.

In a related development, Gupta (1963) studied the above case $\rho = 1/2$ and showed that $P(d) = P(d; \nu)$ satisfies

$$\frac{dP(d;\nu)}{dd} + \nu\left\{P(d;\nu) - P(d;\nu+2)\right\} = 0, \qquad (6.13)$$

which is Hartley's differential-difference equation for the probability integral of a general class of statistics known as Studentized statistics.

Using Hartley's solution (obtained using the theory of characteristics), Gupta obtained an approximation for $P(d;\nu)$ in powers of $1/\nu$ and remarked that it can be computed by using the Gauss-Hermite quadrature. Gupta et al. (1985) extended this result for any $\rho > 0$ and showed that $P(d)$ satisfies (6.13) in this case too. In this case the approximation for $P(d)$ in powers of $1/\nu$ is

$$P(d) = G(d,\ldots,d) + \sum_{k=1}^{m} L_k(d), \tag{6.14}$$

where L_k is the kth correction term and G is the joint cdf of a p-variate normal distribution with zero means, common variance σ^2, and the equicorrelation structure $r_{ij} = \rho$, $i \neq j$. Letting $G^{(k)}(d)$ denote the kth-order derivative of $G(d,\ldots,d)$ with respect to d, the first four correction terms can be written as

$$L_1(d) = \frac{1}{d}\left\{\alpha^{(2)} - \alpha^{(1)}\right\},$$

$$L_2(d) = \frac{1}{6\nu^2}\left\{3\alpha^{(4)} - 10\alpha^{(3)} + 9\alpha^{(2)} - 2\alpha^{(1)}\right\},$$

$$L_3(d) = \frac{1}{6\nu^3}\left\{\alpha^{(6)} - 7\alpha^{(5)} + 17\alpha^{(4)} - 17\alpha^{(3)} + 6\alpha^{(2)}\right\},$$

and

$$\begin{aligned} L_4(d) = \frac{1}{360\nu^4}\Big\{&15\alpha^{(8)} - 180\alpha^{(7)} + 830\alpha^{(6)} - 1848\alpha^{(5)} + 2015\alpha^{(4)} \\ &-900\alpha^{(3)} + 20\alpha^{(2)} + 48\alpha^{(1)}\Big\}, \end{aligned}$$

where

$$\alpha^{(k)} = \frac{1}{2^k}\varphi^{(k)}(d), \qquad k = 1, 2, \ldots, 8$$

and the first eight $\varphi^{(k)}(d)$ are

$$\begin{aligned}
\varphi^{(1)}(d) &= dG^{(1)}(d), \\
\varphi^{(2)}(d) &= d^2G^{(2)}(d) + dG^{(1)}(d), \\
\varphi^{(3)}(d) &= d^3G^{(3)}(d) + 3d^2G^{(2)}(d) + dG^{(1)}(d), \\
\varphi^{(4)}(d) &= d^4G^{(4)}(d) + 6d^3G^{(3)}(d) + 7d^2G^{(2)}(d) + dG^{(1)}(d), \\
\varphi^{(5)}(d) &= d^5G^{(5)}(d) + 10d^4G^{(4)}(d) + 25d^3G^{(3)}(d) + 15d^2G^{(2)}(d) \\
&\quad +dG^{(1)}(d), \\
\varphi^{(6)}(d) &= d^6G^{(6)}(d) + 15d^5G^{(5)}(d) + 65d^4G^{(4)}(d) + 90d^3G^{(3)}(d)
\end{aligned}$$

$$
\begin{aligned}
& +31d^2G^{(2)}(d) + dG^{(1)}(d), \\
\varphi^{(7)}(d) &= d^7G^{(7)}(d) + 21d^6G^{(6)}(d) + 140d^5G^{(5)}(d) + 350d^4G^{(4)}(d) \\
& +301d^3G^{(3)}(d) + 63d^2G^{(2)}(d) + dG^{(1)}(d), \\
\varphi^{(8)}(d) &= d^8G^{(8)}(d) + 28d^7G^{(7)}(d) + 266d^6G^{(6)}(d) + 1050d^5G^{(5)}(d) \\
& +1701d^4G^{(4)}(d) + 966d^3G^{(3)}(d) + 127d^2G^{(2)}(d) \\
& +dG^{(1)}(d). \qquad (6.15)
\end{aligned}
$$

Thus the evaluation of $P(d)$ in (6.14) involves that of $G^{(k)}$ for $k = 0, 1, \ldots, 8$, and we shall discuss in Chapter 8 how the latter can be performed.

6.3 John's Probability Integrals

John (1961) provided alternative formulas for the evaluation of the probability integral. Although the method is discussed in detail only for the bivariate case, it has wider applicability in the sense that it can be adopted to obtain the probability integral of the multivariate t distribution for any dimension.

Let $\mathbf{X}$ be a p-variate vector having the central t distribution with degrees of freedom ν and correlation matrix $\mathbf{R}$. Using the definition that $\mathbf{X}$ can be represented as $(Z_1/S, Z_2/S, \ldots, Z_p/S)$, where $\mathbf{Z}$ is a p-variate normal random vector with correlation matrix $\mathbf{R}$ and $\nu S^2/\sigma^2$ is an independent chi-squared random variable with degrees of freedom ν, one can show that the characteristic function of $\mathbf{X}$ is

$$
\begin{aligned}
E\left(\exp\left(i\mathbf{t}^T\mathbf{X}\right)\right) &= E\left(E\left(\exp(i\mathbf{t}^T\mathbf{Z}/s \mid S = s\right)\right) \\
&= \frac{1}{\Gamma(\nu/2)} \int_0^\infty x^{\nu/2-1} \exp\left(-x - \frac{\nu}{4x}\mathbf{t}^T\mathbf{R}^{-1}\mathbf{t}\right) dx.
\end{aligned}
$$

In the case $p = 2$ with the equicorrelation structure $r_{ij} = \rho$, $i \neq j$, the above expression reduces to

$$
\begin{aligned}
E\left(\exp\left(it_1X_1 + it_2X_2\right)\right) &= \frac{1}{\Gamma(\nu/2)} \int_0^\infty x^{\nu/2-1} \left\{\sum_{i=0}^\infty \frac{1}{i!}\left(-\frac{\nu\rho}{2x}\right)^i t_1^i t_2^i\right\} \\
&\quad \times \exp\left\{-x - \frac{\nu}{4x}\left(t_1^2 + t_2^2\right)\right\} dx.
\end{aligned}
$$

By the inversion theorem, John (1961) derived the corresponding joint pdf as an infinite series of one-dimensional integrals. Integrating the infinite series term by term, the probability integral becomes

$$P(y_1, y_2; \nu, \rho) = y_{\nu,0}(y_1, y_2) + \frac{1}{2\pi} \sum_{i=1}^{\infty} \frac{\rho^i}{i!} y_{\nu,i}(y_1, y_2),$$

where

$$\begin{aligned} y_{\nu,0}(y_1, y_2) &= \frac{1}{\Gamma(\nu/2)} \int_0^{\infty} x^{\nu/2-1} \exp(-x) \\ &\quad \times \Phi\left(\frac{\sqrt{2x}y_1}{\sqrt{\nu}}\right) \Phi\left(\frac{\sqrt{2x}y_2}{\sqrt{\nu}}\right) dx \end{aligned}$$

and

$$\begin{aligned} y_{\nu,i}(y_1, y_2) &= \frac{1}{\Gamma(\nu/2)} \int_0^{\infty} x^{\nu/2-1} \exp\left[-x\left\{1 + \frac{1}{\nu}\left(y_1^2 + y_2^2\right)\right\}\right] \\ &\quad \times H_{i-1}\left(\frac{\sqrt{2x}y_1}{\sqrt{\nu}}\right) H_{i-1}\left(\frac{\sqrt{2x}y_2}{\sqrt{\nu}}\right) dx \end{aligned}$$

for $i = 1, 2, \ldots$. Here, $\Phi(\cdot)$ is the cdf of the standard normal distribution and H_k denotes the Hermite polynomial of order k defined by (6.11). John provided explicit algebraic expressions for $y_{\nu,i}$ for $i = 1, 2, \ldots, 6$. The first three of them are

$$y_{\nu,1}(y_1, y_2) = z^{-\nu/2},$$

$$y_{\nu,2}(y_1, y_2) = y_1 y_2 z^{-(\nu/2+1)},$$

and

$$y_{\nu,3}(y_1, y_2) = \left(1 + \frac{2}{\nu}\right) y_1^2 y_2^2 z^{-(\nu/2+2)} - \left(y_1^2 + y_2^2\right) z^{-(\nu/2+1)} + z^{-\nu/2},$$

where $z = (y_1^2 + y_2^2)/\nu + 1$. In principle, explicit expressions for $y_{\nu,i}$ can be obtained for any $i \geq 1$. To evaluate $y_{\nu,0}$, the integration has to be done numerically. John tabulated values of this quantity for $\nu = 11, 12$ using Gauss' formula for a numerical quadrature (Kopal, 1955, page 371). He also provided several useful recursion relations. For example, values of $y_{\nu,0}(y_1, y_2)$ for y_1 negative or y_2 negative or both negative can be found from the formulas

$$y_{\nu,0}(y_1, y_2) = T_\nu(y_2) - y_{\nu,0}(-y_1, y_2),$$

$$y_{\nu,0}(y_1, y_2) = T_\nu(y_1) - y_{\nu,0}(y_1, -y_2),$$

and

$$y_{\nu,0}(y_1, y_2) = 1 + y_{\nu,0}(-y_1, -y_2) - T_\nu(-y_1) - T_\nu(-y_2),$$

where T_ν is the cdf of the Student's t distribution with ν degrees of freedom.

6.4 Amos and Bulgren's Probability Integrals

In a widely quoted paper, Amos and Bulgren (1969) derived several representations for (6.2) in terms of series and simple one-dimensional quadratures, together with efficient computational procedures for the special functions used in their numerical evaluation. One of the quadrature formulas given is

$$\begin{aligned} P &= \frac{1}{2\pi(\nu+1)(1+\gamma_1^2+\gamma_2^2)^{\nu/2}} \\ &\quad \times \int_{\theta_1}^{\theta_2} {}_2F_1\left(1, \frac{\nu}{2}; \frac{\nu+3}{2}; 1 - c^2\cos^2(\theta-\phi)\right) d\theta \\ &\quad - \frac{\Gamma((\nu+1)/2)}{\sqrt{\pi}\Gamma(\nu/2)(1+\gamma_1^2+\gamma_2^2)^{\nu/2}} \\ &\quad \times \int_{\theta_1}^{\theta_2} \frac{I\{\cos(\theta-\phi)<0\}\cos(\theta-\phi)}{\{1-c^2\cos^2(\theta-\phi)\}^{(\nu+1)/2}} d\theta, \end{aligned}$$

where ${}_2F_1$ is the Gauss hypergeometric function, $I\{\}$ is the indicator function,

$$c = \sqrt{\frac{\gamma_1^2+\gamma_2^2}{1+\gamma_1^2+\gamma_2^2}},$$

$$\gamma_1 = (y_2 + y_1)\sqrt{\frac{\lambda_1}{2\nu}},$$

$$\gamma_2 = (y_2 - y_1)\sqrt{\frac{\lambda_2}{2\nu}},$$

$$\theta_1 = \pi - \arctan\sqrt{\frac{1+\rho}{1-\rho}},$$

$$\theta_2 = \pi + \arctan\sqrt{\frac{1+\rho}{1-\rho}},$$

$$\phi = \begin{cases} \arctan(\gamma_2/\gamma_1), & \text{if } \gamma_1 > 0, \\ \pi + \arctan(\gamma_2/\gamma_1), & \text{if } \gamma_1 < 0, \end{cases}$$

$$\lambda_1 = \frac{1}{1+\rho},$$

and

$$\lambda_2 = \frac{1}{1-\rho}.$$

One of the series formulas given is

$$P = \frac{1}{2\sqrt{\pi}\Gamma(\nu/2)} \sum_{k=0}^{\infty} \frac{(-c)^k}{(1+\gamma_1^2+\gamma_2^2)^{\nu/2}} \frac{\Gamma((\nu+k)/2)}{\Gamma((1+k)/2)} \times \int_{\theta_1}^{\theta_2} \cos^k(\theta-\phi) d\theta. \tag{6.16}$$

For the special case $\nu = 1$, P can be reduced to the closed-form expression

$$P = \frac{1}{\pi} \arctan\left(\frac{2v}{u^2+v^2-1}\right) + I\left\{u^2+v^2<1\right\},$$

where

$$u = \frac{2r\sin\phi}{A(1+r^2+2r\cos\phi)},$$

$$v = \frac{1-r^2}{A(1+r^2+2r\cos\phi)},$$

$$r = \frac{\sqrt{\gamma_1^2+\gamma_2^2}}{1+\sqrt{1+\gamma_1^2+\gamma_2^2}},$$

and

$$A = \tan\left(\frac{\theta_2-\pi}{2}\right).$$

If in addition $\rho = 0$, then the expression for P reduces further to

$$P = \frac{1}{2\pi}\left\{\arctan\left(\frac{y_1 y_2}{\sqrt{1+y_1^2+y_2^2}}\right) + \arctan y_1 + \arctan y_2 + \frac{\pi}{2}\right\}.$$

The advantage of these expressions over the ones given by Dunnett and Sobel (1954) is that these are easier to compute, especially for large degrees of freedom. For instance, the integral in θ in (6.16) can be expressed in terms of incomplete beta functions that are extensively tabulated. Amos and Bulgren (1969) numerically evaluated values of P for all combinations of $\rho = -0.9, -0.5, 0, 0.5, 0.9$ and $\nu = 1, 2, 5, 10, 25, 50$.

6.5 Steffens' Noncentral Probabilities

Consider the p-variate noncentral t distribution defined in (5.1). Motivated by the Studentized maximum and minimum modulus tests, Steffens (1970) studied the particular case for $p = 2$ and $\mathbf{R} = \mathbf{I}_p$. In this case, the joint pdf (5.1) reduces to

$$\begin{aligned} f(x_1, x_2) &= \exp\left(-\frac{\xi_1^2 + \xi_2^2}{2}\right) \frac{1}{\pi \Gamma(\nu/2)} \sum_{k=0}^{\infty} \sum_{l=0}^{\infty} \frac{\Gamma((\nu + k + l)/2 + 1)}{k! l! \nu^{(k+l)/2+1}} \\ &\quad \times \left(\sqrt{2}\xi_1 x_1\right)^k \left(\sqrt{2}\xi_2 x_2\right)^l \left(1 + \frac{x_1^2}{\nu} + \frac{x_2^2}{\nu}\right)^{(\nu+k+l+2)/2}, \end{aligned}$$

where $\xi_j = \mu_j/\sigma$ are the noncentrality parameters and ν denotes the degrees of freedom. The testing procedures involve maximum or minimum values of the components X_1 and X_2 and the computation of the corresponding probabilities. For this reason, Steffens (1970) derived series representations for probabilities of the form $P_1 = \Pr(\mid X_1 \mid \leq A, \mid X_2 \mid \leq A)$ and $P_2 = \Pr(\mid X_1 \mid > A, \mid X_2 > A)$. It is seen that

$$\begin{aligned} P_1 &= 2 \exp\left(-\frac{\xi_1^2 + \xi_2^2}{2}\right) \sum_{k=0}^{\infty} \sum_{l=0}^{\infty} \frac{\left(\xi_1^2/2\right)^k \left(\xi_2^2/2\right)^l}{k! l! B(k + 1/2, l + 1/2)} \\ &\quad \times \int_0^{\pi/4} \left(\sin^{2k} v \cos^{2l} v + \sin^{2l} v \cos^{2k} v\right) \\ &\quad \times I_\alpha\left(k + l + 1, \frac{\nu}{2}\right) dv \end{aligned}$$

and

$$\begin{aligned} P_2 &= 2 \exp\left(-\frac{\xi_1^2 + \xi_2^2}{2}\right) \sum_{k=0}^{\infty} \sum_{l=0}^{\infty} \frac{\left(\xi_1^2/2\right)^k \left(\xi_2^2/2\right)^l}{k! l! B(k + 1/2, l + 1/2)} \\ &\quad \times \int_0^{\pi/4} \left(\sin^{2k} v \cos^{2l} v + \sin^{2l} v \cos^{2k} v\right) \\ &\quad \times \left\{1 - I_\beta\left(k + l + 1, \frac{\nu}{2}\right)\right\} dv, \end{aligned}$$

where I_x denotes the incomplete beta function ratio, $\alpha = A^2 \sec^2 v / (\nu + A^2 \sec^2 v)$, and $\beta = A^2 \mathrm{cosec}^2 v / (\nu + A^2 \mathrm{cosec}^2 v)$. Using these representations, Steffens estimated values of the critical points A for all combinations of $\nu = 1, 2, 5, 10, 20, 50, \infty$ and $\xi_1, \xi_2 = 0(1)5$ for the significance level 0.05. In a more recent development, Bohrer et al. (1982) developed a flexible algorithm to compute probabilities of the form $\Pr(c_{11} \leq X_p \leq c_{21}, \ldots, c_{1p} \leq X_p \leq c_{2p})$ associated with the noncentral p-variate distribution (5.1).

6.6 Dutt's Probability Integrals

Dutt (1975) obtained a Fourier transform representation for the probability integral of a central p-variate t distribution with degrees of freedom ν and correlation matrix $\mathbf{R}$

$$P(y_1, \ldots, y_p) = \int_{-\infty}^{y_1} \cdots \int_{-\infty}^{y_p} f(x_1, \ldots, x_p; \nu) dx_p \cdots dx_1. \quad (6.17)$$

Using the definition of multivariate t, one can rewrite (6.17) as

$$P(y_1, \ldots, y_p) = \frac{2}{2^{\nu/2}\Gamma(\nu/2)} \int_0^{\infty} z^{\nu-1} \exp\left(-z^2/2\right) G\left(\hat{y}_1, \ldots, \hat{y}_p\right) dz, \quad (6.18)$$

where $\hat{y}_k = y_k z/\sqrt{\nu}$, $k = 1, \ldots, p$ and G is the joint cdf of the multivariate normal distribution with zero means and correlation matrix $\mathbf{R}$. In the case $y_k = 0$, one has P independent of ν and

$$P(y_1, \ldots, y_p) = G(0, \ldots, 0).$$

Explicit forms of G for $p = 2, 3, 4$ in terms of the D-functions are given in Dutt (1973). The D-functions are integral forms over $(-\infty, \infty)$ defined by

$$D_k(t_1, \ldots, t_p; \mathbf{R}) = \frac{\left|i^k\right|}{(2\pi)^k} \int_{-\infty}^{\infty} \cdots \int_{-\infty}^{\infty} \frac{d_k}{s_1 \cdots s_k} \times \exp\left(i \sum_{l=0}^{k} t_l s_l - \sum_{l=0}^{k} s_l^2 \Big/ 2\right) ds_k \cdots ds_1,$$

where the first five d_k are

$$\begin{aligned} d_1 &= 1, \\ d_2 &= d_{12}, \\ d_3 &= d_{12+13+23} - \left(d_{12} + d_{13} + d_{23}\right), \\ d_4 &= -d_{12+13+23+14+24+34} + d_{12+13+23} + d_{12+14+24} + d_{13+14+34} \end{aligned}$$

$$
\begin{aligned}
& +d_{23+24+34} - (d_{12} + d_{13} + d_{23} + d_{14} + d_{24} + d_{34}), \\
d_5 \quad = \quad & -d_{12+13+23+24+34+15+25+35+45} + d_{12+13+23+14+24+34} \\
& +d_{12+13+23+15+25+35} + d_{12+14+24+15+25+45} \\
& +d_{13+14+34+15+35+45} + d_{23+24+34+25+35+45} \\
& -\big(d_{12+13+23} + d_{12+14+24} + d_{12+15+25} + d_{13+14+34} \\
& +d_{13+15+35} + d_{14+15+45} + d_{23+24+34} + d_{23+25+35} \\
& +d_{24+25+45} + d_{34+35+45}\big) + d_{12} + d_{13} + \cdots + d_{45},
\end{aligned}
$$

and

$$
d_{p_1 q_1 + \cdots + p_m q_m} = 1 - \exp\left\{-\left(r_{p_1 q_1} s_{p_1} s_{q_1} + \cdots + r_{p_m q_m} s_{p_m} s_{q_m}\right)\right\}.
$$

Using the notation

$$
D_{k:j_1,\ldots,j_k} = D_k\left\{t_{j_1}, \ldots, t_{j_k}; \mathbf{R}\left(t_{j_1}, \ldots, t_{j_k}\right)\right\},
$$

where $\mathbf{R}\left(t_{j_1}, \ldots, t_{j_k}\right)$ is the correlation matrix based on the subscripts $j_1, \ldots, j_k$, Dutt (1973) provided the following explicit forms for G

$$
G(t_1, t_2) = \{1 - \Phi(t_1)\}\{1 - \Phi(t_2)\} + D_{2:1,2},
$$

$$
\begin{aligned}
G(t_1, t_2, t_3) = & \{1 - \Phi(t_1)\}\{1 - \Phi(t_2)\}\{1 - \Phi(t_3)\} \\
& + \{1 - \Phi(t_1)\} D_{2:2,3} + \{1 - \Phi(t_2)\} D_{2:1,3} \\
& + \{1 - \Phi(t_3)\} D_{2:1,2} + D_{3:1,2,3},
\end{aligned}
$$

and

$$
\begin{aligned}
G(t_1, t_2, t_3, t_4) = & \prod_{k=1}^{4} \{1 - \Phi(t_k)\} + \{1 - \Phi(t_1)\}\{1 - \Phi(t_2)\} D_{2:3,4} \\
& + \{1 - \Phi(t_1)\}\{1 - \Phi(t_3)\} D_{2:2,4} \\
& + \{1 - \Phi(t_2)\}\{1 - \Phi(t_3)\} D_{2:1,4} \\
& + \{1 - \Phi(t_1)\}\{1 - \Phi(t_4)\} D_{2:2,3} \\
& + \{1 - \Phi(t_2)\}\{1 - \Phi(t_4)\} D_{2:1,3} \\
& + \{1 - \Phi(t_3)\}\{1 - \Phi(t_4)\} D_{2:1,2} \\
& + \{1 - \Phi(t_1)\} D_{3:2,3,4} \{1 - \Phi(t_2)\} D_{3:1,3,4} \\
& + \{1 - \Phi(t_3)\} D_{3:1,2,4} + \{1 - \Phi(t_4)\} D_{3:1,2,3} \\
& + D_{4:1,2,3,4}.
\end{aligned}
$$

A much simplified representation for G in terms of the error function, erf$(\cdot)$, and integral forms over $(0, \infty)$, denoted as the D^* functions, is given in a later paper by Dutt (1975). These D^*-functions are defined by

$$D_k^*(t_1, \ldots, t_p; \mathbf{R}) = \frac{2}{(2\pi)^k} \int_0^\infty \cdots \int_0^\infty \frac{d_k^*}{s_1 \cdots s_k} \times \exp\left(-\sum_{l=0}^{k} s_l^2 \Big/ 2\right) ds_k \cdots ds_1, \quad (6.19)$$

where for the first few k are

$$\begin{aligned}
d_1^* &= \sin(t_1 s_1), \\
d_2^* &= e_{-12} \cos_{1-2} - e_{12} \cos_{1+2}, \\
d_3^* &= e_{12+13+23+14+24+34} \cos_{1+2+3+4} \\
&\quad + e_{12-13-23-14-24+34} \cos_{-1-2+3+4} \\
&\quad + e_{-12+13-23-14+24-34} \cos_{-1+2-3+4} \\
&\quad + e_{-12-13+23+14-24-34} \cos_{1-2-3+4} \\
&\quad - e_{-12-13+23-14+24+34} \cos_{-1+2+3+4} \\
&\quad - e_{-12+13-23+14-24+34} \cos_{1-2+3+4} \\
&\quad - e_{12-13-23+14+24-34} \cos_{1+2+3+4} \\
&\quad - e_{12+13+23-14-24-34} \cos_{1+2+3-4}
\end{aligned}$$

and for notation

$$\begin{aligned}
e_{p_1 q_1 + \cdots + p_m q_m} &= \exp\{-(r_{p_1 q_1} s_{p_1} s_{q_1} + \cdots + r_{p_m q_m} s_{p_m} s_{q_m})\}, \\
\sin_{p_1 + \cdots + p_m} &= \sin(t_{p_1} s_{p_1} + \cdots + t_{p_m} s_{p_m}), \\
\cos_{p_1 + \cdots + p_m} &= \cos(t_{p_1} s_{p_1} + \cdots + t_{p_m} s_{p_m}).
\end{aligned}$$

(A negative sign on the index $p_1 q_1$ corresponds to $+r_{p_1 q_1} s_{p_1} s_{q_1}$ and $-p_1$ corresponds to $-t_{p_1} s_{p_1}$.) Important special cases of these functions are

$$D_1^*(y) = \frac{1}{2} \operatorname{erf}\left(\frac{y}{\sqrt{2}}\right),$$

$$D_2^*(0, 0; \mathbf{R}) = \frac{1}{2\pi} \arcsin(r_{12}),$$

and

$$D_k^*(\mathbf{0}; \mathbf{R}) \equiv 0, \qquad \text{for } k \text{ odd}.$$

Using the abbreviation that

$$D_{k:j_1,\ldots,j_k}^* = D_k^*\{t_{j_1}, \ldots, t_{j_k}; \mathbf{R}(t_{j_1}, \ldots, t_{j_k})\},$$

Dutt (1975) provided the following representation for G

$$\begin{aligned} G(t_1,\ldots,t_p) &= \left(\frac{1}{2}\right)^p - \left(\frac{1}{2}\right)^{p-1} \sum_{k=1}^{p} D^*_{1:k} + \left(\frac{1}{2}\right)^{p-2} \sum_{k<l=1}^{p} D^*_{2:kl} \\ &\quad + \left(\frac{1}{2}\right)^{p-3} \sum_{k<l<m=1}^{p} D^*_{3:klm} + \cdots + D_{p:1,\ldots,p}. \end{aligned}$$

Hence, by (6.18), the computation of P in (6.17) can be achieved by successive applications of the Gauss-Hermite quadrature formula using only positive Hermite zeros (Abramowitz and Stegun, 1964, page 924). There are several advantages for this approach. First, it is not necessary to invert the correlation matrix. In addition, (6.19) permits the use of Gauss quadrature formula that are remarkably effective in estimating the value of an integral from a few points, provided that the integral excluding the weighting function can be accurately approximated by a polynomial. Moreover, often the integrand separates as a product of two functions, one depending only on correlation coefficients and the other on the original limits of integration.

For selected correlation structures and several values of ν and $y = y_k$, $k = 1,\ldots,p$, Dutt (1975) computed values of P accurate up to six decimal places.

6.7 Amos' Probability Integral

For the equicorrelation structure $r_{ij} = \rho$, $i \neq j$ considered by Gupta and Sobel (1957) and Gupta (1963) – but with the common ρ taken to be any positive real number less than 1 – Amos (1978) derived the following simpler expression for the probability integral

$$\begin{aligned} P(d) &= \frac{2^{\nu-3/2}\Gamma\left((\nu+1)/2\right)}{\sqrt{\pi}\left(1+b^2\right)^{\nu/2}} \int_{-\infty}^{\infty} \exp\left(-\frac{dx^2}{2}\right) \\ &\qquad \times \Phi^p(x) \operatorname{erfc}\left(-\frac{cx}{\sqrt{2}}\right) dx, \end{aligned} \tag{6.20}$$

where erfc$(\cdot)$ is the complementary error function defined by

$$\operatorname{erfc}(x) = \frac{2}{\sqrt{\pi}} \int_x^{\infty} \exp\left(-z^2\right) dz$$

and a, b, c, d are constants given by

$$a = \sqrt{\frac{1-\rho}{\rho}},$$

$$\begin{aligned} b &= \frac{d}{\sqrt{\rho\nu}}, \\ c &= \frac{ab}{\sqrt{1+b^2}}, \\ d &= \frac{a^2}{1+b^2}. \end{aligned}$$

The reduction to (6.20) was obtained by means of a relationship between the parabolic cylinder function and the complementary error function. Amos (1978) suggested computing the integral (6.20) by locating the x_0 for which the derivative of the integrand is zero and then summing quadratures on intervals of length h to the left and right of x_0 until a limit of integration is reached or the truncation error is small enough. The motivation for this procedure comes from the fact that x_0 can vary widely with extreme parameter values, and h, which estimates the spread of the integrand, can be small or large. Thus, x_0 and h accommodate the parameters, producing meaningful results by preventing quadratures over tails that are negligible or preventing gross misjudgments of the scale of integration. Letting $g(x)$ denote the integrand of (6.20), Amos (1978) showed that the derivative of $\log g(x)$ decreases monotonically from ∞ to $-\infty$ as x traverses $(-\infty, \infty)$, guaranteeing a unique root x_0 of $g'(x) = 0$.

6.8 Fujikoshi's Probability Integrals

Fujikoshi (1988) provided asymptotic expansions as well as error bounds for the probability integral (6.17) when the correlation matrix $\mathbf{R} = \mathbf{I}_p$, the $p \times p$ identity matrix. Specifically, letting

$$a_{\delta,j}(y_1, \ldots, y_p) = \frac{d^j}{ds^j}\left\{\Phi\left(s^{-\delta/2}y_1\right)\cdots\Phi\left(s^{-\delta/2}y_p\right)\right\}\Big|_{s=1},$$

where $\delta = -1$, 1, and Φ denotes the cdf of the standard normal distribution, Fujikoshi established the following approximation for the probability integral

$$\begin{aligned} P(y_1, \ldots, y_p) &= \Phi(y_1)\cdots\Phi(y_p) + \sum_{j=1}^{k-1}\frac{1}{j!}a_{\delta,j}(y_1, \ldots, y_p) \\ &\quad \times E\left[\left\{\left(\frac{\chi_\nu^2}{\nu}\right)^\delta - 1\right\}^j\right], \end{aligned} \qquad (6.21)$$

which we shall denote by $A_{\delta,k}(y_1,\ldots,y_p)$. Fujikoshi also derived uniform and nonuniform error bounds for this approximation. Under the assumptions that

$$\bar{a}_{\delta,k} = \sup_{\mathbf{y}} |a_{\delta,k}(y_1,\ldots,y_p)| < \infty,$$

and

$$E\left\{\left(\frac{\chi_\nu^2}{\nu}\right)^k\right\} < \infty, \qquad E\left\{\left(\frac{\nu}{\chi_\nu^2}\right)^k\right\} < \infty,$$

the uniform bound takes the form

$$\begin{aligned} & \sup_{\mathbf{y}} |P(y_1,\ldots,y_p) - A_{\delta,k}(y_1,\ldots,y_p)| \\ \leq\ & \frac{1}{k!}\bar{a}_{\delta,k} E\left[\left\{\left(\frac{\chi_\nu^2}{\nu}\right) \vee \left(\frac{\nu}{\chi_\nu^2}\right) - 1\right\}^k\right]. \end{aligned}$$

Under the assumptions that

$$\bar{a}_{\delta,k}(l) = \sup_{\mathbf{y}} \left(1 + \parallel \mathbf{y} \parallel^l\right) |a_{\delta,k}(y_1,\ldots,y_p)| < \infty$$

and

$$E\left\{\left(\frac{\chi_\nu^2}{\nu}\right)^{k+l/2}\right\} < \infty, \qquad E\left\{\left(\frac{\nu}{\chi_\nu^2}\right)^k\right\} < \infty,$$

the nonuniform bound takes the form

$$\begin{aligned} & |P(y_1,\ldots,y_p) - A_{\delta,k}(y_1,\ldots,y_p)| \\ \leq\ & \frac{1}{k!}\left(1 + \parallel \mathbf{y} \parallel^l\right)^{-1} \bar{a}_{\delta,k}(l) E\left\{\left(\frac{\chi_\nu^2}{\nu}\right)^{l/2} \left|\frac{\chi_\nu^2}{\nu} - 1\right|^k + \left|\frac{\nu}{\chi_\nu^2}\right|^k\right\}. \end{aligned}$$

Clearly the latter bounds are improvements on the uniform bounds in the tail part of the multivariate t distribution. In the case $p = 1$, these results provide useful approximations for the univariate Student's t distribution – see Fujikoshi (1987) and Fujikoshi and Shimizu (1989). The special case of (6.21) for $y_j = y$ has been investigated more recently by Fujikoshi (1988, 1989, 1993), Fujikoshi and Shimizu (1990), and Shimizu and Fujikoshi (1997).

6.9 Probabilities of Cone

Consider the p-dimensional set

$$A_r(c) = \left\{\mathbf{x} : \mathbf{z}^T\mathbf{x} \leq r \parallel \mathbf{z} \parallel, \text{all } \mathbf{z} \text{ in } E(c)\right\}, \tag{6.22}$$

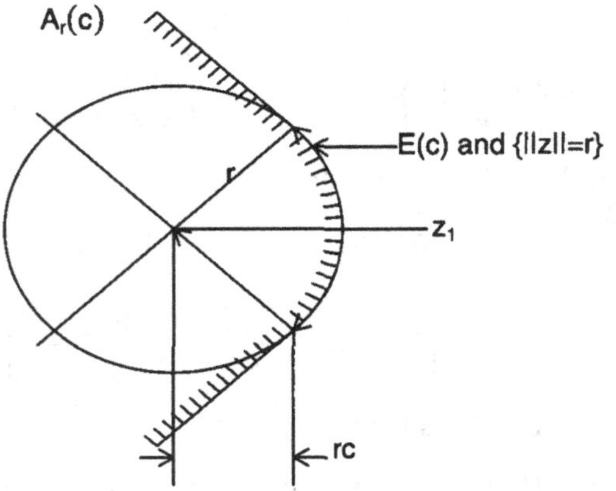

Fig. 6.1. The sets $A_r(c)$ and $E(c) \cap \{\| \mathbf{z} \| = r\}$ in two dimensions

where $E(c) = \{\mathbf{z} : z_1 \geq c \| \mathbf{z} \|\}$, $\| \mathbf{z} \| = \sqrt{\mathbf{z}^T \mathbf{z}}$, and c is a nonnegative constant. The set $E(c)$ is the cone, with vertex at the origin, which intersects origin-centered spheres in spherical caps. This is illustrated in Figure 6.1 for $p = 2$.

Bohrer (1973) studied the analytical shape of $A_r(c)$ and the associated probability

$$p(c, r, p, \nu) = \Pr(\mathbf{X} \in A_r(c))$$

when $\mathbf{X}$ has the p-variate t distribution with mean vector $\mathbf{0}$, covariance matrix $\sigma^2 \mathbf{I}_p$, and degrees of freedom ν. The evaluation of $p(c, r, p, \nu)$ is of statistical interest and use in the construction of confidence bounds (Wynn and Bloomfield, 1971, Section 3; Bohrer and Francis, 1972, equation (2.3)) and in testing multivariate hypotheses (Kudô, 1963, Theorem 3.1, Section 5; Barlow et al., 1972, pages 136ff, 177).

As regards the shape, Bohrer showed that every two-dimensional section of A_r containing the z_1-axis is exactly the two-dimensional version of A_r illustrated in Figure 6.1. Thus, A_r is the solid of revolution about the z_1-axis that is swept out by the A_r in Figure 6.1. To express this more precisely in mathematical terms – for an $p \times 1$ vector $\mathbf{v}$ – define polar coordinates $R_{\mathbf{v}}$ and $\boldsymbol{\mu}_{\mathbf{v}} = \{\theta_{\mathbf{v}i}\}$, with $-\pi < \theta_{\mathbf{v}i} \leq \pi$, by

$$v_1 = R_{\mathbf{v}} \cos\theta_{\mathbf{v}1},$$

$$v_i = R_{\mathbf{v}} \cos\theta_{\mathbf{v}i} \prod_{j=1}^{i-1} \sin\theta_{\mathbf{v}j}, \quad i = 2, \ldots, p-1,$$

and

$$v_p = R_{\mathbf{v}} \prod_{j=1}^{i-1} \sin\theta_{\mathbf{v}j}.$$

Also define

$$\theta^* = \arccos c,$$

$$T_1 = \{\mathbf{x} : |\theta_{\mathbf{t}1}| \leq \theta^*, R_{\mathbf{t}} \leq r\},$$

$$T_2 = \{\mathbf{x} : \theta_{\mathbf{t}1} - \theta^* \in (0, \pi/2], R_{\mathbf{t}} \cos(\theta_{\mathbf{t}1} - \theta^*) \leq r\},$$

$$T_3 = \{\mathbf{x} : \theta_{\mathbf{t}1} + \theta^* \in [-\pi/2, 0), R_{\mathbf{t}} \cos(\theta_{\mathbf{t}1} + \theta^*) \leq r\},$$

and

$$T_4 = \{\mathbf{x} : |\theta_{\mathbf{t}1}| > \theta^* + \pi/2\}.$$

Then the set A_r is the union of the disjoint sets $T_1, \ldots, T_4$. As regards evaluating the probability $p(c, r, p, \nu)$, Bohrer (1973) derived the following expression

$$\begin{aligned} p(c, r, p, \nu) &= \frac{k(\theta^*)}{k(\pi/2)} \Pr\left(F_{p,\nu} \leq \frac{r^2}{\sigma^2}\right) + \frac{k(\pi/2 - \theta^*)}{k(\pi/2)} \\ &\quad + \frac{1}{2k(\pi/2)} \sum_{j=0}^{p-2} B\left(\frac{j+1}{2}, \frac{p-j-1}{2}\right) \binom{p-2}{j} \\ &\quad \times c^j (1-c^2)^{(p-2-j)/2} \Pr\left(F_{p-2-j,\nu} \leq \frac{r^2}{\sigma^2}\right), \end{aligned}$$

where $k(\theta)$ is given by

$$k(\theta) = \frac{(2^m m!)^2}{p!}(1-\cos\theta) - \frac{\cos\theta\sin^{2m}\theta}{p} - \sum_{l=1}^{m-1}\frac{\sin^{2(m-l)}\theta\cos\theta}{p-2l}\prod_{j=1}^{l}\frac{p-2j}{p+1-2j}$$

when $p = 2m+1$ is odd and by

$$k(\theta) = \frac{(p-1)!\theta}{2^{p-1}(m-1)!m!} - \frac{\sin^{p-1-2m}\theta\cos\theta}{p} - \sum_{l=1}^{m-1}\frac{\sin^{p-1-2l}\theta\cos\theta}{p-2l}\prod_{j=1}^{l}\frac{p+1-2j}{p+2-2j}$$

when $p = 2m$ is even. The statistical questions that motivate this work ask what radius r is required so that $p(c,r,p,\nu) = \alpha$ for preassigned values of α. For $p \leq 5$, Bohrer (1973) provided tables of these percentiles for $\alpha = 0.95$ and 0.99 and for a range of (c,ν) pairs.

6.10 Probabilities of Convex Polyhedra

It is well known (Nicholson, 1943; Cadwell, 1951; Owen, 1956) that probabilities of polygons under bivariate normal distributions can be evaluated in terms of probabilities of right-angled triangles with vertices $(0,0)$, $(y_1,0)$, (y_1,y_2), $y_j > 0$, $j = 1,2$ under bivariate normal distributions with zero correlation. John (1964) proved an analogous result that probabilities of polygonal and angular regions for a given bivariate t distribution can be expressed in terms of $V_\nu(y_1,y_2)$, the integral of

$$f(x_1,x_2;\nu) = \frac{\Gamma((\nu+2)/2)}{\nu\pi\Gamma(\nu/2)}\left\{1+\frac{x_1^2+x_2^2}{\nu}\right\}^{-(\nu+2)/2}$$

over the right-angled triangles with vertices $(0,0)$, $(y_1,0)$, and (y_1,y_2). John (1964) also provided several formulas for evaluating $V_\nu(y_1,y_2)$. A formula in terms of the incomplete beta function is

$$V_\nu(y_1,y_2) = \frac{1}{2\pi}\arctan\left(\frac{y_2}{y_1}\right) - \frac{y_1 c^{\nu/2}}{4\pi\sqrt{\nu+y_1^2}}\sum_{k=0}^{\infty}c^k B_u\left(\frac{\nu}{2}+k+\frac{1}{2},\frac{1}{2}\right), \quad (6.23)$$

where

$$c = \frac{\nu}{\nu + y_1^2},$$

$$u = \frac{\nu + y_1^2}{\nu + y_1^2 + y_2^2},$$

and

$$B_x(a, b) = \int_x^1 w^{a-1}(1-w)^{b-1} dw$$

is the incomplete beta function. This series converges slowly unless y_1 is large in relation to ν. In the two cases ν odd and ν even, (6.23) can be reduced considerably. If $\nu = 2m$ for a positive integer m, then

$$V_{2m}(y_1, y_2) = \frac{\sqrt{1-c}}{4\pi} \sum_{k=0}^{m-1} c^k B_u\left(k + \frac{1}{2}, \frac{1}{2}\right) \tag{6.24}$$

while if $\nu = 2m + 1$ for a nonnegative integer m, then

$$\begin{aligned} V_{2m+1}(y_1, y_2) &= \frac{1}{2\pi} \arctan\left(\frac{y_2}{y_1}\right) - \frac{1}{4\pi} B_v\left(\frac{1}{2}, \frac{1}{2}\right) \\ &\quad + \frac{\sqrt{c(1-c)}}{4\pi} \sum_{k=0}^{m-1} c^k B_u\left(k + 1, \frac{1}{2}\right), \end{aligned} \tag{6.25}$$

where

$$v = \frac{\nu\left(\nu + y_1^2 + y_2^2\right)}{\nu\left(\nu + y_1^2 + y_2^2\right) + y_1^2 y_2^2}.$$

An attractive feature of (6.24) and (6.25) is that, when utilizing them for evaluating V_{2m} and V_{2m+1}, they are already evaluated for lower values of m also. If one performs the summations in the order indicated in the formulas, the addition of each term will yield values of V_{2m} or V_{2m+1} for the next higher value of m. This feature makes it particularly suitable for use in preparing tables.

A second formula for $V_\nu(y_1, y_2)$ given in John (1964) is an expansion in powers of $1/\nu$

$$V_\nu(y_1, y_2) = V_\infty(y_1, y_2) - \frac{\exp\left(-y_1^2/2\right)}{2\pi} \sum_{k=1}^{\infty} \frac{\nu^{-k}}{k!} U_k(y_1, y_2), \tag{6.26}$$

where the first three U_k are given by

$$U_1(y_1, y_2) = \frac{y_1^4}{4} W_1(y_1, y_2),$$

$$U_2(y_1, y_2) = y_1^6 \left\{ -\frac{1}{3} W_2(y_1, y_2) + \frac{y_1^2}{16} W_3(y_1, y_2) \right\},$$

and

$$U_3(y_1, y_2) = y_1^8 \left\{ \frac{3}{4} W_3(y_1, y_2) - \frac{y_1^2}{4} W_4(y_1, y_2) + \frac{y_1^4}{64} W_5(y_1, y_2) \right\},$$

where

$$W_\nu(y_1, y_2) = \int_0^{y_2/y_1} \left(1 + t^2\right)^\nu \exp\left(-\frac{y_1^2 t^2}{2}\right) dt.$$

The term V_∞ in (6.26) is the integral of $\exp\{-(y_1^2 + y_2^2)\}/(2\pi)$ over the right-angled triangle with vertices $(0, 0)$, $(y_1, 0)$, and $(0, y_2)$. The method of derivation for (6.26) is similar to the classical method employed by Fisher (1925) for expanding the probability integral of Student's t. Despite the complexity of (6.26) over (6.23), (6.26) should be preferred if ν is sufficiently large. The first two or three terms of (6.26) then can be expected to provide fairly accurate values of V_ν.

John (1964) also provided a recurrence relation and an approximation for $V_\nu(y_1, y_2)$; the latter proved to be satisfactory only when either ν is too small or y_2/y_1 is too large. In a subsequent paper, John (1966) extended this result to higher dimensions, by showing that the probabilities of the p-dimensional convex polyhedra with vertices $(0, 0, 0, 0, \ldots, 0)$, $(y_1, 0, 0, 0, \ldots, 0)$, $(y_1, y_2, 0, 0, \ldots, 0)$, $\ldots$, $(y_1, y_2, y_3, y_4, \ldots, y_p)$, $h_j > 0$, $j = 1, 2, \ldots, p$ under a p-variate t distribution with ν degrees of freedom can be expressed in terms of the function $V_\nu(y_1, y_2, \ldots, y_p)$, the integral of the p-variate t pdf

$$f(x_1, x_2, \ldots, x_p; \nu) = \frac{\Gamma((\nu + p)/2)}{(\nu\pi)^{p/2}\Gamma(\nu/2)} \left\{1 + \frac{x_1^2 + x_2^2 + \cdots + x_p^2}{\nu}\right\}^{-(\nu+p)/2}$$

over the same p-dimensional convex polyhedra. John also provided an important asymptotic expansion in powers of $1/\nu$ connecting $V_\nu(y_1, y_2, \ldots, y_p)$ with $V(y_1, y_2, \ldots, y_p)$, the integral of the p-variate normal pdf

$$f(x_1, x_2, \ldots, x_p; \infty) = (2\pi)^{-p/2} \exp\left\{-\left(x_1^2 + x_2^2 + \cdots + x_p^2\right)/2\right\}$$

over the same polyhedra discussed above. Up to the order of the term $O(1/\nu^2)$, the expansion is

$$
\begin{aligned}
& V_\nu (y_1, y_2, \ldots, y_p) \\
= \; & V(y_1, y_2, \ldots, y_p) + \frac{1}{4\nu}\Big\{ y_1 y_2 f(y_1, y_2) V(y_3, y_4, \ldots, y_p) \\
& \quad - y_1 \left(1 + y_1^2\right) f(y_1) V(y_2, y_3, \ldots, y_p) \Big\} \\
& + \frac{1}{96\nu^2}\Big\{ 3 y_1 y_2 y_3 y_4 f(y_1, y_2, y_3, y_4) V(y_5, y_6, \ldots, y_p) \\
& \quad - y_1 y_2 y_3 \left(2 + 9y_1^2 + 6y_2^2 + 3y_3^2\right) f(y_1, y_2, y_3) V(y_4, \ldots, y_p) \\
& \quad - y_1 y_2 \left(3 + 5y_1^2 + y_2^2 - 9y_1^4 - 9y_1^2 y_2^2 - 3y_2^4\right) \\
& \qquad \times V(y_3, \ldots, y_p) f(y_1, y_2) \\
& \quad + y_1 \left(3 + 5y_1^2 + 7y_1^4 - 3y_1^6\right) f(y_1) V(y_2, \ldots, y_p) \Big\} \\
& + o\left(\frac{1}{\nu^2}\right).
\end{aligned}
$$

In this formula, $V(y_m, y_{m+1}, \cdots, y_p)$ is to be replaced by 0 if $m \geq p+2$ and by 1 if $m = p+1$. In principle, there is no difficulty in determining further terms of this expansion, but the coefficients of higher powers of $(1/\nu)$ have rather complicated expressions. Other useful results given by John (1966) include recursion formulas connecting $V_\nu(y_1, y_2, \cdots, y_p)$ with $V_{\nu\pm2}(y_1, y_2, \cdots, y_p)$.

More recently, several authors have looked into the problem of computing multivariate t probabilities of the form

$$P = \int_A f(\mathbf{x}; \nu)\, d\mathbf{x}, \tag{6.27}$$

where $\mathbf{X}$ has the central multivariate t distribution with correlation matrix $\mathbf{R}$ and A is any convex region. Somerville (1993a, 1993b, 1993c, 1994) developed the first known procedures for evaluating P in (6.27). Let $\mathbf{M}\mathbf{M}^T$ be the Cholesky decomposition of $\mathbf{R}$ (where $\mathbf{M}$ is a lower triangular matrix) and set $\mathbf{X} = \mathbf{M}\mathbf{W}$. Then $\mathbf{W}$ is multivariate t with correlation matrix $\mathbf{I}_p$. If one further sets $r^2 = \mathbf{W}^T\mathbf{W}$, then $F = r^2/p$ has the well known F distribution with degrees of freedom p and ν. Let A be the region bounded by p hyperplanes and described by

$$\mathbf{G}\mathbf{W} \leq \mathbf{d},$$

where $\mathbf{G} = (\mathbf{g}_1, \cdots, \mathbf{g}_p)$ and the jth hyperplane is $\mathbf{g}_j^T\mathbf{W} = d_j$. For a random direction $\mathbf{c}$, let r be the distance from the origin to the boundary of A, that is, the smallest positive distance from the origin to the jth plane, $j = 1, \ldots, p$. Then an unbiased estimate of the integral P in (6.27) is

$$\Pr\left(F \leq r^2/p\right). \tag{6.28}$$

To implement the procedure, Somerville chose successive random directions $\mathbf{c}$ and obtained corresponding estimates of (6.28). The value of P was then taken as the arithmetic mean of the individual estimates.

Somerville (1997, 1998b) provided the following modification of the above procedure. Let r^* be the minimum distance from the origin to the boundary of A, that is, the smallest of the r for all random directions $\mathbf{c}$. Divide A into two regions, the portion inside the hypersphere of radius r^* and centered at the origin, and the region outside. The probability content of the hypersphere is

$$P_1 = \Pr\left(F \leq r^{*2}/p\right),$$

and this can be estimated as in Somerville (1993a, 1993b, 1993c, 1994). If $E(v)$ and $e(v)$, respectively, denote the cdf and the pdf of $v = 1/r$ (the reciprocal distance from the origin from and to the boundary of A), then the probability content of the outer region is

$$P_2 = \int_0^{1/r^*} E(v)e(v)dv.$$

Since $F = r^2/p$, the pdf of v is

$$e(v) = \frac{2\nu^{\nu/2}\Gamma\left((\nu+p)/2\right)}{\Gamma\left(\nu/2\right)\Gamma\left(p/2\right)} \frac{v^{\nu-1}}{\left(1+\nu v^2\right)^{(\nu+p)/2}}.$$

The strategy is to use some numerical method to estimate $E(v)$ and then evaluate the integral P_2 using the Gauss-Legendre quadrature. The approaches of Somerville (1997, 1998b) differ in that Somerville (1997) applied Monte Carlo techniques to estimate $E(v)$ while Somerville (1998b) used a binning procedure. It should be noted, however, that an approach similar to these had been introduced earlier by Deak (1990).

Somerville (1999a) provided an extension of the above methodologies to evaluate P in (6.27) when A is an ellipsoidal region. This has potential applications in the field of reliability (in particular relating to the computation of the tolerance factor for multivariate normal populations) and to the calculation of probabilities for linear combinations of central and noncentral chi-squared and F. In the coordinate system of the transformed variables $\mathbf{W}$, assume, without loss of generality, that the axes of the ellipsoid are parallel to the coordinate axes and the ellipsoid has the equation $(\mathbf{w}-\mathbf{u})^T\mathbf{B}^{-1}(\mathbf{w}-\mathbf{u}) = 1$, where $\mathbf{B}$ is a diagonal matrix with the ith element given by b_i. If the ellipsoid contains the origin, then for each random direction $\mathbf{c}$ there is a unique distance r to the boundary. An unbiased estimate of P is then given by

$$\Pr\left(F \le r^2/p\right).$$

If the ellipsoid does not contain the origin, then, for a random direction, a line from the origin in that direction either intersects the boundary of the ellipsoid at two points (say $r \ge r_*$) or does not intersect it at all. If the line intersects the boundary, then an unbiased estimate of P is given by the difference

$$\Pr\left(F \le r^2/p\right) - \Pr\left(F \le r_*^2/p\right).$$

If the line does not intersect the ellipsoid, an unbiased estimate is 0. As in the first procedure described above, this is repeated for successive random directions $\mathbf{c}$, each providing an unbiased estimate. The value of P is then taken as the arithmetic average. A modification of this procedure along the lines of Somerville (1997, 1998b) is described in Somerville (1999a).

Somerville (1999b) provided an application of his methods for multiple testing and comparisons by taking A in (6.27) to be

$$A = \left\{\mathbf{x} \in \Re^p : \max \mathbf{c}^T\mathbf{x} \le q/\sqrt{2}\right\}, \qquad \mathbf{c} \in B,$$

where B is the set of contrasts corresponding to the different hypotheses and $q > 0$. The purpose is to calculate the value of q for arbitrary $\mathbf{R}$ and ν and arbitrary sets B such that the probability content of A has a preassigned value γ. Somerville and Bretz (2001) have written two *Fortran 90* programs (QBATCH4.FOR and QINTER4.FOR) and two *SAS-IML* programs (QBATCH4.SAS and QINTER4.SAS) for this purpose. QINTER4.FOR and QINTER4.SAS are interactive programs, while the other two are batch programs. A compiled version of the *Fortran 90* programs that should run on any PC with Windows 95 or later can be found at

`http://pegasus.cc.ucf.edu/~somervil/home.html`

These programs implement the methodology described above to evaluate the probability content of A (A *Fortran 90* programs MVI3.FOR used to evaluate multivariate t integrals over any convex region is described in

Somerville (1998a). An extended *Fortran 90* programs MVELPS.FOR to evaluate multivariate t integrals over any ellipsoidal regions is described in Somerville (2001). The average running times for the latter program range from 0.075 and 0.109 second for $p = 2$ and 3, respectively, to 0.379 and 0.843 second for $p = 10$ and 20, respectively.). The so-called "Brent's method," an interactive procedure described in Press (1986), is used to solve for the value of q. The time to estimate the q values (with a standard error of 0.01) using QINTER4 or QBATCH4 range from 10 seconds for Dunnett's multiple comparisons procedure to 52 seconds for Tukey's procedure, using a 486-33 processor.

A problem that frequently arises in statistical analysis is to compute (6.27) when A is a rectangular region, that is,

$$P = \int_{a_1}^{b_1} \int_{a_2}^{b_2} \cdots \int_{a_p}^{b_p} f(x_1, x_2, \ldots, x_p)\, dx_p \cdots dx_2 dx_1. \quad (6.29)$$

Wang and Kennedy (1997) employed numerical interval analysis to compute P. The method is similar to the approaches of Corliss and Rall (1987) for univariate normal probabilities and Wang and Kennedy (1990) for bivariate normal probabilities. The basic idea is to apply the multivariate Taylor expansion to the joint pdf f. Letting $c_j = (a_j + b_j)/2$, the Taylor expansion of f at the mid point $(c_1, c_2, \ldots, c_p)$ is

$$\begin{aligned} & f(x_1, x_2, \ldots, x_p) \\ = & \sum_{k=0}^{m-1} \left\{ \sum_{]k[} \frac{1}{k_1! \cdots k_p!} \frac{\partial^k f(c_1, c_2, \ldots, c_p)}{\partial x_1^{k_1} \partial x_2^{k_2} \cdots \partial x_p^{k_p}} \prod_{j=1}^{p} (x_j - c_j)^{k_j} \right\} \\ & + \sum_{]m[} \frac{1}{m_1! \cdots m_p!} \frac{\partial^m f(\xi_1, \xi_2, \ldots, \xi_p)}{\partial x_1^{m_1} \partial x_2^{m_2} \cdots \partial x_p^{m_p}} \prod_{j=1}^{p} (x_j - c_j)^{m_j}, \end{aligned} \quad (6.30)$$

where ξ_j is contained in the integration region $[a_j, b_j]$ and $]k[$ denotes all possible partitions of k into p parts. For example, in the case $p = 3$, $]2[$ will result in 6 possible partitions of '2' into $\{k_1, k_2, k_3\}$: $\{0, 0, 2\}$, $\{0, 1, 1\}$, $\{0, 2, 0\}$, $\{1, 0, 1\}$, $\{1, 1, 0\}$, and $\{2, 0, 0\}$. The main problem with computing (6.30) is the presence of high-order partial derivatives of f. Defining

$$(f)_{k_1 k_2 \cdots k_p} = \frac{1}{k_1! k_2! \cdots k_p!} \frac{\partial^{k_1 + k_2 + \cdots + k_p} f}{\partial x_1^{k_1} \partial x_2^{k_2} \cdots \partial x_p^{k_p}}, \quad (6.31)$$

Wang and Kennedy derived the following recursive formula

$$
\begin{aligned}
(f)_{k_1 k_2 \cdots k_p} &= -\frac{1}{k_1}\left(1+\frac{\mathbf{x}^T\mathbf{R}^{-1}\mathbf{x}}{\nu}\right)^{-1} \\
&\times \sum_{l_1=0}^{k_1}\sum_{l_2=0}^{k_2}\cdots\sum_{l_p=0}^{k_p}\left\{\frac{p+\nu}{2}(k_1-l_1)+l_1\right\}(f)_{l_1 l_2 \cdots l_p} \\
&\times \left(1+\frac{\mathbf{x}^T\mathbf{R}^{-1}\mathbf{x}}{\nu}\right)_{k_1-l_1, k_2-l_2, \ldots, k_p-l_p}.
\end{aligned}
$$

With regard to the last quadratic term, it should be noted that higher than second-order partial derivatives are all zero. To carry out the computation of (6.31) for a given $(k_1, k_2, \ldots, k_p)$, one can

- first let one l_j be $k_j - 1$ (if this $k_j \neq 1$) and all the other l_j's be their corresponding k_j's;
- next let l_r and l_s be $k_r - 1$ and $k_s - 1$, respectively (if $k_r \neq 1$ and $k_s \neq 1$), while all the other l_j's take their corresponding k_j's;
- finally, let some l_j be $k_j - 2$ (if $k_j \geq 2$) and all other l_j's be the corresponding k_j's.

The total number of terms that contribute to computing $(f)_{k_1 k_2 \cdots k_p}$ is at most $p(p+3)/2$. Compared to the multivariate normal distribution, this number is larger (Wang and Kennedy, 1990). The following table gives the running times and the accuracy for computing (6.29) with $\nu = 10$.

Running time and accuracy for computing P in (6.29)

p	Running time (min)	$a_j = -0.5$ $b_j = 0.5$	$a_j = -0.4$ $b_j = 0.4$	$a_j = -0.3$ $b_j = 0.3$	$a_j = -0.2$ $b_j = 0.2$
10	80			2 sig	4 sig
9	70			3 sig	7 sig
8	85		0 sig	5 sig	10 sig
7	90		3 sig	8 sig	
6	110	3 sig	8 sig		
5	180	10 sig			

Another point to note about Wang and Kennedy's method is that when the integration region is near the origin it works better for larger ν, while when the integration region is off the origin it works better for smaller ν.

The main problem with Wang and Kennedy's (1997) method is that the calculation times required are too large even for low accuracy results (see the table above). Genz and Bretz (1999) proposed a new method for computing (6.29) by transforming the p-variate integrand into a product of univariate integrands. The method is similar to the one used by Genz (1992) for the multivariate normal integral.

Letting $\mathbf{M}\mathbf{M}^T$ be the Cholesky decomposition of $\mathbf{R}$, define the following transformations

$$X_j = \sum_{k=1}^{p} M_{j,k} Y_k,$$

$$Y_j = U_j \sqrt{\frac{\nu + \sum_{k=1}^{j-1} Y_k^2}{\nu + j - 1}},$$

$$U_j = T_{\nu+j-1}\left(Z_j\right),$$

and

$$Z_j = d_j + W_j\left(e_j - d_j\right),$$

where T_τ denotes the cdf of the univariate Student's t distribution with degrees of freedom τ,

$$d_j = T_{\nu+j-1}\left(\hat{a}_j\right),$$

$$e_j = T_{\nu+j-1}\left(\hat{b}_j\right),$$

$$\hat{a}_j = a_j' \sqrt{\frac{\nu + j - 1}{\nu + \sum_{k=1}^{j-1} y_k^2}},$$

$$\hat{b}_j = b_j' \sqrt{\frac{\nu + j - 1}{\nu + \sum_{k=1}^{j-1} y_k^2}},$$

$$a_j' = \frac{a_j - \sum_{k=1}^{j-1} m_{j,k} Y_k}{m_{j,j}},$$

and

$$b'_j = \frac{b_j - \sum_{k=1}^{j-1} m_{j,k} Y_k}{m_{j,j}}.$$

Applying the above transformations successively, Genz and Bretz reduced (6.29) to

$$P = (e_1 - d_1) \int_0^1 (e_2 - d_2) \cdots \int_0^1 (e_p - d_p) \int_0^1 d\mathbf{w} \quad (6.32)$$

$$= \int_0^1 \int_0^1 \cdots \int_0^1 f(\mathbf{w})\, d\mathbf{w}. \quad (6.33)$$

The transformation has the effect of flattening the surface of the original function, and P becomes an integral of $f(\mathbf{w}) = (e_1 - d_1) \cdots (e_p - d_p)$ over the $(p-1)$-dimensional unit hypercube. Hence, one has improved numerical tractability and (6.33) can be evaluated with different multidimensional numerical computation methods. Genz and Bretz considered three numerical algorithms for this: an acceptance-rejection sampling algorithm, a crude Monte Carlo algorithm, and a lattice rule algorithm.

- Acceptance-rejection sampling algorithm: Generate p-dimensional uniform random vectors $\mathbf{w}_1, \mathbf{w}_2, \ldots, \mathbf{w}_N$ and estimate P by

  $$\hat{P} = \frac{1}{N} \sum_{l=1}^{N} h(\mathbf{M}\mathbf{y}_l),$$

  where

  $$h(\mathbf{x}) = \begin{cases} 1 & \text{if } a_j \le x_j \le b_j,\ j = 1, 2, \ldots, p, \\ 0 & \text{otherwise} \end{cases}$$

  and

  $$y_{l,j} = T_{\nu+j-1}^{-1}(w_{l,j}) \sqrt{\frac{\nu + \sum_{k=1}^{j-1} y_k^2}{\nu + j - 1}},$$
  $$j = 1, 2, \ldots, p, \quad l = 1, 2, \ldots, N.$$

- A crude Monte Carlo algorithm: Generate $(p-1)$-dimensional uniform random vectors $\mathbf{w}_1, \mathbf{w}_2, \ldots, \mathbf{w}_N$ and estimate P by

  $$\hat{P} = \frac{1}{N} \sum_{l=1}^{N} f(\mathbf{w}_l),$$

  an unbiased estimator of the integral (6.33).

- A lattice rule algorithm (Joe, 1990; Sloan and Joe, 1994): Generate $(p-1)$-dimensional uniform random vectors $\mathbf{w}_1, \mathbf{w}_2, \ldots, \mathbf{w}_N$ and estimate P by

$$\hat{P} = \frac{1}{Nq} \sum_{l=1}^{N} \sum_{j=1}^{q} f\left(\left| 2\left\{ \frac{j}{q}\mathbf{z} + \mathbf{w}_l \right\} - \mathbf{1}_p \right| \right).$$

  Here N is the simulation size, usually very small, q corresponds to the fineness of the lattice, and $\mathbf{z} \in \Re^{p-1}$ denotes a strategically chosen lattice vector. Braces around vectors indicate that each component has to be replaced by its fractional part. One possible choice of $\mathbf{z}$ follows the good lattice points; see, for example, Sloan and Joe (1994).

For all three algorithms – to control the simulated error – one may use the usual error estimate of the means. Perhaps the most intuitive one of the three is the acceptance-rejection method. However, Deak (1990) showed that, among various methods, it is the one with the worst efficiency. Genz and Bretz (2001) proposed the use of the lattice rule algorithm. Bretz et al. (2001) provided an application of this algorithm for multiple comparison procedures.

The method of Genz and Bretz (1999) described above also includes an efficient evaluation of probabilities of the form

$$P = \int_{\mathbf{a}}^{\mathbf{b}} g(\mathbf{x}) f(\mathbf{x}) d\mathbf{x},$$

where $g(\mathbf{x})$ is some nuisance function. *Fortran* and *SAS-IML* codes to implement the method for $p \leq 100$ are available from the Web sites with URLs

`http://www.bioinf.uni-hannover.de/~betz/`

and

and `http://www.sci.wsu.edu/math/faculty/genz/homepage.`

6.11 Probabilities of Linear Inequalities

Let X be a random variable characterizing the "load," and let Y be a random variable determining the "strength" of a component. Then the probability that a system is "trouble-free" is $\Pr(Y > X)$. In a more complicated situation, the operation of the system may depend on a linear combination of random vectors, say $\mathbf{a}_1^T\mathbf{X}_1 + \mathbf{a}_2^T\mathbf{X}_2 + b$, and the probability of a trouble-free operation will be

$$\Pr\left(\mathbf{a}_1^T\mathbf{X}_1 + \mathbf{a}_2^T\mathbf{X}_2 + b > 0\right), \tag{6.34}$$

where $\mathbf{X}_j$ are independent k_j-dimensional random vectors, $\mathbf{a}_j$ are k_j-dimensional constant vectors, and b is a scalar constant. Absusev and Kolegova (2001) studied the problem of constructing unbiased, maximum likelihood, and Bayesian estimators of the probability (6.34) when $\mathbf{X}_j$ is assumed to have the multivariate t distribution with mean vector $\boldsymbol{\mu}_j$ and correlation matrix $\mathbf{R}_j$. If $\mathbf{x}_{11}, \ldots, \mathbf{x}_{1n_1}$ and $\mathbf{x}_{21}, \ldots, \mathbf{x}_{2n_2}$ are iid samples from the two multivariate t distributions, then – in the where case both $\boldsymbol{\mu}_j$ and $\mathbf{R}_j$ are unknown – it was established that the unbiased and the maximum likelihood estimators are

$$\begin{aligned}\hat{\Pr}\left(\mathbf{a}_1^T\mathbf{X}_1 + \mathbf{a}_2^T\mathbf{X}_2 + b > 0\right) &= \frac{\Gamma(n_1/2)\Gamma(n_2/2)}{\pi\Gamma((n_1-1)/2)\Gamma((n_2-1)/2)} \\ &\quad \times \int_{\Omega_1} \prod_{j=1}^{2} \left(1-\nu_j^2\right)^{(n_j-3)/2} d\nu_1 d\nu_2\end{aligned}$$

and

$$\check{\Pr}\left(\mathbf{a}_1^T\mathbf{X}_1 + \mathbf{a}_2^T\mathbf{X}_2 + b > 0\right) = \Phi\left(\frac{\mathbf{a}_1^T\bar{\mathbf{x}}_{n_1} + \mathbf{a}_2^T\bar{\mathbf{x}}_{n_2} + b}{\sqrt{\mathbf{a}_1^T\mathbf{S}_{n_1+1}\mathbf{a}_1 + \mathbf{a}_2^T\mathbf{S}_{n_2+1}\mathbf{a}_2}}\right),$$

respectively, where

$$\begin{aligned}\Omega_1 &= \Bigg\{\nu_j^2 < 1, j = 1, 2, \\ &\quad \sum_{j=1}^{2} \nu_j \sqrt{n_j \mathbf{a}_j^T \mathbf{S}_{n_j+1}\mathbf{a}_j} + \sum_{j=1}^{2} \mathbf{a}_j^T \bar{\mathbf{x}}_j + b > 0\Bigg\},\end{aligned}$$

$$\bar{\mathbf{x}}_{n_j} = \frac{1}{n_j}\sum_{m=1}^{n_j} \mathbf{x}_m,$$

$$(n_j+1)\,\bar{\mathbf{x}}_j = \sum_{m=1}^{n_j+1} \mathbf{x}_{jm},$$

$$(n_j+1)\,\mathbf{S}_{n_j+1} = \sum_{m=1}^{n_j+1} \left(\mathbf{x}_{jm} - \bar{\mathbf{x}}_j\right)\left(\mathbf{x}_{jm} - \bar{\mathbf{x}}_j\right)^T,$$

and $\mathbf{x}_{n_j+1} = \mathbf{x}$. A Bayesian estimator of (6.34) with unknown parameters $\boldsymbol{\mu}_j$ and $\mathbf{R}_j$ and the Lebesgue measure $p(\boldsymbol{\theta})d\boldsymbol{\theta} = d\boldsymbol{\mu}d\mathbf{R}$ was calculated to be

$$\begin{aligned}\Pr_B\left(\mathbf{a}_1^T\mathbf{X}_1 + \mathbf{a}_2^T\mathbf{X}_2 + b > 0\right) &= \prod_{j=1}^{2}\frac{\Gamma\left((n_j - k_j)/2\right) n_j^{k_j}}{\pi\Gamma\left((n_j-1)/2\right)(n_j+1)^{k_j}} \\ &\quad\times \prod_{j=1}^{2}\frac{\Gamma\left((n_j - k_j - 1)/2\right)}{\Gamma\left((n_j - 2k_j - 1)/2\right)} \\ &\quad\times \int_{\Omega_2}\prod_{j=1}^{2}\left(1 - z_j^2\right)^{\frac{n_j-3}{2}} dz_1 dz_2,\end{aligned}$$

where

$$\Omega_2 = \left\{z_j^2 < 1, j = 1, 2, \sum_{j=1}^{2} z_j\sqrt{n_j\mathbf{a}_j^T\mathbf{S}_{n_j+1}\mathbf{a}_j} + \sum_{j=1}^{2}\mathbf{a}_j^T\bar{\mathbf{x}}_j + b > 0\right\}.$$

This Bayesian estimator is biased and is related to the unbiased estimator via the relation

$$\Pr_B\left(\mathbf{a}_1^T\mathbf{X}_1 + \mathbf{a}_2^T\mathbf{X}_2 + b > 0\right) = A\hat{\Pr}\left(\mathbf{a}_1^T\mathbf{X}_1 + \mathbf{a}_2^T\mathbf{X}_2 + b > 0\right),$$

where

$$A = \prod_{j=1}^{2}\frac{\Gamma\left((n_j - k_j - 1)/2\right)\Gamma\left((n_j - k)/2\right)}{\Gamma\left(n_j - 2k_j - 1)/2\right)\Gamma\left(n_j/2\right)}\frac{(n_j+1)^{k_j}}{n_j^{k_j}}.$$

The coefficient A can be expanded as

$$A = 1 + \frac{k}{n} - \frac{k}{n-k} + O\left(\frac{1}{n^2}\right),$$

where $n = \max(n_1, n_2)$ and $k = \max(k_1, k_2)$. Therefore, the Bayesian estimator is asymptotically unbiased as $n \to \infty$.

Substantial literature is now available on problems concerning probabilities of the form (6.34) for various distributions. For a comprehensive and up-to-date summary, the reader is referred to Kotz et al. (2003).

6.12 Maximum Probability Content

Let $\mathbf{X}$ be a bivariate random vector with the joint pdf of the form

$$f(\mathbf{x}) = g\left((\mathbf{x} - \boldsymbol{\mu})^T\mathbf{R}^{-1}(\mathbf{x} - \boldsymbol{\mu})\right), \tag{6.35}$$

which, of course, includes the bivariate t pdf. Consider the class of rectangles

$$R(a) = \{(x_1, x_2) : |x_1| \leq a, |x_2| \leq \lambda/(4a)\}$$

with the area equal to λ. Kunte and Rattihalli (1984) studied the problem of characterizing the region R in this class for which the probability $P(R(a)) = \Pr(\mathbf{X} \in R(a))$ is maximum. As noted in Rattihalli (1981), the characterizations of such regions is useful for obtaining Bayes regional estimators when (i) the decision space is the class of rectangular regions and (ii) the loss function is a linear combination of the area of the region and the indicator of the noncoverage of the region. It was shown that, for any fixed $\lambda > 0$, the maximal set is

$$\{(x_1, x_2) : |x_1 - \mu_1| \leq c, |x_2 - \mu_2| \leq \lambda/(4c)\},$$

where c is given by

$$c = \sqrt{\frac{\lambda}{4}\sqrt{\frac{r^{22}}{r^{11}}}}.$$

Here, r^{ij} denotes the (i, j)th element of the inverse of $\mathbf{R}$. In particular, if $\boldsymbol{\mu} = \mathbf{0}$, $r^{12} = r^{21} = \rho$ and $|\rho| < 1$ in (6.35), then $P(R(a))$ is increasing for $a < \sqrt{\lambda}/2$ and is decreasing for $a > \sqrt{\lambda}/2$.

6.13 Monte Carlo Evaluation

Let $\mathbf{X}$ be a central p-variate t random vector with correlation matrix $\mathbf{R}$ and degrees of freedom ν. Vijverberg (1996) developed a family of simulators of the multivariate t probability $p = \Pr(\mathbf{X} \leq \mathbf{X}_0)$ based on Monte Carlo simulation and recursive importance sampling. We shall provide the basic steps of this rather complicated but powerful procedure.

Define $\mathbf{Z} = \mathbf{AX}$, where $\mathbf{A}$ is an upper triangular matrix such that $\mathbf{A}^T\mathbf{A} = \mathbf{R}$. Then it is well known that the pdf of $\mathbf{Z}$ can be expressed as a product of univariate Student's t pdfs

$$f(\mathbf{z}) = \prod_{k=1}^{p} f_1\left(z_k; \sigma_k^2, \nu_k\right),$$

where

$$\nu_k = \nu + p - k, \qquad k = 1, 2, \ldots, p,$$

$$\sigma_k^2 = \frac{\nu + y_{k+1}^2 + \cdots + y_p^2}{\nu_k}, \qquad k = 1, 2, \ldots, p - 1,$$

$$\sigma_n^2 = 1,$$

and

$$f_1\left(x;\sigma^2,\nu\right) = \frac{\Gamma\left((\nu+1)/2\right)}{\sqrt{\pi}\sigma\Gamma\left(\nu/2\right)}\left[1+\frac{1}{\nu}\frac{x^2}{\sigma^2}\right]^{-(\nu+1)/2}.$$

We shall denote by $F_1(x;\sigma^2,\nu)$ the cdf corresponding to f_1. For convenience denote $\mathbf{A}^{-1} = \mathbf{B} = b_{ij}$, where $\mathbf{B}$ is an upper triangular matrix with $b_{pp} = 1$ and $b_{jj} > 0$ for all j. Then, since the integral over $\mathbf{X}$ covers the region $\mathbf{X} \leq \mathbf{X}_0$, the integral over $\mathbf{Z}$ is determined by the inequality $\mathbf{BZ} < \mathbf{X}_0$, and the bounds can be written as

$$\begin{aligned} z_p &< x_{p0} \\ &\equiv z_{n0} \end{aligned}$$

and

$$\begin{aligned} z_k &< b_{kk}^{-1}\left(x_{k0} - \sum_{i=k+1}^{p} b_{ki}z_i\right) \\ &\equiv z_{k0}\left(x_{k0}, z_{k+1}, \ldots, z_p\right) \end{aligned}$$

for $k = 1, 2, \ldots, p-1$. Utilizing this transformation, the probability p can be written as $p = J_p$, where

$$\begin{aligned} J_k &= \int_{-\infty}^{y_{k0}} f_1\left(z_k;\sigma_k^2,\nu_k\right) J_{k-1} dz_k \qquad (6.36) \\ &= \int_{-\infty}^{z_{k0}} F_1\left(z_{k_0};\sigma_k^2,\nu_k\right) J_{k-1} f_1^c\left(z_k;\sigma_k^2,\nu_k\right) dz_k \\ &= E_{f_1^c}\left[F_1\left(z_{k_0};\sigma_k^2,\nu_k\right) J_{k-1}\right], \quad k = 2, 3, \ldots, p, \end{aligned}$$

where

$$f_1^c\left(z_k;\sigma_k^2,\nu_k\right) = \frac{f_1\left(z_k;\sigma_k^2,\nu_k\right)}{F_1\left(z_{k0};\sigma_k^2,\nu_k\right)}$$

is the univariate unconditional t pdf for $z_k \leq z_{k0}$ and $J_1 = F_1(z_{10};\sigma_1^2,\nu_1)$. Hence, J_k is the probability over the range of $(z_1, \ldots, z_k)$ conditional on the values for $(z_{k+1}, \ldots, z_p)$.

The Monte Carlo simulation starts off by drawing random values of z_p from the distribution $f_1^c(\cdot; z_{p0}, \sigma_p^2, \nu_p)$, which we shall denote by $\tilde{z}_{p,r}$, $r = 1, \ldots, R$. Each of these yields a different bound $\tilde{z}_{p-1,0,r}$ and parameter value $\tilde{\sigma}_{p-1,r}^2$ for each draw of z_{p-1}; $\tilde{z}_{p-1,r}$ is then drawn from the distribution $f_1^c(\cdot; \tilde{z}_{p-1,0,r}, \tilde{\sigma}_{p-1,r}^2, \nu_{p-1})$. This process continues until $\tilde{z}_{2,r}$ is drawn and $\hat{J}_1 = F_1(\tilde{z}_{10,r}; \tilde{\sigma}_{1,r}^2, \nu_1)$ is computed with a commonly available approximation routine for the univariate Student's t cdf. The simulated estimate $\hat{p}$ of p is then found as the sample average of the $\hat{J}_p$ values across the simulated sample of R elements

$$\hat{p} \equiv \hat{J}_p = \frac{1}{R}\sum_{r=1}^{R} F_1\left(\tilde{z}_{p0,r};\tilde{\sigma}^2_{p,r},\nu_p\right)\hat{J}_{p-1},$$

where $\hat{J}_k = F_1(\tilde{z}_{k0,r};\tilde{\sigma}^2_{k,r},\nu_k)\hat{J}_{k-1}$ for $k = 2,\ldots,p-1$. It is more efficient to estimate J_p by averaging over a large number of elements than to obtain close approximations of its components J_k for $k < p$. Therefore, a better estimate for p is

$$\hat{p} = \frac{1}{R}\sum_{r=1}^{R}\left\{\prod_{k=1}^{p} F_1\left(\tilde{z}_{k0,r};\tilde{\sigma}^2_{k,r},\nu_k\right)\right\}.$$

The right-hand side of (6.36) remains unchanged if the integrand is divided or multiplied by any nonzero function of z. Let g_p be a p-dimensional pdf such that

$$g_p(z;\nu) = \prod_{k=1}^{p} g_1\left(z_k;\tau_k^2\right),$$

where g_1 is a univariate pdf of a type to be mentioned below with $Var(z_k) = \tau_k^2 = \sigma_k^2\nu_k/(\nu_k - 2)$, and σ_k^2 and ν_k are as defined above. Let $G_1(z_k;\tau_k^2)$ be the associated cdf, and let

$$g_1^c\left(z_k;z_{k0},\tau_k^2\right) = \frac{g_1\left(z_k;\tau_k^2\right)}{G_1\left(z_{k0};\tau_k^2\right)}$$

be the conditional pdf. Finally, let

$$g_p^c\left(z;z_0,\nu\right) = \prod_{k=1}^{p} g_1^c\left(z_k;z_{k0},\tau_k^2\right).$$

With these definitions, one can write $p = J_p$ in terms of

$$\begin{aligned} J_k &= \int_{-\infty}^{z_{k0}} \frac{f_1\left(z_k;\sigma_k^2,\nu_k\right)}{g_1^c\left(z_k;z_{k0},\tau_k^2\right)} J_{k-1} g_1^c\left(z_k;z_{k0},\tau_k^2\right) dz_k \\ &= E_{g_1^c}\left[\frac{f_1\left(z_k;\sigma_k^2,\nu_k\right)}{g_1^c\left(z_k;z_{k0},\tau_k^2\right)} J_{k-1}\right], \quad k = 2,\ldots,p \end{aligned}$$

and $J_1 = F(z_{10};\sigma_1^2,\nu_1)$. Clearly, g_p^c, and, more particularly, g_1^c, is an important sampling density (see, for example, Hammersley and Handscomb, 1964). To evaluate p, the procedure is as follows: Generate random drawings $\tilde{z}_{p,r}$ for $r = 1,\ldots,R$ from the distribution $g_1^c(\cdot;z_{n0},\tau_n^2)$;

compute the implied values $\tilde{z}_{p-1,0,r}$ and $\tilde{\tau}^2_{p-1,r}$ for each drawing of z_{p-1}; draw $\tilde{z}_{p-1,r}$ from the distribution $g_1^c(\cdot; \tilde{z}_{p-1,0,r}, \tilde{\tau}^2_{p-1,r}, \nu_p)$; and continue on until $\tilde{z}_{2,0,r}$ is drawn and $\hat{J}_1$ is computed. Based on this procedure, $\hat{p}$ may be written in the form

$$\hat{p} = \frac{1}{R}\sum_{r=1}^{R}\left(F_1\left(\tilde{z}_{10,r}; \tilde{\tau}^2_{1,r}\right)\prod_{k=1}^{p}\frac{f_1\left(\tilde{z}_{k,r}; \tilde{z}_{k0,r}, \tilde{\sigma}^2_{k,r}, \nu_k\right)}{g_1^c\left(\tilde{z}_{k,r}; \tilde{z}_{k0,r}, \tilde{\tau}^2_{k,r}\right)}\right).$$

Three suitable choices for the importance density function are

- the logit with

  $$g_1(x) = \frac{\lambda}{\tau}q(1-q),$$

  where

  $$q = \left[1+\exp\left(-\lambda x/\tau\right)\right]^{-1}$$

  and $\lambda = \pi/\sqrt{3}$;
- transformed beta (2, 2) density (Vijverberg, 1995) with

  $$g_1(x) = 6z^2(1-z)^2,$$

  where

  $$z = \frac{\exp(x/\sigma)}{1+\exp(x/\sigma)};$$
- the normal $N(0,\sigma^2)$ density.

Vijverberg (1997, 2000) has developed a new family of simulators that extends the above research on the simulation of high-order probabilities. For instance, Vijverberg (2000) has reported that the gain in precision using the new family translates into a 40% savings in computational time.

7
Probability Inequalities

Probability inequalities on $\Pr(Y_1 \le y_1, Y_2 \le y_2, \ldots, Y_p \le y_p)$ for multivariate distributions have been a popular topic of investigation since the 1950s. It is well known (Khatri, 1967; Scott, 1967; Šidák, 1965, 1967) that, for arbitrary positive numbers $y_1, y_2, \ldots, y_p$, the inequality

$$\Pr\left(|Y_1| \le y_1, |Y_2| \le y_2, \ldots, |Y_p| \le y_p\right) \quad \ge \quad \prod_{k=1}^{p} \Pr\left(|Y_k| \le y_k\right)$$

holds for any random vector $\mathbf{Y}^T = (Y_1, Y_2, \ldots, Y_p)$ having the multivariate normal distribution with zero means and an arbitrary correlation matrix. A question then arises as to whether there is an analog of this for multivariate t distributions.

7.1 Dunnett and Sobel's Probability Inequalities

Dunnett and Sobel (1955) obtained bounds for the probability integral

$$P \quad = \quad \Pr\left(X_1 \le x_1, X_2 \le x_2, \ldots, X_p \le x_p\right)$$

$x_1 \ge 0$, $x_2 \ge 0$, …, $x_p \ge 0$, when $(X_1, X_2, \ldots, X_p)$ follows the central p-variate t distribution with degrees of freedom ν and the correlation matrix $\mathbf{R}$ taking the special structure $r_{ij} = b_i b_j$ for all $i \ne j$. Using the definition that $\mathbf{X}$ can be represented as $(Z_1/S, Z_2/S, \ldots, Z_p/S)$, where $\mathbf{Z}$ is a p-variate normal random vector with correlation matrix $\mathbf{R}$ and $\nu S^2/\sigma^2$ is an independent chi-squared random variable with degrees of freedom ν, one can rewrite P as

$$P \quad = \quad \Pr\left\{\frac{Z_1}{S} \le x_1, \frac{Z_2}{S} \le x_2, \ldots, \frac{Z_p}{S} \le x_p\right\}$$

$$= \int_0^\infty G(x_1 s, x_2 s, \ldots, x_p s)\, h(s) ds, \tag{7.1}$$

where G is the joint cdf of $\mathbf{Z}$ and h is the pdf of S. If $Y_0, Y_1, Y_2, \ldots, Y_p$ are independent standard normal random variables, then one can represent $Z_j = \sqrt{1-b_j^2}Y_j - b_j Y_0$ for $j = 1, 2, \ldots, p$. Using this result, one can rewrite G as

$$\begin{aligned} G(x_1 s, x_2 s, \ldots, x_p s) &= \Pr\left\{\sqrt{1-b_j^2}Y_j - b_j Y_0 < x_j s, j = 1, 2, \ldots, p\right\} \\ &= \int_{-\infty}^{\infty} \prod_{j=1}^{p} \Phi\left(\frac{x_j s + b_j z}{\sqrt{1-b_j^2}}\right) \phi(z) dz \qquad (7.2) \\ &= E\left\{\prod_{j=1}^{p} \Phi\left(\frac{x_j s + b_j z}{\sqrt{1-b_j^2}}\right)\right\}, \end{aligned}$$

where ϕ and Φ are, respectively, the pdf and the cdf of the standard normal distribution. Using the well known inequality

$$E\left\{\prod_{j=1}^{p} F_j(x)\right\} \geq \prod_{j=1}^{p} E\{F_j(x)\} \tag{7.3}$$

(where F_j denotes a cdf), one can now bound G by

$$\begin{aligned} G(x_1 s, x_2 s, \ldots, x_p s) &\geq \prod_{j=1}^{p} E\left\{\Phi\left(\frac{x_j s + b_j z}{\sqrt{1-b_j^2}}\right)\right\} \\ &= \prod_{j=1}^{p} \Pr\left\{\sqrt{1-b_j^2}Y_j - b_j Y_0 < x_j s\right\} \\ &= \prod_{j=1}^{p} \Phi(x_j s). \end{aligned}$$

Substituting this result into (7.1) and applying (7.3) once more, one obtains the lower bound for P given by Dunnett and Sobel (1955) as

$$\begin{aligned} P &\geq \int_0^\infty \prod_{j=1}^{p} \Phi(x_j s) f(s) ds \\ &\geq \prod_{j=1}^{p} \int_{-\infty}^{\infty} \Phi(x_j s) f(s) ds \end{aligned}$$

$$
\begin{aligned}
&= \prod_{j=1}^{p} \Pr\{Z_j < x_j s\} \\
&= \prod_{j=1}^{p} \Pr\{X_j < x_j\}. \qquad (7.4)
\end{aligned}
$$

This lower bound for P holds more generally for any correlation matrix $\mathbf{R}$ with $r_{ij} \geq 0$ and any arbitrarily fixed $(x_1, \ldots, x_p)$. This is a consequence of the fact that P is an increasing function of each r_{ij} for all $i \neq j$, while other correlations are held fixed. It can be shown further that

$$\Pr(X_1 > x_1, \ldots, X_p > x_p) \geq \prod_{j=1}^{p} \Pr(X_j > x_j)$$

and

$$\Pr(|X_1| \leq x_1, \ldots, |X_p| > x_p) \geq \prod_{j=1}^{p} \Pr(|X_j| > x_j).$$

Since the bound (7.4) does not depend on r_{ij}, it can be calculated easily from a table of the cdf of the univariate Student's t distribution. Dunnett and Sobel (1955) also obtained two sharper bounds by slight modifications of the above arguments: For even $p \geq 2$,

$$P \geq \prod_{j=1}^{p/2} \Pr\{X_{2j-1} < x_{2j-1}, X_{2j} < x_{2j}\}, \qquad (7.5)$$

and for odd $p \geq 3$,

$$P \geq \Pr\{X_1 < x_1\} \prod_{j=1}^{(p-1)/2} \Pr\{X_{2j} < x_{2j}, X_{2j+1} < x_{2j+1}\}. \qquad (7.6)$$

In the case where $r_{ij} = \rho$ for all $i \neq j$ and $x_j = x$ for all j, inequalities that are sharper than (7.5) and (7.6) can be obtained. Let

$$\beta_1(p) = \Pr(X_1 \leq d, X_2 \leq d, \ldots, X_p \leq d)$$

and

$$\beta_2(p) = \Pr(|X_1| \leq d, |X_2| \leq d, \ldots, |X_p| \leq d).$$

It is well known that $\beta_k(p)$, $k = 1, 2$ are monotonically increasing in

r_{ij} $(i \neq j)$ and, if $r_{ij} \geq 0$, then

$$\beta_k(p) \geq \beta_1^k(1), \qquad k = 1, 2. \tag{7.7}$$

Tong (1970) provided the following sharper bounds for β_k

$$\beta_k(p) \geq \{\beta_k(m)\}^{p/m} \geq \beta_k^p(1) + \left\{\beta_k(2) - \beta_k^2(1)\right\}^{p/2}, \tag{7.8}$$

where $p \geq m \geq 2$. These inequalities certainly improve on (7.7), but neither of them are very sharp when p is large. Also observe that the first inequality in (7.8) depends on p and m only through their ratio p/m. Hence, for fairly large p and m as long as p/m is close to 1 (even if the difference $p - m$ is not small), the first inequality is quite adequate. If $\rho = 0$, then a necessary and sufficient condition for $\beta_k(p) \to \beta_k^p(1)$ for every fixed p is that $\nu \to \infty$.

Recently Seneta (1993) pointed out that the "sub-Markov" inequality

$$\beta_1(p) \geq [\Pr\{X_1 < x, X_2 < x\}]^{p-1} / [\Pr\{X_1 < x\}]^{p-2} \tag{7.9}$$

is sharper than the corresponding inequality

$$\beta_1(p) \geq [\Pr\{X_1 < x, X_2 < x\}]^{p/2} \tag{7.10}$$

as given by (7.8). This fact is illustrated in the following table, which is taken from Seneta (1993).

Comparison of the bounds (7.10) and (7.9) for P.
x chosen such that the true value of $P = 0.95$

	ν	x	Bound (7.10)	Bound (7.9)
	10	2.34	0.945	0.946
$p = 3$	15	2.24	0.946	0.947
	20	2.19	0.945	0.946
	60	2.10	0.944	0.945
	10	2.81	0.921	0.921
$p = 9$	15	2.67	0.924	0.924
	20	2.60	0.926	0.927
	60	2.48	0.934	0.936

Actually, (7.9) is a particular case of the following inequalities

$$\begin{aligned}\beta_1(p) \geq\ & \Pr\left(X_j \leq x; j=1,\ldots,m-1\right) \\ & \times \left\{\Pr\left(X_m \leq x \mid X_j \leq x; j=1,\ldots,m-1\right)\right\}^{p-m+1}\end{aligned} \tag{7.11}$$

and

$$\begin{aligned}\beta_2(p) \geq\ & \Pr\left(|X_j| \leq x; j=1,\ldots,m-1\right) \\ & \times \left\{\Pr\left(|X_m| \leq x \mid |X_j| \leq x; j=1,\ldots,m-1\right)\right\}^{p-m+1}\end{aligned}$$

given by Glaz and Johnson (1984), who also provided a formal proof of the fact that (7.11) is sharper than (7.8).

Dunnett (1989) wrote a *Fortran* programs for evaluating the integral (7.2). It uses Simpson's rule to compute an approximation to (7.2) in such a way that a prescribed accuracy is achieved. To approximate the integral of a function $\alpha(z)$, say, over an interval $[a, b]$ using Simpson's rule, the value of the function is computed at the two end points and at the midpoint of the interval; then the approximate value of the integral is given by

$$\left\{\alpha(a) + 4\alpha\left(\frac{a+b}{2}\right) + \alpha(b)\right\}\frac{b-a}{6},$$

with its error bounded by $\alpha_4(a,b)(b-a)^5/2880$, where $\alpha_4(a,b)$ is a bound on the absolute value of the fourth derivative of $\alpha(z)$ over the interval (see, for example, page 66 in Shampine and Allen, 1973). The central processor unit time (on a VAX 8600 computer using single-precision arithmetic) taken to compute (7.2) ranges from 0.01 to 2.37 seconds for cases of equal correlation ($r_{ij} = \rho$) and identical ranges of integration ($x_j = x$). Slightly longer computing times are required for unequal correlations or different limits of integration. Dunnett (1989) suggested that his program can be used along with an appropriate numerical integration routine, such as the Integral Mathematical and Statistical Libraries' (1987, Volume 1, Chapter 4) *QDAGI*, to evaluate the multivariate t probability integral (7.1).

7.2 Dunn's Probability Inequalities

The univariate Student's t distribution has the property that the probability evaluated from $-x$ to $+x$ is an increasing function of the degrees of freedom ν – this also applies to the probability from $-\infty$ to $+x$ (see, for example, Ghosh, 1973, for details). Dunn (1965) pointed out that this monotonicity does not generalize to p dimensions in the usual multivariate t distribution. Specifically, let $\mathbf{X}$ have the central p-variate t distribution with degrees of freedom ν and correlation matrix $\mathbf{R}$. If $F(x)$ is defined by

$$F(x) \quad = \quad \Pr\{\cap_{k=1}^{p} -x < X_k < x\},$$

then $F(x)$ equals the probability mass in the multivariate t distribution evaluated over a p-dimensional hypercube centered at the origin of the half side x and $F(x)$ is the distribution of the maximum of the absolute values of the p X variables. Similarly, if $G(x)$ is defined by

$$G(x) \quad = \quad \Pr\{\cap_{k=1}^{p} -\infty < X_k < x\},$$

then $G(x)$ equals the probability mass evaluated from $-\infty$ to x in each direction and $G(x)$ is the distribution of the maximum of the p X variables. Dunn showed that, for any given $x > 0$ and degrees of freedom $\nu_1 > \nu_2$, there exists an integer K such that, for all $p \geq K$,

$$F_{p,\nu_1}(x) \quad < \quad F_{p,\nu_2}(x)$$

and

$$G_{p,\nu_1}(x) \quad < \quad G_{p,\nu_2}(x).$$

Here, $F_{p,\nu}$ and $G_{p,\nu}$ are F and G as defined above, with dimension p and degrees of freedom ν. This result covers the case of all correlations equal to 0. When all correlations are equal to 1, the distribution is the same as the univariate Student's t distribution, so that, for all dimensions, $F(x)$ and $G(x)$ are monotonically increasing functions of ν. Other correlation matrices may be considered in some sense to lie between these two extremes. In various unpublished tables of $F(x)$, the change is found to occur at a dimension where $F(x)$ is approximately 0.25 or 0.30.

7.3 Halperin's Probability Inequalities

Halperin (1967) extended the inequality (7.4) for generalized bivariate t distributions as follows. Let (Y_{i1}, Y_{i2}), $i = 0, 1, 2, \ldots, r$, $r \geq 1$ be independent samples from a bivariate normal distribution with zero means, variances σ_1^2, σ_2^2, and covariances $\sigma_1\sigma_2\rho_i$, $\mid \rho_i \mid \leq 1$. Let Y_{i1}, $i = r+1, \ldots, r+n$ and Y_{i2}, $i = r+1, \ldots, r+m$ be independent normal samples with zero means and variances σ_1^2 and σ_2^2, respectively, and independent of (Y_{i1}, Y_{i2}), $i = 0, 1, 2, \ldots, r$. Define

$$(X_1, X_2) = \left(\frac{Y_{10}}{S_1}, \frac{Y_{20}}{S_2}\right),$$

where

$$S_1 = \frac{1}{r+n}\sum_{i=1}^{r+n} Y_{i1}^2$$

and

$$S_2 = \frac{1}{r+m}\sum_{i=1}^{r+m} Y_{i2}^2.$$

Halperin (1967) then showed that the probability integral of (X_1, X_2) satisfies the inequality

$$\Pr(|X_1| \leq x_1, |X_2| \leq x_2) \geq \Pr(|X_1| \leq x_1)\Pr(|X_2| \leq x_2)$$

for all real numbers x_1 and x_2.

7.4 Šidák's Probability Inequalities

In the bivariate case considered above, it is assumed that the correlation between Y_{i1} and Y_{i2} may be different for different i's. For a general p, but for a special correlation structure of Y's, Šidák (1967) established the following result. Let $\mathbf{Y}^T = (Y_1, Y_2, \ldots, Y_p)$ have a p-variate normal distribution with zero means and an arbitrary correlation matrix. Let $\mathbf{Z}_i^T = (Z_{i1}, Z_{i2}, \ldots, Z_{ip})$, $i = 1, \ldots, n$ be a p-variate normal random sample, which is mutually independent and independent of $\mathbf{Y}$, each of which has zero means, unit variances, and the decomposable correlation structure given by

$$Corr(Z_{ki}, Z_{kj}) = b_i b_j$$

for $i, j = 1, \ldots, p$; $i \neq j$; $k = 1, \ldots, n$ with $0 \leq b_j \leq 1$, $i = 1, \ldots, p$. Then,

$$\Pr\left(\frac{|Y_1|}{\sqrt{Z_{11}^2 + \cdots + Z_{n1}^2}} \leq x_1, \ldots, \frac{|Y_p|}{\sqrt{Z_{1p}^2 + \cdots + Z_{np}^2}} \leq x_p\right)$$

$$\geq \prod_{i=1}^{p} \Pr\left(\frac{|Y_i|}{\sqrt{Z_{1i}^2 + \cdots + Z_{ni}^2}} \leq x_i\right). \tag{7.12}$$

Essentially the same result, assuming more generally only $| b_i | \leq 1$, follows by an easy specialization of Corollary 8 in Khatri (1967).

A general proof of the inequality (7.12) under the assumption that $\mathbf{Y}$ and all $\mathbf{Z}_i$'s have the same normal distribution with zero means and an arbitrary covariance matrix was provided by Scott (1967). Unfortunately, this proof is correct only for $p = 2$, and Šidák (1971) produced a counterexample showing its incorrectness for $p > 2$. Šidák (1971) went on to show that, if

$$Corr\,(Y_i, Y_j) \;=\; c_i c_j r_{ij}$$

for $i, j = 1, 2, \ldots, p$; $i \neq j$ with $| c_j | \leq 1$ $(j = 1, 2, \ldots, k)$ and $\{r_{ij}\}$ any fixed correlation matrix, and if

$$Corr\,(Z_{li}, Z_{lj}) \;=\; b_{li} b_{lj}$$

for $i, j = 1, 2, \ldots, p$; $i \neq j$; $l = 1, 2, \ldots, n$ with $| b_{li} | < 1$ $(i = 1, 2, \ldots, p$, $l = 1, 2, \ldots, n)$, then the left-hand side probability in (7.12) as a function of c_j is nonincreasing for $-1 \leq c_j < 0$ and nondecreasing for $0 < c_j \leq 1$, so that it has a minimum for $c_j = 0$ and, as a function of b_{li}, is nonincreasing for $-1 < b_{li} < 0$ and nondecreasing for $0 < b_{li} < 1$, so that it has a minimum for $b_{li} = 0$. Hence, (7.12) is also true for this more general correlation structure.

Šidák (1973) obtained an inequality using exchangeability when $\mathbf{X}$ is a central p-variate t random vector with the equicorrelation structure $r_{ij} = \rho$, $i \neq j$. He showed that

$$\begin{aligned} & \Pr\,(b \leq X_1 \leq a, \ldots, b \leq X_p \leq a) \\ > \; & \{\Pr\,(b \leq X_1 \leq a, \ldots, b \leq X_r \leq a)\}^{p/r} \\ > \; & \{\Pr\,(b \leq X_i \leq a)\}^p \end{aligned}$$

for all $p > r \geq 2$ and $a > b$. In an earlier paper, Tong (1970) obtained similar results for a much larger class of random vectors.

7.5 Tong's Probability Inequalities

We noted in Chapter 1 that the multivariate t density is Schur-concave in the particular case $r_{ij} = \rho$, $i \neq j$. Tong (1982) used this property to derive certain probability inequalities. He showed that if $f : \Re^p \rightarrow [0, \infty)$ is Borel-measureable and Schur-concave, then, provided that the integral exists, $\int_{A(\mathbf{x})} f(\mathbf{y}) d\mathbf{y}$ is also a Schur-concave function of $(x_1, \ldots, x_p)$, where $A(\mathbf{x})$ denotes the rectangular set

$$A(\mathbf{x}) = \{\mathbf{y} \mid \mathbf{y} \in \Re^p, |y_j| \le x_j, j = 1, \ldots, p\}.$$

Taking f to be the pdf of the multivariate t, it follows that

$$\Pr(|X_j| \le x_j, j = 1, \ldots, p) \le \Pr(|X_j| \le \bar{x}, j = 1, \ldots, p),$$

where $\bar{x} = -(x_1 + \cdots + x_p)$.

8
Percentage Points

From the 1950s numerous authors have tried to compute the percentage points of multivariate t distributions. It is an indication of the interest in problems leading to applications of this "new" distribution. This research continued well into the 1990s and is still going strong. Although some of the results have by now lost their practical importance – in our opinion – it is essential for historical reasons to describe a majority of these contributions. This will certainly assist historians and experts in multivariate distributions to gain a better perspective of the developments in the area. Moreover, many of the techniques involved in these calculations are ingenious and worthy of emulation and further investigation.

8.1 Dunnett and Sobel's Percentage Points

Let (X_1, X_2) have the bivariate t distribution with joint pdf (6.1). For given ν, ρ and probability level γ, let d denote the equicoordinate percentage point satisfying

$$\int_{-\infty}^{d} \int_{-\infty}^{d} f(x_1, x_2; \nu, \rho) dx_1 dx_2 = \gamma.$$

The value of d can be determined for any ν by trial and error using (6.3) and (6.4). However, this procedure becomes more involved as ν increases. Dunnett and Sobel (1954) derived an asymptotic expansion in powers of $1/\nu$ that expresses d in terms of the corresponding quantity e for the bivariate normal distribution: e is defined by

$$\int_{-\infty}^{e} \int_{-\infty}^{e} f(x_1, x_2; \infty, \rho)\, dx_1 dx_2 = \gamma$$

and can be obtained by interpolation in the classical tables of Pearson (1931), for example. Their expansion yields a good approximation of d even for moderately small values of ν. The method of derivation is essentially the same as that used by Fisher (1941) in deriving an asymptotic expansion for the percentage points of the univariate Student's t distribution. Up to the terms of $O(1/\nu)$, Dunnett and Sobel obtained

$$d = e - \frac{e}{4\nu}\left\{e^2 + 1 - \frac{t\Phi'(t)}{\Phi(t)}\right\}$$

and by inverting

$$e = d + \frac{d}{4\nu}\left\{d^2 + 1 - \frac{d\Phi'(d)}{\Phi(d)}\right\}.$$

They also tabulated numerical values of the coefficients of the terms $1/\nu^i$ for $i = 0, 1, 2, 3, 4$.

8.2 Krishnaiah and Armitage's Percentage Points

Consider $\mathbf{X}$ having a central p-variate t distribution with degrees of freedom ν and with the equicorrelation structure $r_{ij} = \rho$, $i \neq j$. Krishnaiah and Armitage (1966) evaluated the multivariate percentage point d by solving the integral

$$P(d) = \int_{-\infty}^{d} \cdots \int_{-\infty}^{d} f(x_1, \ldots, x_p; \nu, \rho) dx_1 \cdots dx_p = \gamma \quad (8.1)$$

and produced extensive tables for all combinations of $p = 1(1)10$; $\nu = 5(1)35$; $\rho = 0.05(0.05)0.9$ and $\gamma = 0.90, 0.95, 0.975, 0.99$ (see also Armitage and Krishnaiah, 1965). These computations use the approximation that

$$P(d) \approx 2\int_0^c \frac{x^{\nu-1}\exp\left(-x^2\right)}{\sqrt{\pi}\Gamma\left(\nu/2\right)} \int_{-\infty}^{\infty} \exp\left(-y^2\right) \times \Phi^p\left(\sqrt{\frac{2}{\nu(1-\rho)}}dx + \sqrt{\frac{\rho}{1-\rho}}y\right) dydx,$$

where Φ is the cdf of the standard normal distribution and the upper limit c is chosen large enough to make the error of approximation as small as desired. Krishnaiah and Armitage (1966) took $c = 10$.

8.3 Gupta et al.'s Percentage Points

Gupta et al. (1985) solved equation (8.1) by setting

$$|P(d) - \gamma| \leq 10^{-8}$$

with $P(d)$ computed by the approximation (6.14). The numerical evaluation of (6.14) involves the evaluation of the derivatives $G^{(k)}(d)$ given by (6.15) for $k = 0, 1, \ldots, 8$. But it is easily seen from (7.2) that

$$G^{(k)}(d) = \frac{1}{\rho^{k/2}} \int_{-\infty}^{\infty} \Phi^p \left(\frac{\sqrt{\rho} z + d}{\sqrt{1-\rho}} \right) H_k(z) \phi(z) dz, \qquad (8.2)$$

where $H_k(z)$ is the Hermite polynomial of degree k given by (6.11). By letting $A = \sqrt{\rho}/\sqrt{1-\rho}$ and $B = 1/\sqrt{1-\rho}$ and changing the variable by the transformation $u = Az + Bd$, (8.2) can be rewritten as

$$G^{(k)}(d) = \frac{1}{A} \int_{-\infty}^{\infty} \left(\frac{B}{A} \right)^k \Phi^p(u) H_k \left(\frac{u - Bd}{A} \right) \phi \left(\frac{u - Bd}{A} \right) du.$$

Gupta et al. (1985) approximated this integral by

$$G^{(k)}(d) \approx \frac{1}{A} \int_{-9}^{9A+4B} \left(\frac{B}{A} \right)^k \Phi^p(u) H_k \left(\frac{u - Bd}{A} \right) \phi \left(\frac{u - Bd}{A} \right) du,$$

and the integration was carried out by Gauss' method over intervals of length $D = 0.5$ starting from -9 until $9A + 4B$ was included. They provided tabulations of the percentage point d for all combinations of

- $p = 1(1)9(2)19$; $\nu = 15(1)20, 24, 30, 36, 48, 60, 120, \infty$; $\gamma = 0.75, 0.9, 0.95, 0.99$ and $\rho = 0.1, 0.2(0.1)0.6$;
- $p = 1(1)9(2)15$; $\nu = 15, 17, 20, 24, 36, 60, 120, \infty$; $\gamma = 0.9, 0.95$ and $\rho = 0.7(0.1)0.9$.

8.4 Rausch and Horn's Percentage Points

Rausch and Horn (1988) considered the particular case of (8.1) when the common $\rho = 0$. They used the approximation

$$P(d) \approx \sum_{i=1}^{n} w_i \Phi^p \left(d \sqrt{\frac{2 z_i}{\nu}} \right).$$

The weights w_i are calculated according to the formula

$$w_i = \frac{\Gamma(n+\beta)}{\Gamma(1+\beta)\, n!\, (n+\beta)} z_i \left\{ L_{n-1}^{\beta}(z_i) \right\}^{-2},$$

where $\beta = \nu/2 - 1$,

$$L_m^\beta(z) = \sum_{l=0}^{m} \binom{m+\beta}{m-l} \frac{(-z)^l}{l!}$$

are the Laguerre polynomials and $z_1, \ldots, z_n$ are the zeros of $L_m^\beta(z)$. Rausch and Horn computed d for all combinations of $3 \leq p \leq 100$; $5 \leq \nu \leq 120, \nu = \infty$; and $0.5 \leq \gamma \leq 0.99$.

8.5 Hahn and Hendrickson's Percentage Points

Hahn and Hendrickson (1971) computed percentage points d by solving the equation

$$P(d) = \int_{-d}^{d} \cdots \int_{-d}^{d} f(x_1, \ldots, x_p; \nu, \rho) dx_1 \cdots dx_p = \gamma \qquad (8.3)$$

for all combinations of $p = 1(1)6$, 8, 10, 12, 15, 20; $\nu = 3(1)12$, 15, 20, 25, 30, 40, 60; $\rho = 0$, 0.2, 0.4, 0.5; and $\gamma = 0.90$, 0.95, 0.99. As one would expect, these values are comparable to the positive square root of the values given by Krishnaiah and Armitage (1966, Section 8.2). Hahn and Hendrickson's computations use the approximation that

$$P(d) \approx \int_0^\infty \left[\int_{-\infty}^{\infty} \left\{ \Phi\left(\frac{dx + \sqrt{d}y}{\sqrt{1-\rho}} \right) - \Phi\left(\frac{-dx + \sqrt{\rho}y}{\sqrt{1-\rho}} \right) \right\}^p \phi(y) dy \right] h(x) dx,$$

where ϕ and Φ are, respectively, the pdf and the cdf of the standard normal distribution and h is the pdf of $\sqrt{\chi_\nu^2/\nu}$.

8.6 Siotani's Percentage Points

Siotani (1964) suggested two interesting approximations for computing d in (8.3). The first approximation is the value d_1 satisfying

$$p \Pr\left(X_i^2 > d_1^2\right) = 1 - \gamma;$$

this approximation had been suggested previously by Dunn (1958, 1961). By Bonferroni's inequalities, one notes that

$$1 - \gamma - \epsilon_1(\gamma, p) \leq 1 - P(d_1) \leq 1 - \gamma,$$

where

$$\epsilon_1(\gamma, p) = \sum_{i<j} \Pr\left(X_i^2 > d_1^2, X_j^2 > d_1^2\right).$$

Thus, if $\epsilon_1(\gamma, p)$ is sufficiently small, then one can use d_1 as a good estimate. A modified second approximation is the value d_2 satisfying

$$2p\Pr\left(X_i^2 > d_2^2\right) = 1 - \gamma + \epsilon_1(\gamma, p).$$

This time, one notes that

$$-\epsilon_2(\gamma, p) \leq \gamma - P(d_2) \leq \epsilon_3(\gamma, p),$$

where

$$\epsilon_2(\gamma, p) = \sum_{i<j} \Pr\left(X_i^2 > d_2^2, X_j^2 > d_2^2\right) - \epsilon_1(\gamma, p)$$

and

$$\epsilon_3(\gamma, p) = \sum_{i<j<k} \Pr\left(X_i^2 > d_2^2, X_j^2 > d_2^2, X_k^2 > d_2^2\right).$$

Since both $\epsilon_2(\gamma, p) > 0$ and $\epsilon_3(\gamma, p) > 0$, the absolute value of $\gamma - P(d_2)$ may be expected to be sufficiently small for the tail of the p-variate t distribution to correspond to $1-\gamma$ for values of $\gamma \geq 0.95$. For the particular case $p = 2$, $\mu_1 = \mu_2 = 0$, and the equicorrelation structure $r_{ij} = \rho$, $i \neq j$, Siotani (1964) tabulated estimates of the probability in (8.3) for all combinations of $d = 2.0(0.5)4.5$; $\nu = 10(2)50(5)90, 100, 120, 150, 200, \infty$; and $\mid \rho \mid = 0.0(0.1)0.9, 0.95$. He also illustrated applications to interval estimation of the parameters in the model of a randomized block design and for coefficients in a normal regression equation.

8.7 Graybill and Bowden's Percentage Points

Graybill and Bowden (1967) derived bounds for d satisfying (8.3) for the special case $p = 2$ and $\rho = 0$. In this special case (8.3) becomes

$$\Pr\left\{X_1^2 \leq d^2, X_2^2 \leq d^2\right\} = \gamma$$

or, equivalently,

$$\Pr\left\{\max\left(X_1^2, X_2^2\right) \leq d^2\right\} = \gamma.$$

But

$$\Pr\left\{\max\left(X_1^2, X_2^2\right) \leq d^2\right\} = \Pr\left\{X_1^2 + X_2^2 \leq 2F_{\alpha,2,\nu-2}\right\}$$

and

$$\begin{aligned}\Pr\left\{\max\left(X_1^2, X_2^2\right) \le F_{\gamma,2,\nu-2}\right\} &< \Pr\left\{X_1^2 + X_2^2 \le 2F_{\gamma,2,\nu-2}\right\} \\ &< \Pr\left\{\max\left(X_1^2, X_2^2\right) \le 2F_{\gamma,2,\nu-2}\right\},\end{aligned}$$

where $F_{\gamma,2,\nu-2}$ is percentage point of the F distribution with degrees of freedom 2 and $\nu - 2$ corresponding to γ. Hence, one obtains

$$F_{\gamma,2,\nu-2} < d^2 < 2F_{\gamma,2,\nu-2},$$

the bounds given by Graybill and Bowden. In a related development, McCann and Edwards (1996) obtained the following lower bound for the left-hand side of (8.3) when the underlying correlation matrix $\mathbf{R}$ is of rank r

$$\begin{aligned}P(d) \;\ge\; 1 - \int_0^{1/d} \min\Bigg[\frac{\Lambda}{\pi} F_{r-2,2}\left\{\frac{2(ds)^{-2} - 1}{r-2}\right\} \\ + F_{r-1,1}\left\{\frac{(ds)^{-2} - 1}{r-1}\right\}, 1\Bigg] q(s)ds \qquad (8.4)\end{aligned}$$

with

$$\Lambda = \sum_{k=1}^{p-1} \arccos\left(r_{j,j+1}\right), \qquad (8.5)$$

where $F_{m,n}$ is the cdf of an F distribution with degrees of freedom m and n, and q denotes the pdf of $\sqrt{F_{\nu,r}/r}$. This inequality requires only the evaluation of a one-dimensional integral and depends on $\mathbf{R}$ through its rank r and also through the constant Λ. If one writes $\mathbf{R} = \mathbf{A}\mathbf{A}^T$ for a $p \times r$ matrix $\mathbf{A}$ of rank r with rows $\mathbf{a}_k^T$, then it is interesting to note that the terms $\arccos(r_{j,j+1})$ in (8.5) are the angles between consecutive $\mathbf{a}_k$ vectors, which are points on an r-dimensional sphere. It is also straightforward to show that the d that sets the right-hand side of (8.4) equal to γ is strictly increasing in Λ. This implies that, as $\Lambda \to \infty$, one has $d \to \sqrt{rF_{1-\gamma,r,\nu}}$, which is the percentage point given by Scheffé's method. On the other hand, as $\Lambda \to 0$, one has $d \to t_{(1-\gamma)/2,\nu}$, the percentage point of the univariate Student's t distribution corresponding to $(1 - \gamma)/2$. This is intuitively pleasing because $\Lambda \to 0$ implies that correlations in $\mathbf{R}$ approach 1, in which case the p-dimensional distribution becomes one-dimensional for all practical purposes.

8.8 Pillai and Ramachandran's Percentage Points

Pillai and Ramachandran (1954) tabulated solutions of (8.1) and (8.3) for

- $p = 1(1)8$; $\nu = 3(1)10, 12, 14(1)16(2)20, 24, 30, 40, 60, 120, \infty$; $\gamma = 0.05$ and $\rho = 0$;
- $p = 1(1)8$; $\nu = 5(5)20, 24, 30, 40, 60, 120, \infty$; $\gamma = 0.05$ and $\rho = 0$,

respectively. For computing these percentage points, Pillai and Ramachandran used the pdfs of

$$U_p = \max(X_1, X_2, \ldots, X_p)$$

and $|U_p|$, which were derived as

$$\begin{aligned} f(u_p) &= \frac{p(\nu/2)^{\nu/2}}{\Gamma(\nu/2)\sqrt{2\pi}} \sum_{k=0}^{\infty} u_p^k \left\{ \frac{6}{(p+2)u_p^2 + 3\nu} \right\}^{(k+\nu+1)/2} \\ &\quad \times \Gamma\left(\frac{k+\nu+1}{2}\right) a_k^{p-1} \end{aligned}$$

and

$$\begin{aligned} f(|u_p|) &= \frac{p\nu^{\nu/2}}{\Gamma(\nu/2)\pi^{\nu/2}} \sum_{k=0}^{\infty} |u_p|^{p+2k-1} \left\{ \frac{6}{(p+2)u_p^2 + 3\nu} \right\}^{(p+2k+\nu)/2} \\ &\quad \times \Gamma\left(\frac{p+2k+\nu}{2}\right) b_k^{p+2k-1}, \end{aligned}$$

respectively, where the a's and b's are the coefficients of the expansions

$$\begin{aligned} &\left\{ \frac{1}{\sqrt{2\pi}} \int_{-\infty}^{y} \exp\left(-\frac{t^2}{2}\right) dt \right\}^k \\ &= \exp\left(-\frac{ky^2}{6}\right) \left[a_0^{(k)} + a_1^{(k)} y + a_2^{(k)} y^2 + \cdots \right] \end{aligned}$$

and

$$\left\{ \int_0^y \exp\left(-\frac{t^2}{2}\right) dt \right\}^k = y^k \exp\left(-\frac{ky^2}{6}\right) \left[1 + b_2^{(k)} y^4 + b_3^{(k)} y^6 + \cdots \right],$$

respectively. Note that $a_0^{(k)} = (1/2)^k$ and $b_0^{(k-1)} = 1$.

8.9 Dunnett's Percentage Points

Dunnett (1955) tabulated solutions of (8.1) and (8.3) for $p = 1(1)9$; $\nu = 5(1)20, 24, 30, 40, 60, 120, \infty$; $\gamma = 0.01, 0.05$; and $\rho = 0.5$. For solving (8.1), Dunnett evaluated the integral in (7.1) by using tables of the multivariate normal cdf computed by the National Bureau of Standards. For (8.3), Dunnett bounded $P(d)$ by

$$P(d) \geq [\Pr(-d < X_1 < d, -d < X_2 < d)]^{p/2}$$

(Dunnett and Sobel, 1955) and evaluated the probability integral of the bivariate t distribution using expressions (6.3)–(6.4). In a latter paper, Dunnett (1964) obtained approximations for d in (8.3) for all ρ's lying between 0 and 0.5.

8.10 Gupta and Sobel's Percentage Points

Gupta and Sobel (1957) solved (8.1) for the special case $\rho = 1/2$. Note from (6.8) that

$$P(d) = \Pr\left(Z < \sqrt{2}d\right)$$

and that $Z = (M_p - Y)/S$ is asymptotically normal as both ν and p tend to infinity. This allows for the use of a technique developed by Cornish and Fisher (1950) for computing the percentage points d directly without first computing a table of probability integral values. Applying their result, Gupta and Sobel arrived at

$$d = y_\gamma + \alpha_3 I_c + \alpha_4 I_d + \alpha_3^2 I_{c^2} + \alpha_5 I_e + \alpha_3\alpha_4 I_{cd} + \alpha_3^2 I_{c^3} + \cdots,$$

where y_γ is the percentage point of the standard normal distribution corresponding to γ, α_k is the standardized cumulant defined in (6.12), and I_c, I_d, I_{c^2}, ... are tabulated in Table I of Cornish and Fisher (1950) for the probability levels $\gamma = 0.75$, 0.90, 0.95, 0.975, 0.99, 0.995, 0.9975, 0.999 and 0.9995. Gupta and Sobel tabulated d for all combinations of $p + 1 = 2$, 5, 10(1)16, 18, 20(5)40, 50; $\nu = 15(1)20$, 24, 30, 36, 40, 48, 60, 80, 100, 120, 360, ∞; and $\gamma = 0.75$, 0.9, 0.95, 0.975, 0.99.

Gupta and Sobel (1957) also obtained several bounds for the percentage point d satisfying (8.3). An upper bound for d is obtained by setting (6.8) to be equal to $(1+\gamma)/2$ while a lower bound is obtained by setting (6.8) to be equal to γ. These bounds are best for large γ's. For smaller values of γ, Gupta and Sobel provided the following lower bound

$$d \geq \frac{(1+p)^{1/(2p)}}{4\left\{\Gamma\left(1+\frac{p}{2}\right)\right\}^{1/p}} \sqrt{2\pi\chi^2_{p,\gamma}},$$

where $\chi^2_{p,\gamma}$ is the percentage point of the chi-squared distribution (with p degrees of freedom) corresponding to γ.

8.11 Chen's Percentage Points

Chen (1979) provided an alternative formulation of percentage points d by solving the equation

$$F(d,\ldots,d;p,\nu) - F(-d,\ldots,-d;p,\nu) = \gamma \tag{8.6}$$

for the special case $\rho = 0$, where

$$F(x,\ldots,x;p,\nu) = \int_{-\infty}^{x}\cdots\int_{-\infty}^{x} f(x_1,\ldots,x_p;\nu,\rho)dx_p\cdots dx_1$$

and f is the joint pdf of a central p-variate t distribution with degrees of freedom ν and the equicorrelation structure $r_{ij} = \rho$, $i \neq j$. Chen noted that (8.6) can be rewritten as

$$\int_0^{\infty}\left\{\Phi^k(dy) - \Phi^k(-dy)\right\} g(y)dy = \gamma, \tag{8.7}$$

where $\Phi(\cdot)$ is the cdf of the standard normal distribution and g denotes the pdf of $\sqrt{\chi^2_\nu}$; further, by a change of variable, $z = \sqrt{\nu/2}y$, (8.7) becomes

$$2\int_0^{\infty}\left\{\Phi^k\left(\frac{\sqrt{2}dz}{\sqrt{\nu}}\right) - \Phi^k\left(\frac{-\sqrt{2}dz}{\sqrt{\nu}}\right)\right\}\frac{z^{\nu-1}\exp\left(-z^2\right)}{\Gamma(\nu/2)}dz = \gamma. \tag{8.8}$$

Using the fact that the tail integral

$$\begin{aligned} & 2\int_{10}^{\infty}\left\{\Phi^k\left(\frac{\sqrt{2}dz}{\sqrt{\nu}}\right) - \Phi^k\left(\frac{-\sqrt{2}dz}{\sqrt{\nu}}\right)\right\}\frac{z^{\nu-1}\exp\left(-z^2\right)}{\Gamma(\nu/2)}dz \\ \leq\ & 2\int_{10}^{\infty}\frac{z^{\nu-1}\exp\left(-z^2\right)}{\Gamma(\nu/2)}dz \\ =\ & \gamma \\ <<\ & 10^{-8} \end{aligned}$$

for all $\nu \leq 60$, Chen approximated the left-hand side of (8.8) by

$$P(d) = 2\int_0^{10}\left\{\Phi^k\left(\frac{\sqrt{2}dz}{\sqrt{\nu}}\right) - \Phi^k\left(\frac{-\sqrt{2}dz}{\sqrt{\nu}}\right)\right\}\frac{z^{\nu-1}\exp\left(-z^2\right)}{\Gamma(\nu/2)}dz$$

and found the value $\hat{d}_0$ such that $P(\hat{d}_0) = \gamma$. Tables of d_0 were given for all combinations of $\gamma = 0.8, 0.9, 0.95, 0.99$; $p = 2(1)20$; $\nu = 2(1)30(5)60$; and $\rho = 0$.

8.12 Bowden and Graybill's Percentage Points

Bowden and Graybill (1966) presented percentage points of the bivariate t distribution when the percentage points are not necessarily equal, that is, the case where

$$P(d, g) = \int_{-d}^{d} \int_{-g}^{g} f(x_1, x_2)\, dx_2 dx_1 = \gamma.$$

Setting $D = d$ and $A = g/d$, one can rewrite

$$P(d, g) = \int_{-D}^{D} \int_{-AD}^{AD} f(x_1, x_2)\, dx_2 dx_1,$$

which can be solved for D using the expressions (6.3) and (6.4). Bowden and Graybill computed D for all combinations of $p - 2 = 4$ (2) 16 (4) 24 (6) 30 (10) 50; $A = 0.5(0.1)1.5$; $\mid \rho \mid = 0.0(0.1)0.1(0.2)0.9$; and $\gamma = 0.90$, 0.95. Trout and Chow (1972) extended this development for trivariate t distributions: Setting

$$\int_{-d}^{d} \int_{-g}^{g} \int_{-h}^{h} f(x_1, x_2, x_3)\, dx_3 dx_2 dx_1 = \gamma$$

and using the transformations $D = d$, $A = g/d$ and $B = h/d$, the following expression is obtained

$$\int_{-D}^{D} \int_{-AD}^{AD} \int_{-BD}^{BD} f(x_1, x_2, x_3)\, dx_3 dx_2 dx_1 = \gamma.$$

Tables for D were given for all combinations of $\nu = 5(1)9(2)29$; $A = 0.5(0.1)1.5$; $B = 0.5(0.5)1.5$; $r_{ii} = 0.1(0.4)0.9$; $r_{12} = r_{13} = r_{23} = 0$; and $\gamma = 0.05$.

8.13 Dunnett and Tamhane's Percentage Points

More recently, Dunnett and Tamhane (1992) extended Bowden and Graybill's (1966) calculations for multivariate t distributions of any dimension with zero means and the equicorrelation structure $r_{ij} = \rho$, $i \neq j$. Consider iid standard normal random variables Z_j, $j = 0, 1, \ldots, p$ and let S be a $\sqrt{\chi_\nu^2/\nu}$ random variable independent of the Z_j. Then the random variables defined by

$$X_j = \frac{\sqrt{1-\rho} Z_j - \sqrt{\rho} Z_0}{S}, \qquad 1 \leq j \leq p$$

have the desired multivariate t distribution. Thus

$$\begin{aligned} P &= \Pr(X_1 \leq d_1, \ldots, X_p \leq d_p) \\ &= \int_0^\infty \int_{-\infty}^\infty \Pr(Z_1 < e_1, \ldots, Z_p \leq e_p)\, \phi(z)\, h(s) dz ds, \end{aligned} \tag{8.9}$$

where $e_j = (d_j s + \sqrt{\rho} z)/\sqrt{1-\rho}$, ϕ is the pdf of the standard normal distribution and h is the pdf of S. Dunnett and Tamhane (1992) obtained the following recursive formula for evaluating $\Phi_{1\cdots p} = \Pr(Z_1 < e_1, \ldots, Z_p < e_p)$ in the integrand of (8.9)

$$\begin{aligned} \Phi_{1\cdots p} &= \Pr(Z_1 < e_2, Z_2 < e_3, \ldots, Z_{p-1} < e_p)\, \Phi(e_1) \\ &\quad + \Pr(Z_1 < e_1, Z_2 < e_3, \ldots, Z_{p-1} < e_p) \\ &\qquad \times \{\Phi(e_2) - \Phi(e_1)\} \\ &\quad \vdots \\ &\quad + \Pr(Z_1 < e_1, Z_2 < e_2, \ldots, Z_{p-1} < e_{p-1}) \\ &\qquad \times \{\Phi(e_p) - \Phi(e_{p-1})\}, \end{aligned} \tag{8.10}$$

where $\Phi(\cdot)$ denotes the cdf of the standard normal distribution. They also suggested the following algorithm for computing $\Phi_{1\cdots p}$

Step 1: Calculate $\Phi_j = \Phi(e_j)$, for $j = 1, \ldots, p$, a total of p terms.
Step 2: Calculate $\Phi_{jk} = \Phi_j \Phi_k + \Phi_j(\Phi_k - \Phi_l)$, for $1 \leq j < k \leq p$, a total of $\binom{p}{2}$ terms.
Step 3: Calculate $\Phi_{jkl} = \Phi_{kl}\Phi_j + \Phi_{jl}(\Phi_k - \Phi_j) + \Phi_{jk}(\Phi_l - \Phi_k)$, for $1 \leq i < j < k \leq p$, a total of $\binom{p}{3}$ terms.

$\vdots$

Step p: Calculate $\Phi_{12\cdots p} = \Phi_{2\cdots p}\Phi_1 + \Phi_{13\cdots p}(\Phi_2 - \Phi_1) + \cdots + \Phi_{1\cdots p-1}(\Phi_p - \Phi_{p-1})$.

The computational details of this algorithm can be found in Dunnett and Tamhane (1990). Kwong and Liu (2000) – using Kwong's (2001b) lemma – proposed the following modification of (8.10)

$$\Phi_{1\cdots m} = \sum_{j=1}^{m} (-1)^{j-1} \binom{m}{m-j} \{\Phi(e_{m-j+1})\}^j J_{m-j}, \tag{8.11}$$

where $m \geq 2$, $J_1 = \Phi(e_1)$, and $J_0 = 1$ (see also Kwong, 2001a).

There are three commonly known approaches for determining the percentage points $d_1, \ldots, d_p$ in (8.9) (after setting $P = \gamma$): the step-up procedure, the step-down procedure, and the simulation approach (Dunnett and Tamhane, 1995). We shall describe them below by means of recursive algorithms. Throughout we shall let $X_{(1)} < \cdots < X_{(p)}$ denote the order statistics of $X_1, \ldots, X_p$ and let $c_1, \ldots, c_p$ denote the corresponding ordering of $d_1, \ldots, d_p$.

- Step-up procedure
  (i) Take c_1 to be the 100γ percentage point of the univariate Student's t distribution with degrees of freedom ν.
  (ii) Solve the equation

$$\begin{aligned} & \Pr\left(X_{(1)} < c_1, X_{(2)} < c_2\right) \\ = & \Pr\left(X_1 < c_1, X_2 < c_2\right) + \Pr\left(c_1 < X_1 < c_2, X_2 < c_1\right) \\ = & \gamma \end{aligned}$$

  for c_2 by evaluating the two bivariate probabilities using (8.9) and the value for c_1 defined in (i).
  (iii) Solve the equation

$$\begin{aligned} & \Pr\left(X_{(1)} < c_1, X_{(2)} < c_2, X_{(3)} < c_3\right) \\ = & \Pr\left(X_1 < c_1, X_2 < c_2, X_3 < c_3\right) \\ & \quad + \Pr\left(X_1 < c_1, c_2 < X_2 < c_3, X_3 < c_2\right) \\ & \quad + \Pr\left(c_1 < X_1 < c_2, X_2 < c_1, X_3 < c_3\right) \\ & \quad + \Pr\left(c_1 < X_1 < c_2, c_1 < X_2 < c_3, X_3 < c_1\right) \\ & \quad + \Pr\left(c_2 < X_1 < c_3, X_2 < c_1, X_3 < c_2\right) \\ & \quad + \Pr\left(c_2 < X_1 < c_3, c_1 < X_2 < c_2, X_3 < c_1\right) \\ = & \gamma \end{aligned}$$

  for c_3 by evaluating the six trivariate probabilities using (8.9) and the values for c_1 and c_2 defined in (i) and (ii), respectively.
  (iv) In general, the recursive formula below defines how the region over which the probability must be evaluated can be subdivided to obtain probability expressions

$$\begin{aligned} & \left[X_{(1)} < c_1, \ldots, X_{(p)} < c_p\right] \\ = & \left\{X_1 < c_1, \left[X_{(2)} < c_2, \ldots, X_{(p)} < c_p\right]\right\} \\ & + \left\{c_1 < X_1 < c_2, \left[X_{(2)} < c_1, X_{(3)} < c_3, \ldots, X_{(p)} < c_p\right]\right\} \\ & \vdots \\ & + \left\{c_{p-1} < X_1 < c_p, \left[X_{(2)} < c_1, \ldots, X_{(p)} < c_{p-1}\right]\right\}, \end{aligned} \tag{8.12}$$

where $X_{(2)} < \cdots < X_{(p)}$ denote the order statistics of $X_2, \ldots, X_p$ with X_1 separated out. Formula (8.12) is applied recursively to the terms enclosed within the square brackets. This leads to a division of the region into $p!$ subregions that have rectangular boundaries, making it possible to evaluate the individual probabilities (using (8.9)).

- Step-down procedure: See Dunnett and Tamhane (1991) for a lucid description.
- Simulation approach: This approach is feasible provided that p is not so large that sampling errors in the values of $c_3, \ldots, c_{p-1}$ accumulate and render the estimated value of c_p too uncertain to be of practical use. The procedure for estimating c_m given the values of $c_1, \ldots, c_{m-1}$ (Edwards and Berry, 1987) is as follows.

    (i) Let N_T denote the total number of simulations to be performed and choose N_T so that $N_0 = (1-\gamma)(1+N_T)$ is an integer.

    (ii) Initialize a counter, $N_c = N_0$.

    (iii) For each simulation, draw m standard normal deviates $Z_1, \ldots, Z_m$ having the desired correlation structure and, if ν is finite, a random χ^2_ν/ν variate S^2.

    (iv) Set $X_i = Z_i/S$ if ν is finite or $X_i = Z_i$ if $\nu = \infty$ and order the X values to obtain the order statistics $X_{(1)} \leq \cdots \leq X_{(m)}$.

    (v) Check whether $X_{(1)} < c_1, \ldots, X_{(m-1)} < c_{m-1}$. If this is the case, store the value of $X_{(m)}$ and return to step (iii). Otherwise, decrease N_c by 1 and return to step (iii).

    (vi) After completing the N_T simulations, find the estimate of c_m by counting down N_c from the top of the ordered values of the stored $X_{(m)}$. Note that this approach is general and does not impose restrictions on the correlation structure r_{ij}.

Dunnett and Tamhane (1992) computed values of $d_1, \ldots, d_p$ using the step-up procedure for all combinations of $p = 2(1)8$; $\nu = 10, 20, 30, \infty$; $\rho = 0, 0.1(0.2)0.5$; and $\gamma = 0.95$. Kwong and Liu (2000), using the same procedure but with the modification (8.11), computed values of $d_1, \ldots, d_p$ for all combinations of $p = 9(1)20$; $\nu = 10, 20, 30, \infty$; $\rho = 0.1, 0.3, 0.5$; and $\gamma = 0.95$.

8.14 Kwong and Liu's Percentage Points

In the case $r_{ij} = b_i b_j$, Kwong and Liu (2000) pointed out that (8.9) can be generalized to

$$P = \int_0^\infty \int_{-\infty}^\infty H_{1\cdots p} \phi(z) h(s) dz ds,$$

where $H_{1\cdots p} = \Pr(X_1 < e_1, \ldots, X_p < e_p)$, $e_j = d_j x$ and X_j are iid normal random variables with means $b_j z$ and variances $1 - b_j^2$. The recursive formula (8.10) and the algorithm for computing it can be generalized in a natural manner. Hence the step-up, step-down, or the simulation-based procedure can be used to compute the percentage points $d_1, \ldots, d_p$. A fourth procedure not discussed above is one based on approximation.

(i) c_1 and c_2 can be determined as in the step-up procedure.

(ii) To determine c_3, replace r_{12}, r_{13}, and r_{23} by $\bar{\rho}_3 = (r_{12} + r_{13} + r_{23})/3$. Taking this as the common ρ and using the previous values c_1, c_2, apply the step-up procedure to estimate c_3.

(iii) To determine c_4, replace r_{12}, r_{13}, r_{14}, r_{23}, r_{24}, and r_{34} by $\bar{\rho}_4 = (r_{12} + r_{13} + r_{14} + r_{23} + r_{24} + r_{34})/6$. Taking this as the common ρ and using the previous values c_1, c_2, c_3, apply the step-up procedure to estimate c_4.

(iv) In general, replace the $\binom{m}{2}$ correlation coefficients by their arithmetic average $\bar{\rho}_m$ and use the previously calculated values of $c_1, \ldots, c_{m-1}$ to obtain an estimate for c_m using the step-up procedure.

This procedure is similar to the ones presented in Dunnett (1985), Hochberg and Tamhane (1987, page 146), and Iyengar (1988).

8.15 Other Results

Some other tabulations of percentage points of the multivariate t distributions with the equicorrelation structure are contained in the following references.

- Paulson (1952) for $p = 3, 6$ and $\rho = 0$.
- Dunnett and Sobel (1955) used the lower bounds (7.4), (7.5), (7.6), and (7.10) for $p = 3$, 9; $\nu = 5, \infty$; $\rho = 1/2$; and $\gamma = 0.50$, 0.75, 0.95, 0.99.
- Halperin et al. (1955) for $p = 3(1)10$, 15, 20, 30, 40, 60; $\nu = 3(1)10$, 15, 20, 30, 40, 60, 120, ∞; $\rho = 1 - 1/p$; and $\gamma = 0.95, 0.99$.

- Gupta (1963) for $p = 1(1)50$; $\nu = \infty$; $\rho = 1/2$; and $\gamma = 0.75, 0.9, 0.95, 0.975$.
- Milton (1963) for extensions of Gupta's (1963) tables for γ ranging from 0.5 to 0.9999.
- Steffens (1969b) for $p = 2$.
- Dunn and Massey (1965) for $p = 2, 6, 10, 20$; $\nu = 4, 10, 30, \infty$; $\rho = 0.0(0.1)1.0$; and $\gamma = 0.5(0.1)0.9, 0.95, 0.975, 0.99$.
- Tong (1970) for a procedure to calculate conservative estimates of the percentage points for $p > 20$ using tabulated values for $p = 20$.
- Freeman and Kuzmack (1972) for $p = 6, 8, 10(5)30$; $N_0 = 10, 20$, mean (40–70), 50, 100, mean (90–500), 500; $\rho = 0$; and $\gamma = 0.90, 0.95, 0.99$, where $\nu = p(N_0 - 1)$.
- Gupta et al. (1973) for $p = 1(1)10(2)50$; $\nu = \infty$; $\rho = 0.1, 0.125, 0.2, 1/3, 0.375, 0.4, 1/2, 0.64, 0.625, 2/3, 0.7, 0.75, 0.8, 0.875, 0.9$; and $\gamma = 0.75, 0.9, 0.95, 0.975, 0.99$.
- Amos (1978) for $p = 100$; $\nu = 100$; $\rho = 0.01$; and $\gamma = 0.05, 0.1, \ldots, 0.95$.
- Ahner and Passing (1983) for $1 \leq p \leq 20$; $2 \leq \nu \leq 120$, $\nu = \infty$; $\rho = 0$; and $\gamma = 0.95, 0.99$.
- Bechhofer and Dunnett (1988) for the most comprehensive table to date for $p = 2(1)16, 18, 20$; $\nu = 2(1)30(5)50, 60(20)120, 200, \infty$; $\rho = 0.0(0.1)0.9, 1/(1 + \sqrt{p})$; and $\gamma = 0.80, 0.90, 0.95, 0.99$.
- Kwong and Iglewicz (1996) for $p = 4, 5$; $\nu = p(1)20, 24, 30(10)100, 120, \infty$; $\rho = -1/(p - 1)$; and $\gamma = 0.90, 0.95, 0.99$ (note that the correlation matrix is singular).

There has been relatively little work concerned with the percentage points of multivariate t distributions when the correlations are not equicorrelated. Apart from those mentioned above, four results known to us are

- In the case that the (i, j)th element of the inverse of $\mathbf{R}$ is

$$r^{ij} = \begin{cases} 2p/(p+1), & \text{if } i = j, \\ -2/(p+1), & \text{if } i \neq j, \end{cases}$$

  Goldberg and Levine (1946) computed the percentage point d satisfying

$$\int_{-\infty}^{d} \cdots \int_{-\infty}^{d} f(x_1, \ldots, x_p; \nu)\, dx_1 \cdots dx_p = \gamma \qquad (8.13)$$

  for combinations of $p = 3$; $\nu = 1(1)30(3)60(15)120, 150, 300, 600, \infty$;

and $\gamma = 0.50, 0.75, 0.90, 0.95$. This seems to be the earliest paper on this topic.

- In the case

$$r^{ij} = 2\min(i,j)\left\{1 - \frac{\max(i,j)}{p}\right\},$$

Freeman et al. (1967) computed the percentage point d satisfying (8.13) for combinations of $p = 3(1)5$; $(\nu/p) + 1 = 10(10)100, 200, 500$; and $\gamma = 0.95$. See also Bechhofer et al. (1954) and Table 4 in Dunnett and Sobel (1954).

- In the case that the (i, j)th element of $\mathbf{R}$ is

$$r_{ij} = \sqrt{\frac{n_i n_j}{(n_0 + n_i)(n_0 + n_j)}},$$

where n_k denotes some treatment sample size, Dutt et al. (1976) computed d in (8.13) for all combinations of $\gamma = 0.95, 0.99$ and $3 \le n_i \le 12$, $i = 0, 1, 2, 3$ with $p = 3$ and the degrees of freedom $\nu = \sum(n_i - 1)$. See also Dutt et al. (1975).

- In the case $r_{ij} = 0$ for all $i \neq j$ except that

$$r_{i,i+1} = \frac{-\sqrt{n_i n_{i+2}}}{\sqrt{n_i + n_{i+1}}\sqrt{n_{i+1} + n_{i+2}}}$$

for some treatment sample sizes n_k, Lee and Spurrier (1995) provided tables of one-sided and two-sided percentage points – of the form (8.13) – for $3 \le p \le 6$ and $n_k = n$ (the balanced case). Liu et al. (2000) extended these tables for $3 \le p \le 10$; $\nu = 5(1)8(2)20, 25, 30, 40, 60, 120, \infty$; and $\gamma = 0.90, 0.95, 0.99$. See also Somerville et al. (2001).

Calculations of percentage points for singular correlation structures of the form $r_{ij} = -b_i b_j$ are discussed in Spurrier and Isham (1985) and Kwong and Iglewicz (1996). The former provided tabulations of the percentage points for $p = 3$, $3 \le n_1 \le n_2 \le n_3$, $10 \le N \le 29$, and $\gamma = 0.90, 0.95, 0.99$ when $b_k = \sqrt{n_k/(N - n_k)}$ with $N = n_1 + n_2 + n_3$ (where n_k denotes some treatment sample size).

Calculations of percentage points for correlation structures more general than the decomposable structure $r_{ij} = b_i b_j$ are quite difficult and challenging. Even the generalization to quasi-decomposable structures of the form $r_{ij} = b_i b_j + \gamma_{ij}$ (Yang and Zhang, 1997), where the γ_{ij}'s are nonzero deviations for some i and j, is a rather restrictive assumption. A solution is to find the "closest" $\hat{\mathbf{R}}$ for a given matrix $\mathbf{R}$, which still possesses the decomposable correlation structure (Hsu, 1992; Hsu and Nelson, 1998). One also has the choice of adopting the simulation approach or one of the general approaches due to Somerville (1997, 1998b) and Genz and Bretz (1999) – described in Section 6.10.

9
Sampling Distributions

Here, we shall consider sampling distributions of certain statistics associated with multivariate t distributions.

9.1 Wishart Matrix

Suppose $\mathbf{X}_1, \ldots, \mathbf{X}_n$ is a random sample from a p-variate t distribution with the common pdf

$$\begin{aligned} f(\mathbf{x}_i) &= \frac{\Gamma((\nu+p)/2)}{(\pi\nu)^{p/2}\Gamma(\nu/2)|\mathbf{R}|^{1/2}} \\ &\quad \times \left[1+\frac{1}{\nu}(\mathbf{x}_i-\boldsymbol{\mu})^T \mathbf{R}^{-1}(\mathbf{x}_i-\boldsymbol{\mu})\right]^{-(\nu+p)/2}. \end{aligned}$$

The joint pdf of the n independent observations is given by

$$f(\mathbf{x}_1, \ldots, \mathbf{x}_n) = f(\mathbf{x}_1)\cdots f(\mathbf{x}_n). \tag{9.1}$$

However, it is more instructive to consider dependent but uncorrelated t distributions. Joarder and Ahmed (1996) suggested the model

$$\begin{aligned} f(\mathbf{x}_1, \ldots, \mathbf{x}_n) &= \frac{\Gamma((\nu+p)/2)}{(\pi^n\nu)^{p/2}\Gamma(\nu/2)|\mathbf{R}|^{n/2}} \\ &\quad \times \left[1+\frac{1}{\nu}\sum_{i=1}^{n}(\mathbf{x}_i-\boldsymbol{\mu})^T \mathbf{R}^{-1}(\mathbf{x}_i-\boldsymbol{\mu})\right]^{-(\nu+np)/2}, \end{aligned} \tag{9.2}$$

which they referred to as the *multivariate t model.* Joarder and Ali (1997) remarked that this model can also be written as a scale mixture of multivariate normal distributions given by

$$\begin{aligned} & f(\mathbf{x}_1, \ldots, \mathbf{x}_n) \\ = & \int_0^\infty \frac{\left|\tau^2 \mathbf{R}\right|^{-n/2}}{(2\pi)^{np/2}} \exp\left\{-\frac{1}{2}\sum_{i=1}^n (\mathbf{x}_i - \boldsymbol{\mu})^T \left(\tau^2 \mathbf{R}\right)^{-1} (\mathbf{x}_i - \boldsymbol{\mu})\right\} h(\tau)\, d\tau, \end{aligned}$$

where τ has the inverted gamma distribution with the pdf

$$h(\tau) = \frac{2\tau^{-(1+\nu)}}{(2/\nu)^{\nu/2}\Gamma(\nu/2)} \exp\left(-\frac{\nu}{2\tau^2}\right) \tag{9.3}$$

for $\tau > 0$. Equivalently, $\mathbf{X}_j \mid \tau$ has the multivariate normal distribution $N_p(\boldsymbol{\mu}, \tau^2 \mathbf{R})$. Among others, Zellner (1976) and Sutradhar and Ali (1986) considered (9.2) in the context of stock market problems. By successive integration, one can show that the marginal distribution of $\mathbf{X}_j$ in the multivariate t model (9.2) is p-variate t. It also follows from (9.2) that $E(\mathbf{X}_j - \boldsymbol{\mu})(\mathbf{X}_l - \boldsymbol{\mu}) = 0$ for $j \neq l$. Thus, in (9.2), although $\mathbf{X}_1, \ldots, \mathbf{X}_n$ are pairwise uncorrelated, they are not necessarily independent. More specifically, $\mathbf{X}_1, \ldots, \mathbf{X}_n$ in (9.2) are not independent if $\nu < \infty$, since independence would imply that $\mathbf{X}_1, \ldots, \mathbf{X}_n$ are normally distributed. The case of independent normally distributed random vectors can be included in (9.2) by letting $\nu \to \infty$. In the case $\nu = 1$, (9.2) is the multivariate Cauchy distribution for which neither the mean nor the variance exists. Kelejian and Prucha (1985) proved that (9.2) is better able to capture heavy-tailed behavior than an independent t model given by (9.1).

The sampling quantities of interest are the mean vector and the sum of product matrix (Wishart matrix) given by

$$\bar{\mathbf{X}} = \frac{1}{n}\sum_{i=1}^n \mathbf{X}_i$$

and

$$\mathbf{A} = \sum_{i=1}^n \left(\mathbf{X}_i - \bar{\mathbf{X}}\right)\left(\mathbf{X}_i - \bar{\mathbf{X}}\right)^T, \tag{9.4}$$

respectively. Sutradhar and Ali (1989) derived the corresponding pdfs, which are

$$\begin{aligned} f(\bar{\mathbf{x}}) = & \frac{\nu^{\nu/2}\Gamma((\nu+p)/2)}{\pi^{p/2}\Gamma(\nu/2)} \left|\mathbf{R}/n\right|^{-1/2} \\ & \times \left[\nu + (\bar{\mathbf{x}} - \boldsymbol{\mu})^T \mathbf{R}^{-1} (\bar{\mathbf{x}} - \boldsymbol{\mu})\right]^{-(\nu+p)/2} \end{aligned}$$

and

$$f(\mathbf{A}) = \frac{\Gamma((\nu+p)/2)}{\nu^{p/2}\Gamma(\nu/2)\Gamma_p(n/2)}|\mathbf{R}|^{-(n-1)/2}|\mathbf{A}|^{-(n-p-2)/2} \times \left[\nu + \mathrm{tr}\mathbf{R}^{-1}\mathbf{A}\right]^{-(\nu+p(n-1))/2}, \tag{9.5}$$

respectively, where $\mathbf{A} > 0$, $n > 1+p$ and $\Gamma_p(x)$ is the generalized gamma function defined by

$$\Gamma_p(x) = \pi^{p(p-1)/4}\prod_{i=1}^{p}\Gamma\left(\frac{2x-i+1}{2}\right). \tag{9.6}$$

The distribution of the Wishart matrix, (9.5), has its applications in factor analysis. More specifically, in practice, one may be confronted with the situation where the observed data have a symmetrical distribution with tails that are fatter than that predicted by the normal distribution. In such cases, one could explicitly account for the observed "fat tails" by using the multivariate t model (9.2). Consequently, in factor analysis, analogous to the Wishart distribution, one may use the distribution of the sum of products matrix under (9.2).

It is easily checked that, as $\nu \to \infty$, the pdf (9.5) converges to

$$f(\mathbf{A}) = \frac{|\mathbf{R}|^{-(n-1)/2}|\mathbf{A}|^{-(n-p-2)/2}}{2^{p(n-1)/2}\Gamma_p((n-1)/2)}\exp\left[-\frac{1}{2}\mathrm{tr}\mathbf{R}^{-1}\mathbf{A}\right],$$

which is the pdf of the usual Wishart distribution $\mathcal{W}_p(\mathbf{R}, n)$. The pdf (9.5) can also be written as the mixture of distributions

$$f(\mathbf{A}) = \int_0^\infty \frac{|\tau^2\mathbf{R}|^{-(n-1)/2}}{2^{p(n-1)/2}\Gamma_p((n-1)/2)}|\mathbf{A}|^{-(n-p-2)/2} \times \exp\left[-\frac{1}{2}\mathrm{tr}\left(\left(\tau^2\mathbf{R}\right)^{-1}\mathbf{A}\right)\right] f(\tau)d\tau,$$

where τ has the inverted gamma pdf (9.3). This is equivalent to saying that $\mathbf{A} \mid \tau = \tau^2\mathbf{W}$ has the usual Wishart distribution $\mathcal{W}_p(\tau^2\mathbf{R}, n)$. Joarder and Ali (1992) and Joarder (1998) derived various expectations of the Wishart matrix $\mathbf{A}$. Specifically, one has the following expressions

$$E(\mathbf{A}) = \frac{n-1}{1-2/\nu}\mathbf{R},$$

$$E\left(\mathbf{A}^2\right) = \frac{(n-1)^2\mathbf{R}^2 + (n-1)\left\{\mathbf{R}\mathrm{tr}\mathbf{R} + \mathbf{R}^2\right\}}{(1-2/\nu)(1-4/\nu)}, \quad \nu > k,$$

$$E\left(|\mathbf{A}|^k\right) = \frac{\Gamma(\nu/2 - kp)}{\nu^{-kp}\Gamma(\nu/2)} \frac{\Gamma_p((n-1)/2+k)}{\Gamma_p((n-1)/2)} |\mathbf{R}|^k, \quad \nu > k,$$

$$E\left(|\mathbf{A}|^k \mathbf{A}\right) = \frac{((n-1)/2+k)\Gamma(\nu/2 - kp - 1)}{\nu^{-(1+kp)}\Gamma(\nu/2)} \times \frac{\Gamma_p((n-1)/2+k)}{\Gamma_p((n-1)/2)} |\mathbf{R}|^k \mathbf{R}, \quad n + 2k > 1, \quad \nu > 2(kp+1),$$

$$E\left(|\mathbf{A}|^k \mathbf{A}^{-1}\right) = \frac{2\Gamma(\nu/2 - kp + 1)\Gamma_p((n-1)/2+k)}{\nu^{1-kp}(n+2k-p)\Gamma(\nu/2)\Gamma_p((n-1)/2)} |\mathbf{R}|^k \mathbf{R}^{-1}, \quad n + 2k > p + 2, \quad \nu > 2(kp-1),$$

$$E\left[(\mathrm{tr}\mathbf{A})^2\right] = \frac{n-1}{(1-2/\nu)(1-4/\nu)} \left[(n-1)(\mathrm{tr}\mathbf{R})^2 + 2\mathrm{tr}\left(\mathbf{R}^2\right)\right], \quad \nu > 4$$

and

$$E\left[\mathrm{tr}\left(\mathbf{A}^2\right)\right] = \frac{n-1}{(1-2/\nu)(1-4/\nu)} \left[n\mathrm{tr}\left(\mathbf{R}^2\right) + (\mathrm{tr}\mathbf{R})^2\right], \quad \nu > 4,$$

where k is any real number and $\nu > 0$. These expectations are important tools in developing estimation theories for the correlation matrix, inverted correlation matrix, trace of the correlation matrix, and other characteristics of the correlation matrix, of the multivariate t model under quadratic loss functions. Extensions of these expectations to the class of scale mixtures of normal distributions – which may be useful in inferential works having a t distribution or the scale mixture of normal distributions as the parent population – are discussed below.

Sutradhar and Ali (1989) derived an elementwise expression for the variance-covariance matrix of $\mathbf{A}$. Letting m_{ij} denote the (i,j)th element of $\mathbf{R}^{1/2}$, they showed that

$$\begin{aligned} Cov(A_{ij}, A_{kl}) = \frac{\nu-2}{\nu-4}\Bigg[& 3(n-1)\sum_{u=1}^{p} m_{iu}m_{ju}m_{ku}m_{lu} \\ & + (n-1)(n-2)\sum_{u=1}^{p} m_{iu}m_{ju}m_{ku}m_{lu} \\ & + (n-1)^2 \sum_{u<v} m_{iu}m_{ju}m_{kv}m_{lv} \end{aligned}$$

$$
\begin{aligned}
&+(n-1)^2 \sum_{u<v} m_{iv} m_{jv} m_{ku} m_{lu} \\
&+(n-1) \sum_{u<v} m_{iu} m_{jv} \left(m_{ku} m_{lv} + m_{kv} m_{lu}\right) \\
&+(n-1) \sum_{u<v} m_{iv} m_{ju} \left(m_{ku} m_{lv} + m_{kv} m_{lu}\right)\Bigg] \\
&-(n-1)^2 \left(\sum_{u=1}^{p} m_{iu} m_{ju}\right) \left(\sum_{u=1}^{p} m_{ku} m_{lu}\right)
\end{aligned}
$$

for $i \neq k$, $j \neq l$, $i, j, k, l = 1, \ldots, p$, and

$$
\begin{aligned}
Var\left(A_{ij}\right) &= \frac{\nu-2}{\nu-4} \left[(n-1)^2 \left(\sum_{u=1}^{p} m_{iu} m_{ju}\right)^2 + 2(n-1) \sum_{u=1}^{p} m_{iu}^2 m_{ju}^2 \right. \\
&\left. +(n-1) \sum_{u<v} \left(m_{iu} m_{jv} + m_{iv} m_{ju}\right)^2 \right] \\
&-(n-1)^2 \left(\sum_{u=1}^{p} m_{iu} m_{ju}\right)^2
\end{aligned}
$$

for $i, j = 1, \ldots, p$.

Let $\mathbf{A} = \mathbf{T}\mathbf{T}^T$ and $\mathbf{A} = \mathbf{S}\mathbf{M}\mathbf{S}^T$ be the triangular and spectral decompositions of $\mathbf{A}$. Let $\mathbf{W} = \mathbf{U}\mathbf{U}^T$ be the triangular decomposition of $\mathbf{W} \sim \mathcal{W}_p(\mathbf{R}, n)$. Let $l_1, \ldots, l_p$ and $m_1, \ldots, m_p$ denote the latent roots of $\mathbf{W}$ and $\mathbf{A}$, respectively. Also define

$$
d_i = \frac{1}{n+p+1-2i}
$$

and

$$
d_i^* = \frac{\nu-2}{\nu} \frac{1}{n+p+1-2i}
$$

with $\mathbf{D} = \mathrm{diag}(d_1, \ldots, d_p)$ and $\mathbf{D}^* = \mathrm{diag}(d_1^*, \ldots, d_p^*)$. Then, some further expectation identities involving $\mathbf{A}$ useful in the estimation of $\mathbf{R}$ are

$$
E\left[\log\left(|\mathbf{A}|\right)\right] = E\left[\log\left(|\mathbf{W}|\right)\right] + 2pE\left(\log \tau\right),
$$

$$
E\left[\log\left(\left|\mathbf{R}^{-1}\mathbf{A}\right|\right)\right] = \sum_{i=1}^{p} E\left[\log\left(\chi^2_{n+1-i}\right)\right] + 2pE\left(\log \tau\right),
$$

and

$$E\left[\operatorname{tr}\left(\mathbf{R}^{-1}\mathbf{T}\boldsymbol{\Delta}\mathbf{T}^{T}\right)\right] = \frac{\nu}{\nu-2}E\left[\operatorname{tr}\left(\mathbf{R}^{-1}\mathbf{U}\boldsymbol{\Delta}\mathbf{U}^{T}\right)\right]$$

(Joarder and Ali, 1997), where $\boldsymbol{\Delta}$ is a positive definite diagonal matrix and τ has the inverted gamma distribution given by (9.3).

Joarder and Ahmed (1998) considered a generalization of the multivariate t model in (9.2) when the random sample $\mathbf{X}_1, \ldots, \mathbf{X}_n$ is assumed to come from a p-variate elliptical distribution with the joint pdf

$$f(\mathbf{x}_i) = \int_0^\infty \frac{\left|\tau^2\mathbf{R}\right|^{-1/2}}{(2\pi)^{p/2}} \exp\left\{-\frac{1}{2}(\mathbf{x}_i-\boldsymbol{\mu})^T\left(\tau^2\mathbf{R}\right)^{-1}(\mathbf{x}_i-\boldsymbol{\mu})\right\} h(\tau)d\tau,$$

where $h(\cdot)$ is the pdf of a nondiscrete random variable τ. Many multivariate distributions having a constant pdf on the hyperellipse $(\mathbf{x}-\boldsymbol{\mu})^T$ $\mathbf{R}^{-1}$ $(\mathbf{x}-\boldsymbol{\mu}) = c^2$ may be generated by varying $h(\cdot)$. In this general case, the model corresponding to (9.2) has the joint pdf

$$f(\mathbf{x}_1, \ldots, \mathbf{x}_n) = \int_0^\infty \frac{\left|\tau^2\mathbf{R}\right|^{-n/2}}{(2\pi)^{np/2}} \exp\left\{-\frac{1}{2}\sum_{i=1}^n(\mathbf{x}_i-\boldsymbol{\mu})^T\left(\tau^2\mathbf{R}\right)^{-1}(\mathbf{x}_i-\boldsymbol{\mu})\right\} h(\tau)d\tau. \tag{9.7}$$

The observations $\mathbf{X}_1, \ldots, \mathbf{X}_n$ are independent only if τ is degenerate at the point unity, in which case the joint pdf (9.7) denotes the pdf of the product of n independent p-variate normal distributions each being $N_p(\boldsymbol{\mu}, \mathbf{R})$. Furthermore, if ν/τ^2 has the chi-squared distribution with degrees of freedom ν, then (9.7) reduces to (9.2). The pdf of the Wishart matrix $\mathbf{A}$ under the generalized model (9.7) takes the form

$$f(\mathbf{A}) = \frac{\left|\tau^2\mathbf{R}\right|^{-n/2}|\mathbf{A}|^{(n-p-1)/2}}{2^{np/2}\Gamma_p(n/2)} \exp\left[-\frac{1}{2}\operatorname{tr}\left(\left(\tau^2\mathbf{R}\right)^{-1}\mathbf{A}\right)\right] h(\tau)d\tau,$$

where $\mathbf{A} > 0$, $n > 1+p$, and $\Gamma_p(\cdot)$ is as defined in (9.6). Some expectations of $\mathbf{A}$ useful for estimating $\mathbf{R}$ are

$$E\left(|\mathbf{A}|^r\right) = 2^{pr}\frac{\Gamma(n/2+r)}{\Gamma(n/2)}|\mathbf{R}|^r\gamma_{2pr},$$

$$E\left[|\mathbf{A}|^k\mathbf{A}\right] = 2^{kp}(n+2k)\frac{\Gamma_p(n/2+k)}{\Gamma(n/2)}|\mathbf{R}|^k\mathbf{R}\gamma_{2kp+2},$$

and

$$E\left[(\operatorname{tr}\mathbf{A})^2\right] = n\gamma_4\left[n(\operatorname{tr}\mathbf{R})^2 + 2\operatorname{tr}\left(\mathbf{R}^2\right)\right],$$

where $\gamma \in \Re$, $k \in \Re$, $n + 2k > 0$, $\gamma_{2pr} = E(\tau^{2pr}) > 0$, $\gamma_{2kp+2} = E(\tau^{2kp+2}) > 0$, and $\gamma_4 = E(\tau^4)$ (all assumed to exist).

The pdfs of $\mathbf{A}$ in the real and complex cases – under the independence model (9.1) – were originally studied by Cornish (1955) and Gupta (1964), respectively. Nagarsenker (1975) provided a very detailed study of the distribution of $\mathbf{A}$ and its quadratic forms. He investigated both the noncentral real and the noncentral complex cases.

Let $\mathbf{Y}$ be a $p \times n$ matrix of iid normal random variables with means $E(Y_{ij}) = \mu_{ij}$ and covariance matrix $\sigma^2\mathbf{R}$. Assume that S is an independent random variable having the $\sqrt{\sigma'^2\chi^2_{2\nu}/(2\nu)}$ distribution. Then the noncentral version of $\mathbf{A}$ is defined by

$$a_{ij} = \frac{1}{S^2}\sum_{k=1}^{n}\left(Y_{ik} - \bar{Y}_i\right)\left(Y_{jk} - \bar{Y}_j\right),$$

where

$$\bar{Y}_l = \frac{1}{n}\sum_{j=1}^{n} Y_{lj}.$$

In the real case, $S^2\mathbf{A}$ has the noncentral Wishart distribution. In the complex case, it should be interpreted as a Hermitian positive definite matrix having the noncentral complex Wishart distribution (James, 1964). In the noncentral real case, Nagarsenker (1975) established that the pdf of $\mathbf{A}$ is given by the complicated expression involving zonal polynomials

$$\begin{aligned} & f(\mathbf{A}) \\ = & \frac{(\sigma')^{p(n-1)} \mid \mathbf{R} \mid^{-(n-1)/2} \mid \mathbf{A} \mid^{(n-p)/2} \exp\left\{-\mathrm{tr}\mathbf{R}^{-1}\mu\mu^T / \left(2\sigma^2\right)\right\}}{\sigma^{p(n-1)}(2\nu)^{p(n-1)/2}\Gamma_p\left((n-1)/2\right)\Gamma(\nu)} \\ & \times \sum_{k=0}^{\infty}\sum_{\kappa} \frac{\nu\left(\sigma'\right)^{2k}\Gamma\left(\nu + k + p(n-1)/2\right) C_\kappa\left(\mathbf{R}^{-1}\mu\mu^T\mathbf{R}^{-1}\mathbf{A}\right)}{k!((n-1)/2)_\kappa\left(4\nu\sigma^4\right)^k\left\{1 + \left(\sigma'^2\mathrm{tr}\mathbf{R}^{-1}\mathbf{A}\right)/2\nu\sigma^2\right\}} \\ & + \frac{p(n-1)}{2} + k, \end{aligned} \tag{9.8}$$

where $\kappa = \{k_1, \ldots, k_m\}$, $k_1 \geq k_2 \geq \cdots \geq k_m \geq 0$, $k_1 + k_2 + \cdots + k_m = k$,

$$(x)_\kappa = \frac{\Gamma_m(x, \kappa)}{\Gamma_m(x)},$$

$$\Gamma_r(x, \kappa) = \pi^{r(r-1)/4}\Gamma\left(x + k_1\right)\Gamma\left(x + k_2 - \frac{1}{2}\right)\cdots\Gamma\left(x + k_r - \frac{r-1}{2}\right),$$

and $C_\kappa(\mathbf{T})$ are symmetric homogeneous polynomials of degree k in the latent roots of $\mathbf{T}$. In the particular case $\sigma = \sigma'$ and $\mu = \mathbf{0}$, (9.8) reduces to the expression given in Cornish (1955). If $\mathbf{B}$ is an $(n-1) \times (n-1)$ symmetric positive definite matrix of full rank, then Nagarsenker (1975) established further that the quadratic form $\mathbf{Q} = \mathbf{A}\mathbf{B}\mathbf{A}^T$ has the formidable pdf

$$\begin{aligned} f(\mathbf{Q}) &= \frac{(\sigma')^{p(n-1)} \mid \mathbf{R} \mid^{-(n-1)/2} \mid \mathbf{Q} \mid^{(n-p)/2}}{\sigma^{p(n-1)} (2\nu)^{p(n-1)/2} \Gamma_p((n-1)/2) \Gamma(\nu) \mid \mathbf{B} \mid^{p/2}} \\ &\quad \times \sum_{k=0}^{\infty} \sum_{\kappa} \frac{C_\kappa\left(-\mathbf{R}^{-1}\mathbf{Q}/2\nu\right) C_\kappa\left(\mathbf{B}^{-1}\right)}{k! \left(\sigma/\sigma'\right)^{2k} C_\kappa\left(\mathbf{I}_{n-1}\right)} \Gamma\left(\frac{p(n-1)}{2} + k + \nu\right). \end{aligned} \tag{9.9}$$

In the particular case $\sigma = \sigma'$ and $\mathbf{B} = \mathbf{I}_{n-1}$, (9.9) reduces to equation (14) in Cornish (1955). Nagarsenker also provided the joint cdf and the moment generating function of $\mathbf{Q}$ as well as the corresponding expressions for the noncentral complex case (which generalizes those given in Gupta, 1964).

9.2 Multivariate t Statistic

A random variable X with iid copies $X_1, X_2, \ldots$ is said to be in the domain of attraction of the normal law if there exists $a_n \to \infty$ such that

$$\frac{1}{a_n} \sum_{i=1}^{n} (X_i - \mu) \quad \to \quad N(0,1)$$

as $n \to \infty$. It is well known that, for X in the domain of attraction of the normal law, the t statistic defined by

$$T_n = \frac{\sum_{i=1}^{n} (X_i - \mu)}{\sqrt{\sum_{i=1}^{n} (X_i - \bar{X})^2}} \quad \to \quad N(0,1) \tag{9.10}$$

as $n \to \infty$, where, as usual, $\bar{X} = (1/n) \sum_{i=1}^{n} X_i$.

Sepanski (1994, 1996) provided two multivariate analogs f (9.10). Let $\mathbf{X}$ be a p-variate random vector with mean vector $\boldsymbol{\mu}$ and covariance matrix $\boldsymbol{\Sigma}$. Also let $\mathbf{X}_1, \mathbf{X}_2, \ldots$ be iid copies of $\mathbf{X}$. Then $\mathbf{X}$ is said to be in the domain of attraction of a p-variate normal law if there exists

$a_n \to \infty$ such that

$$\frac{1}{a_n}\sum_{i=1}^{n}(\mathbf{X}_i - \boldsymbol{\mu}) \quad \to \quad N(\mathbf{0}, \mathbf{C}) \tag{9.11}$$

for some nonsingular matrix $\mathbf{C}$. Sepanski (1994) defined the multivariate t statistic by

$$\mathbf{T}_n \quad = \quad \mathbf{D}_n^{-1/2}\sum_{i=1}^{n}(\mathbf{X}_i - \boldsymbol{\mu}), \tag{9.12}$$

where, for some sequence $b_n > 0$, $\mathbf{D}_n = \mathbf{C}_n + b_n\mathbf{I}$ and

$$\mathbf{C}_n \quad = \quad \sum_{i=1}^{n}(\mathbf{X}_i - \bar{\mathbf{X}})(\mathbf{X}_i - \bar{\mathbf{X}})^T.$$

Note that $\mathbf{C}_n$ is symmetric nonnegative definite while $\mathbf{D}_n$ is symmetric positive definite. Under the assumption that $\mathbf{X}$ satisfies (9.11), Sepanski (1994) showed that $\mathbf{T}_n \to N(\mathbf{0}, \mathbf{I})$ as $n \to \infty$. Sepanski (1996) established the same limiting result under weaker conditions by taking

$$\mathbf{T}_n \quad = \quad \mathbf{C}_n^{-1/2}\sum_{i=1}^{n}(\mathbf{X}_i - \boldsymbol{\mu}) \tag{9.13}$$

and considering its behavior when $\mathbf{X}$ is in the *generalized domain of attraction of a normal law*, which means that there exist matrices $\mathbf{A}_n$ and vectors $\boldsymbol{\mu}_n$ such that

$$\mathbf{A}_n\sum_{i=1}^{n}\mathbf{X}_i - \boldsymbol{\mu}_n \quad \to \quad N(\mathbf{0}, \mathbf{I}) \tag{9.14}$$

as $n \to \infty$. See Hahn and Klass (1980a, 1980b) for several examples of random vectors satisfying this condition and for an algorithm for constructing the normalizing matrices $\mathbf{A}_n$.

9.3 Hotelling's T^2 Statistic

A customary approach to the estimation/testing problem is based on the so-called Hotelling's T^2 statistic. It is defined by

$$H_n^2 \quad = \quad n^2(\bar{\mathbf{X}} - \boldsymbol{\mu})^T \mathbf{C}_n^{-1}(\bar{\mathbf{X}} - \boldsymbol{\mu}).$$

Under the normality, it is well known that $(n-p)H_n^2/(p(n-1))$ is distributed as an F distribution with degrees of freedom p and $n-p$ (see, for example, Anderson, 1984). The distribution of H_n^2 has been studied under a mixture of two normal distributions by Srivastava and Awan (1982) and Kabe and Gupta (1990). Iwashita (1997) investigated the asymptotics of H_n^2 under an elliptical distribution. Unfortunately, there are no direct results for the specific case of multivariate t distributions. For completeness, however, we shall survey the results when $\mathbf{X}$ has an elliptical distribution. In this case, the characteristic function of $\mathbf{X}$ can be written as

$$\phi(\mathbf{t}) = \exp\left\{i\mathbf{t}^T\mathbf{m}\right\} \Psi\left(\mathbf{m}^T\mathbf{Q}^{-1}\mathbf{m}\right) \qquad (9.15)$$

for some nonnegative function Ψ (Kelker, 1970) and the parameter

$$\kappa = \frac{\Psi''(0)}{\{\Psi'(0)\}^2} - 1,$$

which controls the kurtosis of the distribution. Iwashita (1997) provided an asymptotic distribution of H_n^2 under the null hypothesis that $\mathbf{m} = \boldsymbol{\mu}$ and a local alternative of it. Up to the order of $1/n$, the asymptotic null pdf of H_n^2 is given by

$$f(x) = g_p(x) + \frac{1}{n}\left\{\sum_{j=0}^{2} c_j g_{p+2j}(x)\right\} + o\left(\frac{1}{n}\right),$$

where

$$c_0 = -\frac{1}{4}p\{p + \kappa(p+2)\},$$

$$c_1 = -\frac{1}{2}p\{1 - \kappa(p+2)\},$$

$$c_2 = \frac{1}{4}p(p+2)(1-\kappa),$$

and $g_k(\cdot)$ denotes the pdf of a chi-squared distribution with degrees of freedom k. Iwashita (1997) also derived the percentiles and approximate powers of the H_n^2 statistic. An asymptotic expansion of the cdf of H_n^2 under the two assumptions

(i) $E(\| \mathbf{Y} \|^8) < \infty$, where $\mathbf{Y} = \boldsymbol{\Sigma}^{-1/2}(\mathbf{X} - \boldsymbol{\mu})$ and $\mathbf{X}$ is a $p \times 1$ random vector with mean vector $\boldsymbol{\mu}$ and covariance matrix $\boldsymbol{\Sigma}$;

(ii) the distribution of $\mathbf{Y} = (Y_1, \ldots, Y_p)$ has an absolutely continuous component with a positive density on some nonempty open set U such that $1, y_1, \ldots, y_p, y_1^2, y_1y_2, \ldots, y_p^2$ are linearly independent on U

is given by Fujikoshi (1997). It takes the form

$$\Pr\left(H_n^2 \le x\right) = G_p(x) + \frac{1}{n}\sum_{j=0}^{3}\beta_j G_{p+2j}(x) + o\left(\frac{1}{n}\right) \tag{9.16}$$

uniformly for all positive real numbers x, where $G_k(\cdot)$ denotes the cdf of a chi-squared distribution with degrees of freedom k. The coefficients β_j's are given by

$$\beta_0 = -\frac{1}{4}p^2 + \frac{1}{6}\left(\kappa_3^{(1)}\right)^2 - \frac{1}{4}\kappa_4^{(1)},$$

$$\beta_1 = -\frac{1}{2}p - \frac{1}{2}\left(\kappa_3^{(1)}\right)^2 + \frac{1}{2}\kappa_4^{(1)},$$

$$\beta_2 = \frac{1}{4}p(p+2) - \frac{1}{2}\left(\kappa_3^{(2)}\right)^2 - \frac{1}{4}\kappa_4^{(1)},$$

and

$$\beta_3 = \frac{1}{3}\left(\kappa_3^{(1)}\right)^2 + \frac{1}{2}\left(\kappa_3^{(2)}\right)^2,$$

where

$$\kappa_3^{(1)} = \sqrt{\sum_{i,j,k}\left(\kappa^{(i,j,k)}\right)^2},$$

$$\kappa_3^{(2)} = \sqrt{\sum_{i,j,k}\kappa^{(i,i,k)}\kappa^{(j,k,k)}},$$

$$\kappa_4^{(1)} = \sum_{i,j}\kappa^{(i,i,j,j)},$$

and $\kappa^{(i_1,\ldots,i_j)}$ are the jth cumulants of $\mathbf{Y}$. If $\mathbf{X}$ has the elliptical distribution given by (9.15), then β_3 vanishes to zero and β_k reduces to c_k for $k = 0, 1, 2$. Kano (1995) obtained the same asymptotic expansion as in (9.16), using a different method.

It is well known that, for large samples, H_n^2 has a limiting chi-squared distribution with degrees of freedom p. The usual underlying assumption for this result is simply that $E \parallel \mathbf{X} \parallel < \infty$. More general limiting behavior of H_n^2 has been studied by Eaton and Efron (1970), Sepanski (1994), and Fujikoshi (1997). Sepanski (1994) showed that, under the assumption of the generalized domain of attraction (defined in (9.14)), the modified Hotelling's T^2 statistic

$$H_n'^2 = n^2 \left(\bar{\mathbf{X}} - \boldsymbol{\mu}\right)^T \mathbf{D}_n^{-1} \left(\bar{\mathbf{X}} - \boldsymbol{\mu}\right)$$

still has an asymptotic chi-squared distribution with degrees of freedom p. Eaton and Efron (1970) studied the distribution H_n^2 when $\mathbf{X}$ has orthant symmetry, that is, $\mathbf{X}$ has the same distribution as $\mathbf{DX}$ for any choice of the diagonal matrix $\mathbf{D}$ with diagonal elements equal to ± 1.

We shall now consider the Hotelling's T^2 statistic in the context of testing equality of means. Suppose $\mathbf{X}_1 = (\mathbf{X}_{11}, \ldots, \mathbf{X}_{1n_1})^T$ and $\mathbf{X}_2 = (\mathbf{X}_{21}, \ldots, \mathbf{X}_{2n_2})^T$ are two samples of size n_1 and n_2, respectively. In analogy with (9.2), assume that $\mathbf{X}_1$ and $\mathbf{X}_2$ have the joint pdf given by

$$f\left(\mathbf{x}_1, \mathbf{x}_2\right) \propto |\mathbf{R}|^{-n/2} \left[\nu - 2 + \sum_{i=1}^{2} \sum_{j=1}^{n_i} \left(\mathbf{x}_{ij} - \boldsymbol{\mu}_i\right)^T \mathbf{R}^{-1} \left(\mathbf{x}_{ij} - \boldsymbol{\mu}_i\right)\right]^{-(\nu+np)/2},$$

where $n = n_1 + n_2$. It is immediate that $\mathbf{X}_{ij}$ is p-variate t with mean vector $\boldsymbol{\mu}_i$, correlation matrix $\mathbf{R}$, and degrees of freedom ν. Also, the elements of the combined sample of size $n = n_1 + n_2$ are pairwise uncorrelated. The Hotelling's T^2 for testing equality of means takes the form

$$T^2 = \frac{n_1 n_2}{n_1 + n_2} \left(\bar{\mathbf{X}}_1 - \bar{\mathbf{X}}_2\right)^T \mathbf{S}_{\text{pooled}}^{-1} \left(\bar{\mathbf{X}}_1 - \bar{\mathbf{X}}_2\right),$$

where

$$\mathbf{S}_{\text{pooled}} = \frac{1}{n_1 + n_2 - 2} \sum_{i=1}^{2} \sum_{j=1}^{n_i} \left(\bar{\mathbf{X}}_{ij} - \bar{\mathbf{X}}_i\right) \left(\bar{\mathbf{X}}_{ij} - \bar{\mathbf{X}}_i\right)^T.$$

Sutradhar (1990) derived the nonnull distribution of the T^2 statistic, given by the pdf

$$f\left(t^2\right) = \frac{2}{\nu - 2} \left[\beta_{\delta/(\nu-2)} \left(r + 1, \frac{\nu}{2} - 1\right) \times \beta_f \left(r + \frac{p}{2}, \frac{n_1 + n_2 - p - 1}{2}\right)\right], \qquad (9.17)$$

where

$$\beta_x(k, m) = \frac{\Gamma(k)\Gamma(m) x^{k-1}}{\Gamma(k+m)(1+x)^{k+m}}$$

and

$$\delta = \frac{n_1 n_2}{n_1 + n_2} (\mu_1 - \mu_2)^T \mathbf{R}^{-1} (\mu_1 - \mu_2).$$

Note that, under $H_0 : \mu_1 = \mu_2$, where $\delta = 0$, the pdf of T^2 in (9.17) reduces to

$$f(t^2) = \beta_f \left(\frac{p}{2}, \frac{n_1 + n_2 - p - 1}{2} \right),$$

which implies that, under H_0, $T^2(n_1 + n_2 - p - 1)/p$ has the usual F distribution with degrees of freedom p and $n_1 + n_2 - p - 1$. Thus the null distribution remains the same as in the normal case. Furthermore, the power of the Hotelling's T^2 test can be computed by using the nonnull pdf in (9.17).

Kozumi (1994) considered testing equality of means when the two samples $\mathbf{X}_1$ and $\mathbf{X}_2$ have mutually independent t distributions with equal correlation matrices and equal degrees of freedom. When the sample sizes are equal (say, $n_1 = n_2 = n$) the T^2 statistic is given by

$$T_d^2 = n \bar{\mathbf{y}}^T \mathbf{S}_d^{-1} \bar{\mathbf{y}},$$

where $\bar{\mathbf{y}}$ and $\mathbf{S}_d$ are, respectively, the sample mean and the sample covariance matrix of the differences $y_j = x_{1j} - x_{2j}$. For unequal sample sizes, assuming without loss of generality that $n_1 < n_2$, the T^2 statistic is given by

$$T_s^2 = n_1 \bar{\mathbf{z}}^T \mathbf{S}_s^{-1} \bar{\mathbf{z}},$$

where

$$\mathbf{z}_j = \mathbf{x}_{1j} - \sqrt{\frac{n_1}{n_2}} \mathbf{x}_{2j} + \frac{1}{\sqrt{n_1 n_2}} \sum_{s=1}^{n_1} \mathbf{x}_{2s} - \frac{1}{n_2} \sum_{s=1}^{n_2} \mathbf{x}_{2s}$$

and $\bar{\mathbf{z}}$ and $\mathbf{S}_s$ are the sample mean and the sample covariance matrix of the $\mathbf{z}_j$'s. It should be noted that T_s^2 reduces to T_d^2 in the case $n_1 = n_2 = n$. Under the $H_0 : \mu_1 = \mu_2$, $(n_1 - p) T_s^2 / (p(n_1 - 1))$ has the usual F distribution with degrees of freedom p and $n_1 - p$. The nonnull pdf of T_s^2 is given by the infinite sum involving Student's t pdfs

$$\begin{aligned} f(t_s^2) = {} & \frac{\nu^\nu (n_2/n_1)^{\nu/2}}{(n_1 - 1) \Gamma^2 (\nu/2)} \sum_{k=0}^{\infty} \frac{(n_2 \delta)^k \Gamma(k + \nu)}{k! B (p/2 + k, (n_1 - p)/2)} \\ & \times \left(\frac{t_s^2}{n_1 - 1} \right)^{p/2+k-1} \left(1 + \frac{t_s^2}{n_1 - 1} \right)^{-(n_1/2+k)} \\ & \times J (x; n_1, n_2, k, \delta, \nu), \end{aligned}$$

where $\delta = (\boldsymbol{\mu}_1 - \boldsymbol{\mu}_2)^T \mathbf{R}^{-1} (\boldsymbol{\mu}_1 - \boldsymbol{\mu}_2)$ and the integral

$$
\begin{aligned}
J(x; n_1, n_2, k, \delta, \nu) &= \int_0^1 \{x(1-x)\}^{k+\nu/2-1} \left\{ n_2 \delta x (1-x) \right. \\
&\qquad \left. + \nu \left(1 - \frac{n_2}{n_1}\right) x + \nu \frac{n_2}{n_1} \right\}^{-(k+\nu)} dx.
\end{aligned}
$$

Kozumi (1994) also provided an expression for the cdf of T_s^2 and calculated the powers of T_s^2 corresponding to the sizes $\alpha = 0.01, 0.05$, $p = 5$ and for various values of n_1, n_2, δ, and ν.

9.4 Entropy and Kullback-Leibler Number

The forms of entropy and Kullback-Leibler number for the multivariate t distribution were discussed earlier in Chapter 1 (see equations (1.27), (1.29), and (1.31)). Here, we shall discuss the corresponding sampling properties.

The entropy for the central p-variate t involves the correlation matrix $\mathbf{R}$, and it is known that the maximum likelihood estimator of $\mathbf{R}$ for a sample of n observations is based on the Wishart matrix $\mathbf{A}$ in (9.4). Hence it is of interest to consider the sampling properties of the difference $\delta = H(\mathbf{X}; \mathbf{A}) - H(\mathbf{X}; \mathbf{R})$. Guerrero-Cusumano (1996a) derived the corresponding moment generating function, mean, variance, and some asymptotics. Specifically,

$$
M(t) = \nu^{pt/2} \Gamma_p \left(\frac{n-1+t}{2} \right) \Gamma \left(\frac{\nu - tp}{2} \right) \Big/ \Gamma_p \left(\frac{n-1}{2} \right) \Gamma \left(\frac{\nu}{2} \right),
$$

$$
E(\delta) = \frac{1}{2} \sum_{i=1}^{p} \psi \left(\frac{n-i}{2} \right) - p \psi \left(\frac{\nu}{2} \right) + \frac{p}{2} \log \nu,
$$

$$
Var(\delta) = \frac{1}{4} \sum_{i=1}^{p} \psi' \left(\frac{n-i}{2} \right),
$$

and

$$
\begin{aligned}
\sqrt{\frac{n-\nu}{2}} \delta^* &= \sqrt{\frac{n-\nu}{2}} \left[2\delta - p \log \{ \nu (p - \nu) \} \right] \\
&\to N(0, p)
\end{aligned}
$$

as $n \to \infty$, where $\psi(\cdot)$ denotes the digamma function. Note that

$H_u(\mathbf{X};\mathbf{R}) = H(\mathbf{X};\mathbf{A}) - E(\delta)$ is an unbiased estimator for $H(\mathbf{X};\mathbf{R})$ with $E(H_u) = 0$ and $Var(H_u) = Var(\delta)$. Also note that, as $\nu \to \infty$,

$$E(\delta) \quad \to \quad \frac{1}{2}\sum_{i=1}^{p} \psi\left(\frac{n-i}{2}\right) + p\log 2,$$

coinciding with the result given in Ahmed and Gokhale (1989) for the multivariate normal distribution. The expression for $Var(\delta)$ given above is also valid for the multivariate normal distribution since it is independent of ν.

The Kullback-Leibler number for the central p-variate t is given by (1.31). The corresponding maximum likelihood estimator for a sample of n observations is

$$\widehat{T}(\mathbf{X};\mathbf{R}) \quad = \quad \Omega - \frac{1}{2}\log\left|\frac{\mathbf{A}}{n-1}\right| - \frac{p}{2}\log\left(\frac{\nu-2}{\nu}\right).$$

Thus, the sampling quantity of interest is the difference $\delta = \widehat{T}(\mathbf{X};\mathbf{R}) - T(\mathbf{X};\mathbf{R})$. Guerrero-Cusumano (1996b, 1998) derived the corresponding moment generating function, cumulant generating function, cumulants, mean, variance, and some asymptotics. Specifically,

$$M(t) \quad = \quad \left(\frac{n-1}{\nu-2}\right)^{pt/2} \Gamma_p\left(\frac{n-1+t}{2}\right)\Gamma\left(\frac{\nu+tp}{2}\right) \Big/ \Gamma_p\left(\frac{n-1}{2}\right)\Gamma\left(\frac{\nu}{2}\right),$$

$$K(t) \quad = \quad \frac{pt}{2}\log\left(\frac{n-1}{\nu-2}\right) + \log\Gamma\left(\frac{\nu+pt}{2}\right) - \log\Gamma\left(\frac{\nu}{2}\right) + \sum_{i=1}^{p}\left\{\log\Gamma\left(\frac{n-i-t}{2}\right) - \log\Gamma\left(\frac{n-i}{2}\right)\right\},$$

$$\kappa_j \quad = \quad \frac{1}{2^j}\sum_{i=1}^{p}\left\{p^{j-1}\psi^{(j-1)}\left(\frac{\nu+p}{2}\right) + (-1)^j\psi^{(j-1)}\left(\frac{n-i}{2}\right)\right\},$$

$$E(\delta) \quad = \quad \frac{1}{2}\sum_{i=1}^{p}\left\{\psi\left(\frac{\nu}{2}\right) - \psi\left(\frac{n-i}{2}\right)\right\} + \frac{p}{2}\log\left(\frac{n-1}{\nu-2}\right),$$

$$Var(\delta) \quad = \quad \frac{1}{4}\sum_{i=1}^{p}\left\{p\psi^{(1)}\left(\frac{\nu+p}{2}\right) - \psi^{(1)}\left(\frac{n-i}{2}\right)\right\},$$

and

$$\frac{\nu + p}{p}\delta \quad \rightarrow \quad \chi^2_{(\nu+p)^2}$$

as $n \rightarrow \infty$ and $\nu \rightarrow \infty$. Furthermore,

$$(n-1)\widehat{T}\left(\mathbf{X};\mathbf{R}\right) \quad \rightarrow \quad \chi^2_{p(p-1)/2}$$

and

$$\sqrt{n-1}\delta \quad \rightarrow \quad N\left(0, \frac{\mathrm{tr}\left(\mathbf{B}^2\right) - p}{2}\right)$$

as $n \rightarrow \infty$, where $\mathbf{B} = \mathbf{A}_d^{-1/2}\mathbf{A}\mathbf{A}_d^{-1/2}$ with $\mathbf{A}_d$ denoting the diagonal matrix of $\mathbf{A}$. In the latter limit, it is assumed that ν is known. When ν is unknown, the limit still holds for n sufficiently large. The exact distribution of δ is quite complicated to obtain in a closed form.

10
Estimation

The material in this chapter is of special interest to researchers attempting to model various phenomena based on multivariate t distributions. We shall start with a popular result in the bivariate case.

10.1 Tiku and Kambo's Estimation Procedure

In Chapter 4, we studied a bivariate t distribution due to Tiku and Kambo (1992) given by the joint pdf

$$\begin{aligned} f(x_1, x_2) &= \frac{1}{\sigma_1 \sigma_2 \sqrt{k(1-\rho^2)}} \left\{ 1 + \frac{(x_2 - \mu_2)^2}{k\sigma_2^2} \right\}^{-\nu} \\ & \quad \times \exp\left[-\frac{1}{2\sigma_1^2 (1-\rho^2)} \left\{ x_1 - \mu_1 - \rho \frac{\sigma_1}{\sigma_2} (x_2 - \mu_2)^2 \right\} \right]. \end{aligned} \tag{10.1}$$

Here, we discuss estimation of the parameters μ_1, μ_2, σ_1, σ_2, and ρ when ν is known. The method for estimating the location and scale parameters developed by Tiku and Suresh (1992) is used for this problem. For a random sample $\{(X_{1i}, X_{2i}), i = 1, \ldots, n\}$ from (10.1), the likelihood function is

$$\begin{aligned} L &\propto \left\{ \sigma_1^2 \sigma_2^2 (1-\rho^2) \right\}^{-n/2} \prod_{i=1}^{n} \left\{ 1 + \frac{X_{(2:i)} - \mu_2}{k\sigma_2^2} \right\}^{-\nu} \\ & \quad \times \exp\left[-\frac{1}{2\sigma_1^2 (1-\rho^2)} \sum_{i=1}^{n} \left\{ X_{[1:i]} - \mu_1 - \frac{\rho\sigma_1}{\sigma_2} \left(X_{(2:i)} - \mu_2 \right) \right\}^2 \right], \end{aligned}$$

where $k = 2\nu - 3$, $X_{(2:i)}$, $i = 1, \ldots, n$ are the order statistics of X_{2i} and $X_{[1:i]}$, $i = 1, \ldots, n$ are the corresponding concomitant X_1 observations. Consider the following three situations:

(i) Complete samples are available and ν is not too small ($\nu > 3$).

(ii) Complete samples are available but ν is small ($\nu \leq 3$).

(iii) A few smallest or a few largest X_{2i} observations and the corresponding concomitant $X_{[1:i]}$ are censored due to the constraints of an experiment. This situation arises in numerous practical situations. In a time mortality experiment, for example, n mice are inoculated with a uniform culture of human tuberculosis. What is recorded is X_{2i}: the time to death of the first $A(< n)$ mice, and X_{1i}: the corresponding weights at the time of death.

These situations also arise in the context of ranking and selection (David, 1982). We provide some details of the inference for situation (i) as described in Tiku and Kambo (1992). Using a linear approximation of the likelihood based on the expected values of order statistics, it is shown that the maximum likelihood estimators are

$$\widehat{\mu}_1 = \bar{x}_1 - \frac{\widehat{\rho}\widehat{\sigma}_1}{\widehat{\sigma}_2}\left(\bar{x}_2 - \mu_2\right),$$

$$\widehat{\sigma}_1 = \sqrt{s_1^2 + \frac{s_{12}^2}{s_2^2}\left(\frac{\widehat{\sigma}_2^2}{s_2^2} - 1\right)},$$

$$\widehat{\mu}_2 = \bar{x}_2 - \frac{\widehat{\rho}\widehat{\sigma}_2}{\widehat{\sigma}_1}\left(\bar{x}_1 - \mu_1\right),$$

$$\widehat{\sigma}_2 = \sqrt{s_2^2 + \frac{s_{12}^2}{s_1^2}\left(\frac{\widehat{\sigma}_1^2}{s_1^2} - 1\right)},$$

and

$$\widehat{\rho} = \frac{s_{12}}{s_2^2}\frac{\widehat{\sigma}_2}{\widehat{\sigma}_1},$$

where $(\bar{x}_1, \bar{x}_2)$ are the usual sample means, (s_1^2, s_2^2) are the usual sample variances, and s_{12} is the sample covariance. The estimators $\widehat{\mu}_1$, $\widehat{\mu}_2$, $\widehat{\sigma}_1$, $\widehat{\sigma}_2$, and $\widehat{\rho}$ are asymptotically unbiased and minimum variance bound estimators. The estimator $\widehat{\sigma}_1^2$ is always real and positive while the estimator $\widehat{\rho}$ always assumes values between -1 and 1. The asymptotic variances and covariances of the estimators can be written as

$$\mathbf{V} = \begin{pmatrix} \mathbf{V}_1 & \mathbf{0} \\ \mathbf{0} & \mathbf{V}_2 \end{pmatrix},$$

where

$$\mathbf{V}_1 = \frac{1}{n}\begin{pmatrix} \sigma_1^2 & \rho\sigma_1\sigma_2 \\ \rho\sigma_2\sigma_1 & \sigma_2^2 \end{pmatrix} - \frac{2m\nu - nk}{2\nu m n\sigma_2^2}\begin{pmatrix} \rho^2\sigma_1^2\sigma_2^2 & \rho\sigma_1\sigma_2^3 \\ \rho\sigma_2^3\sigma_1 & \sigma_2^4 \end{pmatrix} \tag{10.2}$$

is positive definite and is the asymptotic variance-covariance matrix of $(\widehat{\mu}_1, \widehat{\mu}_2)$ while

$$\mathbf{V}_2 = \frac{1}{2n}\begin{pmatrix} \sigma_1^2 & \rho\sigma_1\sigma_2 & \rho\sigma_1(1-\rho^2) \\ \rho\sigma_1\sigma_2 & \sigma_2^2 & \rho\sigma_2(1-\rho^2) \\ \rho\sigma_1(1-\rho^2) & \rho\sigma_2(1-\rho^2) & 2(1-\rho^2)^2 \end{pmatrix} - \frac{\delta}{2n(2+\delta)}\begin{pmatrix} \rho^4\sigma_1^2 & \rho^2\sigma_1\sigma_2 & \rho^3\sigma_1(1-\rho^2) \\ \rho^2\sigma_1\sigma_2 & \sigma_2^2 & \rho\sigma_2(1-\rho^2) \\ \rho^3\sigma_1(1-\rho^2) & \rho\sigma_2(1-\rho^2) & \rho^2(1-\rho^2)^2 \end{pmatrix} \tag{10.3}$$

is positive definite and is the asymptotic variance-covariance matrix of $(\widehat{\sigma}_1, \widehat{\sigma}_2, \widehat{\rho})$. The parameters m and δ are determined by the linear approximation of the likelihood. Interestingly, $Var(\widehat{\mu}_1)$ and $Var(\widehat{\mu}_2)$ decrease with increasing ρ^2 unless $\nu = \infty$. The first component on the right of (10.2) is the variance-covariance matrix of $\widehat{\mu}_1$, and $\widehat{\mu}_2$ under bivariate normality, and the first component on the right of (10.3) is the asymptotic variance-covariance matrix of $\widehat{\sigma}_1$, $\widehat{\sigma}_2$, and $\widehat{\rho}$ under bivariate normality. The second components in (10.2) and (10.3) represent the effect of nonnormality due to the family (10.1). The asymptotic distribution of $\sqrt{n}(\widehat{\mu}_1 - \mu_1, \widehat{\mu}_2 - \mu_2)$ is bivariate normal with zero means and variance-covariance matrix $n\mathbf{V}_1$. For testing $H_0 : (\mu_1, \mu_2) = (0, 0)$ versus $H_1 : (\mu_1, \mu_2) \neq (0, 0)$, a useful statistic is $T_p^2 = (\widehat{\mu}_1, \widehat{\mu}_2)^T \widehat{\mathbf{V}}_1^{-1} (\widehat{\mu}_1, \widehat{\mu}_2)$, the asymptotic null distribution of which is chi-squared with degrees of freedom 2. The asymptotic nonnull distribution is noncentral chi-squared with degrees of freedom 2 and noncentrality parameter

$$\lambda_\nu = \lambda_\infty + \left(\frac{2m\nu}{kn} - 1\right)\left(\frac{\mu_2}{\sigma_2}\right)^2,$$

where

$$\lambda_\infty = \frac{n}{1-\rho^2}\left\{\left(\frac{\mu_1}{\sigma_1}\right)^2 + \left(\frac{\mu_2}{\sigma_2}\right)^2 - 2\rho\left(\frac{\mu_1}{\sigma_1}\right)\left(\frac{\mu_2}{\sigma_2}\right)\right\}.$$

Note that λ_∞ is the noncentrality parameter of the asymptotic nonnull distribution of the Hotelling's T^2 statistic based on the sample means

$(\bar{x}_1, \bar{x}_2)$, sample variances (s_1^2, s_2^2), and the sample correlation coefficient $\widehat{\rho} = s_{12}/(s_1 s_2)$. Tiku and Kambo (1992) also provided evidence to the fact that the use of T_p^2 in place of the Hotelling's T^2 statistic can result in a substantial gain in power.

10.2 ML Estimation via EM Algorithm

Consider fitting a p-variate t distribution to data $\mathbf{x}_1, \ldots, \mathbf{x}_n$ with the log-likelihood function

$$L(\boldsymbol{\mu}, \mathbf{R}, \nu) = -\frac{n}{2} \log|\mathbf{R}| - \frac{\nu + p}{2} \sum_{i=1}^{n} \log(\nu + s_i), \qquad (10.4)$$

where $s_i = (\mathbf{x} - \boldsymbol{\mu})^T \mathbf{R}^{-1} (\mathbf{x} - \boldsymbol{\mu})$ and ν is assumed to be fixed. Differentiating (10.4) with respect to $\boldsymbol{\mu}$ and $\mathbf{R}$ leads to the estimating equations

$$\boldsymbol{\mu} = \text{ave}\{w_i \mathbf{x}_i\} / \text{ave}\{w_i\} \qquad (10.5)$$

and

$$\mathbf{R} = \text{ave}\left\{w_i (\mathbf{x} - \boldsymbol{\mu})(\mathbf{x} - \boldsymbol{\mu})^T\right\}, \qquad (10.6)$$

where $w_i = (\nu + p)/(\nu + s_i)$ and "ave" stands for the arithmetic average over $i = 1, 2, \ldots, n$. Note that equations (10.5)–(10.6) can be viewed as an adaptively weighted sample mean and sample covariance matrix where the weights depend on the Mahalanobis distance between $\mathbf{x}_i$ and $\boldsymbol{\mu}$. The weight function $w(s) = (\nu + p)/(\nu + s)$, where $s = (\mathbf{x} - \boldsymbol{\mu})^T \mathbf{R}^{-1} (\mathbf{x} - \boldsymbol{\mu})$, is a decreasing function of s, so that the outlying observations are downweighted. Maronna (1976) proved, under general assumptions, the existence, uniqueness, consistency, and asymptotic normality of the solutions of (10.5)–(10.6). For instance, if there exists $a > 0$ such that, for every hyperplane H, $\Pr(H) \leq p/(\nu + p) - a$, then (10.5)–(10.6) has a unique solution. Also, every solution satisfies the consistency property that $\lim_{n \to \infty} (\widehat{\boldsymbol{\mu}}, \widehat{\mathbf{R}}) = (\boldsymbol{\mu}, \mathbf{R})$ with probability 1.

The standard approach for solving (10.5)–(10.6) for $\boldsymbol{\mu}$ and $\mathbf{R}$ is the popular EM algorithm because of its simplicity and stable convergence (Dempster et al., 1977; Wu, 1983). The EM algorithm takes the form of iterative updates of (10.5)–(10.6), using the current estimates of $\boldsymbol{\mu}$ and $\mathbf{R}$ to generate the weights. The iterations take the form

$$\boldsymbol{\mu}^{(m+1)} = \text{ave}\left\{w_i^{(m)} \mathbf{x}_i\right\} / \text{ave}\left\{w_i^{(m)}\right\}$$

and

$$\mathbf{R}^{(m+1)} = \text{ave}\left\{w_i^{(m)}\left(\mathbf{x}_i - \boldsymbol{\mu}^{(m+1)}\right)\left(\mathbf{x}_i - \boldsymbol{\mu}^{(m+1)}\right)^T\right\},$$

where

$$w_i^{(m)} = (\nu + p) \Big/ \left\{\nu + \left(\mathbf{x}_i - \boldsymbol{\mu}^{(m)}\right)^T \left(\mathbf{R}^{(m)}\right)^{-1} \left(\mathbf{x}_i - \boldsymbol{\mu}^{(m)}\right)\right\}.$$

This is known as the direct EM algorithm and is valid for any $\nu > 0$. For details of this algorithm see the pioneering papers of Dempster et al. (1977, 1980), Rubin (1983), and Little and Rubin (1987). Several variants of the above have been proposed in the literature, as summarized in the table below.

Algorithm	Primary References
Extended EM	Kent et al. (1994), Arsian et al. (1995)
Restricted EM	Arsian et al. (1995)
MC-ECM1	Liu and Rubin (1995)
MC-ECM2	Liu and Rubin (1995), Meng and van Dyk (1997)
ECME1	Liu and Rubin (1995), Liu (1997)
ECME2	Liu and Rubin (1995)
ECME3	Liu and Rubin (1995), Meng and van Dyk (1997)
ECME4	Liu and Rubin (1995), Liu (1997)
ECME5	Liu (1997)
PXEM	Liu et al. (1998)

Consider the maximum likelihood (ML) estimation for a g-component mixture of t distributions given by

$$f(\mathbf{x}; \Psi) = \sum_{i=1}^{g} \pi_i f\left(\mathbf{x}; \boldsymbol{\mu}_i, \mathbf{R}_i, \nu_i\right),$$

where

$$f\left(\mathbf{x}; \boldsymbol{\mu}_i, \mathbf{R}_i, \nu_i\right) = \frac{\Gamma\left(\left(\nu_i + p\right)/2\right)}{\left(\pi \nu_i\right)^{p/2} \Gamma\left(\nu_i/2\right) \left|\mathbf{R}_i\right|^{1/2}} \times \left[1 + \frac{\left(\mathbf{x} - \boldsymbol{\mu}_i\right)^T \mathbf{R}_i^{-1} \left(\mathbf{x} - \boldsymbol{\mu}_i\right)}{\nu_i}\right]^{-\left(\nu_i + p\right)/2},$$

$\boldsymbol{\Psi} = (\pi_1, \ldots, \pi_{g-1}, \boldsymbol{\theta}^T, \boldsymbol{\nu}^T)^T$, $\boldsymbol{\theta} = ((\boldsymbol{\mu}_1, \mathbf{R}_1)^T, \ldots, (\boldsymbol{\mu}_g, \mathbf{R}_g)^T)^T$, and $\boldsymbol{\nu} = (\nu_1, \ldots, \nu_g)^T$. The application of the EM algorithm for this model in a clustering context has been considered by McLachlan and Peel (1998) and Peel and McLachlan (2000). The iteration updates now take the form

$$\boldsymbol{\mu}_i^{(m+1)} = \sum_{j=1}^{n} \tau_{ij}^{(m)} u_{ij}^{(m)} \mathbf{x}_j \Big/ \sum_{j=1}^{n} \tau_{ij}^{(m)} u_{ij}^{(m)}$$

and

$$\mathbf{R}_i^{(m+1)} = \sum_{j=1}^{n} \tau_{ij}^{(m)} u_{ij}^{(m)} \left(\mathbf{x}_j - \boldsymbol{\mu}_i^{(m+1)}\right) \left(\mathbf{x}_j - \boldsymbol{\mu}_i^{(m+1)}\right)^T \Big/ \sum_{j=1}^{n} \tau_{ij}^{(m)},$$

where

$$u_{ij}^{(m)} = \frac{\nu_i^{(m)} + p}{\nu_i^{(m)} + \left(\mathbf{x}_j - \boldsymbol{\mu}_i^{(m)}\right)^T \mathbf{R}_i^{(m)^{-1}} \left(\mathbf{x}_j - \boldsymbol{\mu}_i^{(m)}\right)}$$

and

$$\tau_{ij}^{(m)} = \frac{\pi_i^{(m)} f\left(\mathbf{x}_j; \boldsymbol{\mu}_i^{(m)}, \mathbf{R}_i^{(m)}, \nu_i^{(m)}\right)}{f\left(\mathbf{x}_j; \boldsymbol{\Psi}^{(m)}\right)}.$$

The EMMIX program of McLachlan et al. (1999) for the fitting of normal mixture models has an option that implements the above procedure for the fitting of mixtures of t-components. The program automatically generates a selection of starting values for the fitting if they are not provided by the user. The user only has to provide the data set, the restrictions on the component-covariance matrices (equal, unequal, diagonal), the extent of the selection of the initial groupings to be used to determine the starting values, and the number of components that are to be fitted. The program is available from the software archive StatLib or from Professor Peel's homepage at the Web site address

`http://www.maths.uq.edu.au/~gjm/`

10.3 Missing Data Imputation

When a data set contains missing values, multiple imputation for missing data (Rubin, 1987) appears to be an ideal technique. Most importantly, it allows for valid statistical inferences. In contrast, any single imputation method, such as filling in the missing values with either their marginal means or their predicted values from linear regression, typically leads to biased estimates of parameters and thereby often to an invalid inference (Rubin, 1987, pages 11–15).

The multivariate normal distribution has been a popular statistical model in practice for rectangular continuous data sets. To impute the missing values in an incomplete normal data set, Rubin and Schafer (1990) (see also Schafer, 1997, and Liu, 1993) proposed an efficient method, called monotone data augmentation (MDA), and implemented it using the factorized likelihood approach. A more efficient technique to implement the MDA than the factorized likelihood approach is provided by Liu (1993) using Bartlett's decomposition, which is the extension of the Bayesian version of Bartlett's decomposition of the Wishart distribution with complete rectangular normal data to the case with monotone ignorable missing data.

When a rectangular continuous data set appears to have longer tails than the normal distribution, or it contains some values that are influential for statistical inferences with the normal distribution, the multivariate t distribution becomes useful for multiple imputation as an alternative to the multivariate normal distribution. First, when the data have longer tails than the normal distribution, the multiply imputed data sets using the t distribution allow more valid statistical inferences than those using the normal distribution with some "influential" observations deleted. Second, it is well known that the t distribution is widely used in applied statistics for robust statistical inferences. Therefore, when an incomplete data set contains some influential values or outliers, the t distribution allows for a robust multiple imputation method. Furthermore, the multiple imputation appears to be more useful than the asymptotic method of inference since the likelihood functions of the parameters of the t distribution given the observed data can have multiple modes. For a complete description of the MDA using the multivariate t distribution, see Liu (1995). See also Liu (1996) for extensions in two aspects, including covariates in the multivariate t models (as in Liu and Rubin, 1995), and replacing the multivariate t distribution with a more general class of distributions, that is, the class of normal/independent distributions (as in Lange and Sinsheimer, 1993). These extensions provide a flexible class of models for robust multivariate linear regression and multiple imputation. Liu (1996) described methods to implement the MDA for these models with fully observed predictor variables and possible missing values from outcome variables.

10.4 Laplacian T-Approximation

The Laplacian T-approximation (Sun et al., 1996) is a useful tool for Bayesian inferences for variance component models. Let $p(\boldsymbol{\theta} \mid \mathbf{y})$ be the posterior pdf of $\boldsymbol{\theta} = (\theta_1, \ldots, \theta_p)^T$ given data $\mathbf{y}$, and let $\eta = g(\boldsymbol{\theta})$ be the parameter of interest. Leonard et al. (1994) introduced a Laplacian T-approximation for the marginal posterior of η of the form

$$p^*(\eta|\mathbf{y}) \propto |\mathbf{T}_\eta|^{-1/2} p(\boldsymbol{\theta}_\eta|\mathbf{y}) \lambda_\eta^{-w/2} f(\eta | w, \boldsymbol{\theta}_\eta^*, \mathbf{T}_\eta) \tag{10.7}$$

to be the marginal posterior pdf of η, where

$$\mathbf{T}_\eta = \frac{w}{(w+p)\lambda_\eta}\mathbf{Q}_\eta,$$

$$\lambda_\eta = 1 - \frac{\mathbf{1}_\eta^T \mathbf{Q}_\eta^{-1} \mathbf{1}_\eta}{w+p-1},$$

$$\mathbf{Q}_\eta = \mathbf{U}_\eta + \frac{2\mathbf{1}_\eta \mathbf{1}_\eta^T}{w+p-1},$$

$$\mathbf{1}_\eta = \left.\frac{\partial \log p(\boldsymbol{\theta} \mid \mathbf{y})}{\partial \boldsymbol{\theta}}\right|_{\boldsymbol{\theta}=\boldsymbol{\theta}_\eta},$$

$$\mathbf{U}_\eta = \left.\frac{\partial^2 \log p(\boldsymbol{\theta} \mid \mathbf{y})}{\partial \left(\boldsymbol{\theta}\boldsymbol{\theta}^T\right)}\right|_{\boldsymbol{\theta}=\boldsymbol{\theta}_\eta},$$

$$\boldsymbol{\theta}_\eta^* = \boldsymbol{\theta}_\eta + \mathbf{Q}_\eta^{-1}\mathbf{1}_\eta,$$

and $f(\eta \mid w, \boldsymbol{\theta}_\eta^*, \mathbf{T}_\eta)$ denotes the pdf of $\eta = g(\boldsymbol{\theta})$ when $\boldsymbol{\theta}$ possesses a multivariate t distribution with mean vector $\boldsymbol{\theta}_\eta^*$, covariance matrix $\mathbf{T}_\eta$, and degrees of freedom w. Here, $\boldsymbol{\theta}_\eta$ represents some convenient approximation to the conditional posterior mean vector of $\boldsymbol{\theta}$, given η, and w should be taken to roughly approximate the degrees of freedom of a generalized multivariate T-approximation to the conditional distribution of $\boldsymbol{\theta}$ given η.

When $\boldsymbol{\theta}_\eta$ is the conditional posterior mode vector of $\boldsymbol{\theta}$, given η, (10.7) reduces to the Laplacian approximation introduced by Leonard (1982) and shown by Tierney and Kadane (1986) and Leonard et al. (1989) to possess saddlepoint accuracy as well as an excellent finite-sample accuracy, in many special cases. It was previously used for hierarchical models by Kass and Steffey (1989).

In the special case where $\eta = \mathbf{a}^T\boldsymbol{\theta}$ is a linear combination of the θ's, the approximation (10.7) is equivalent to

$$p^*(\eta\,|\mathbf{y}) \quad \propto \quad |\mathbf{T}_\eta|^{-1/2}\, p(\boldsymbol{\theta}_\eta|\,\mathbf{y})\,\lambda_\eta^{-(w+p)/2} t_\eta\left(w, \mathbf{a}^T\boldsymbol{\theta}_\eta^*, \left(\mathbf{a}^T\mathbf{T}_\eta\mathbf{a}\right)^{-1}\right),$$

where $t_\eta(w, \mu, \tau)$ denotes a generalized t pdf.

10.5 Sutradhar's Score Test

Consider a random sample $\mathbf{X}_1, \ldots, \mathbf{X}_n$ from a p-variate t distribution with the pdf

$$\begin{aligned} f(\mathbf{x}_j) \quad &= \quad \frac{(\nu-2)^{\nu/2}\Gamma((\nu+p)/2)}{\pi^{p/2}\Gamma(\nu/2)\,|\mathbf{R}|^{1/2}} \\ &\quad \times \left[\nu - 2 + (\mathbf{x}_j - \boldsymbol{\mu})^T \mathbf{R}^{-1}(\mathbf{x}_j - \boldsymbol{\mu})\right]^{-(\nu+p)/2}. \end{aligned}$$

Note this is a slight reparameterization of the usual t pdf. The log-likelihood

$$G \quad = \quad \sum_{j=1}^{n} \log f(\mathbf{x}_j)$$

is a function of the parameters $\mathbf{R}$, $\boldsymbol{\mu}$, and ν.

Frequently in social sciences, and particularly in factor analysis, one of the main inference problems is to test the null hypothesis $\mathbf{R} = \mathbf{R}_0$ when $\boldsymbol{\mu}$ and ν are known. Sutradhar (1993) proposed Neymann's (1959) score test for this test for large n. Le $\mathbf{r} = (r_{11}, \ldots, r_{hl}, \ldots, r_{pp})^T$ be the $p(p+1)/2 \times 1$ vector formed by stacking the distinct elements of $\mathbf{R}$, with r_{hl} being the (h, l)th element of the $p \times p$ matrix $\mathbf{R}$. Also let

$$\left(\lambda_1, \ldots, \lambda_i, \ldots, \lambda_{p(p+1)/2}\right)^T \quad \equiv \quad b(\mathbf{r}_0, \widehat{\boldsymbol{\mu}}, \widehat{\nu})$$

and

$$(\gamma_1, \ldots, \gamma_j, \ldots, \lambda_{p+1})^T \quad \equiv \quad \begin{bmatrix} \xi(\mathbf{r}_0, \widehat{\boldsymbol{\mu}}, \widehat{\nu}) \\ \eta(\mathbf{r}_0, \widehat{\boldsymbol{\mu}}, \widehat{\nu}) \end{bmatrix},$$

where $b(\mathbf{r}_0, \widehat{\boldsymbol{\mu}}, \widehat{\nu})$, $\xi(\mathbf{r}_0, \widehat{\boldsymbol{\mu}}, \widehat{\nu})$, and $\eta(\mathbf{r}_0, \widehat{\boldsymbol{\mu}}, \widehat{\nu})$ are the score functions obtained under the null hypothesis $\mathbf{r} = \mathbf{r}_0$, by replacing $\boldsymbol{\mu}$ and ν with their consistent estimates $\widehat{\boldsymbol{\mu}}$ and $\widehat{\nu}$ in

$$b(\mathbf{r}_0, \widehat{\boldsymbol{\mu}}, \widehat{\nu}) \quad = \quad \frac{\partial G}{\partial \mathbf{r}}, \tag{10.8}$$

$$\xi(\mathbf{r}_0, \widehat{\boldsymbol{\mu}}, \widehat{\nu}) = \frac{\partial G}{\partial \boldsymbol{\mu}}, \tag{10.9}$$

and

$$\eta(\mathbf{r}_0, \widehat{\boldsymbol{\mu}}, \widehat{\nu}) = \frac{\partial G}{\partial \nu}, \tag{10.10}$$

respectively. Furthermore, let $T_i(\mathbf{r}_0, \widehat{\boldsymbol{\mu}}, \widehat{\nu}) = \lambda_i - \sum_{j=1}^{p+1} \beta_{ij}\gamma_j$, where β_{ij} is the partial regression coefficient of λ_i on γ_j. Then, Neyman's partial score test statistic is given by

$$\begin{aligned} & W(\widehat{\boldsymbol{\mu}}, \widehat{\nu}) \\ = \; & \mathbf{T}^T \left[\widehat{\mathbf{M}}_{11} - \left(\widehat{\mathbf{M}}_{12}\widehat{\mathbf{M}}_{13} \right) \begin{pmatrix} \widehat{\mathbf{M}}_{22} & \widehat{\mathbf{M}}_{23} \\ & \widehat{M}_{33} \end{pmatrix}^{-1} \begin{pmatrix} \widehat{\mathbf{M}}_{12}^T \\ \widehat{\mathbf{M}}_{13}^T \end{pmatrix} \right]^{-1} \mathbf{T}, \end{aligned} \tag{10.11}$$

where $\mathbf{T} \equiv [T_1(\mathbf{r}_0, \widehat{\boldsymbol{\mu}}, \widehat{\nu}), \ldots, T_{p(p+1)/2}(\mathbf{r}_0, \widehat{\boldsymbol{\mu}}, \widehat{\nu})]^T$ for $i, r = 1, 2, 3$; $\widehat{\mathbf{M}}_{ir}$ are obtained from $\mathbf{M}_{ir} = E(-D_{ir})$ by replacing $\boldsymbol{\mu}$ and ν with their consistent estimates; and $\mathbf{D}_{ir}$ for $i, r = 1, 2, 3$ are the derivatives given by

$$\mathbf{D}_{11} = \frac{\partial^2 G}{\partial \mathbf{r} \partial \mathbf{r}'},$$

$$\mathbf{D}_{12} = \frac{\partial^2 G}{\partial \mathbf{r} \partial \boldsymbol{\mu}'},$$

$$\mathbf{D}_{13} = \frac{\partial^2 G}{\partial \mathbf{r} \partial \nu},$$

$$\mathbf{D}_{22} = \frac{\partial^2 G}{\partial \boldsymbol{\mu} \partial \boldsymbol{\mu}'},$$

$$\mathbf{D}_{23} = \frac{\partial^2 G}{\partial \boldsymbol{\mu} \partial \nu},$$

and

$$D_{33} = \frac{\partial^2 G}{\partial \nu^2}.$$

Under the null hypothesis $\mathbf{r} = \mathbf{r}_0$, the test statistic $W(\widehat{\boldsymbol{\mu}}, \widehat{\nu})$ has an approximate chi-squared distribution with degrees of freedom $p(p+1)/2$. The test based on (10.11) is asymptotically locally most powerful. Clearly the implementation of this test requires consistent estimates of

$\widehat{\mu}$, $\widehat{\nu}$ as well as expressions for the score functions and the information matrix. The maximum likelihood estimates of μ and ν are obtained by simultaneously solving

$$\widehat{\mu} = \sum_{j=1}^{n} q_j^{-1}\mathbf{X}_j \Big/ \sum_{j=1}^{n} q_j^{-1}$$

and

$$\eta\left(\widehat{\mu}, \mathbf{r}_0, \widehat{\nu}\right) = 0,$$

where $q_j = \widehat{\nu} - 2 + (\mathbf{X}_j - \widehat{\mu})^T\mathbf{R}_0(\mathbf{X}_j - \widehat{\mu})$ and $\mathbf{R}_0$ is specified by the null hypothesis. The moment estimates of μ and ν (which also turn out to be consistent) are

$$\widehat{\mu} = \frac{1}{n}\sum_{j=1}^{n}\mathbf{X}_j$$

and

$$\widehat{\nu} = \frac{2\left\{2\widehat{\beta}_2 - f\left(\mathbf{r}_0\right)\right\}}{\widehat{\beta}_2 - f\left(\mathbf{r}_0\right)},$$

where

$$\widehat{\beta}_2 = \frac{1}{n}\sum_{j=1}^{n}\left[\left(\mathbf{X}_j - \bar{\mathbf{X}}\right)^T \mathbf{R}_0 \left(\mathbf{X}_j - \bar{\mathbf{X}}\right)\right]^2$$

is a consistent estimator of the multivariate measure of skewness (see, for example, Mardia, 1970b), and

$$f\left(\mathbf{r}_0\right) = 3\sum_{h=1}^{p}\left(r_0^{hh}\right)^2\left\{r_{hh}^{(0)}\right\}^2 + \sum_{h\neq h'}^{p}\left\{r_{hh}^{(0)}\right\}^2\left\{r_0^{hh}r_0^{h'h'} + \left(r_0^{hh'}\right)^2\right\},$$

where $r_{hh'}^{(0)}$ and $r_0^{hh'}$ denote the (h,h')th element of $\mathbf{R}_0$ and $\mathbf{R}_0^{-1}$, respectively.

10.5.1 Score Functions

The score functions defined in (10.8), (10.9), and (10.10) are given by

$$b\left(\mathbf{r}, \widehat{\mu}, \widehat{\nu}\right) = -\frac{1}{2}\left[n\mathbf{I}_p - (\nu + p)\mathbf{R}^{-1}\sum_{j=1}^{n} q_j^{-1}\mathbf{A}_j\right]\mathbf{R}^{-1},$$

$$\xi(\mathbf{r},\widehat{\mu},\widehat{\nu}) = (\nu+p)\mathbf{R}^{-1}\sum_{j=1}^{n} q_j^{-1}(\mathbf{X}_j-\mu),$$

and

$$\eta(\mathbf{r},\widehat{\mu},\widehat{\nu}) = n\left[\frac{\nu}{2(\nu-2)}+\frac{1}{2}\log(\nu-2)+\frac{1}{2}\psi\left(\frac{\nu+p}{2}\right)-\frac{1}{2}\psi\left(\frac{\nu}{2}\right)\right] \\ -\frac{1}{2}\left[(\nu+p)\sum_{j=1}^{n}\frac{1}{q_j}+\sum_{j=1}^{n}\log q_j\right],$$

respectively, where $\psi(\cdot)$ denotes the digamma function and q_j is a non-homogeneous quadratic form given by $q_j = \nu - 2 + \mathrm{tr}\mathbf{R}^{-1}\mathbf{B}_j$ with $\mathbf{B}_j = (\mathbf{X}_j-\mu)(\mathbf{X}_j-\mu)^T$.

10.5.2 Information Matrix

By taking the second derivatives and then applying expectations, one can derive the elements of the information matrix. The first element takes the complicated form

$$\mathbf{M}_{11} = [m^*(1,1), m^*(1,2),\ldots,m^*(h,l),\ldots,m^*(p,p)],$$

where, for $l \geq h$, $h,l = 1,\ldots,p$, $m^*(h,l)$ is the $p(p+1)/2$-dimensional vector, formed by stacking the distinct elements of the $p\times p$ symmetric matrix

$$\mathbf{M}^*_{h,l} = \frac{n}{2}\left[r^h\otimes\left(r^l\right)^T\right]-\frac{n(\nu+p)}{2}\mathbf{R}^{-1}\mathbf{Q}_{h,l}\mathbf{R}^{-1}.$$

Here, r^k denotes the kth column of the $\mathbf{R}^{-1}$ matrix, and the (u,v)th element of the $p\times p$ matrix $\mathbf{Q}_{h,l}$ is given by

$$\frac{(\nu+2)^2}{(\nu+4)^2(\nu+p)(\nu+p+2)}\sum_{i=1}^{p}\sum_{j=1}^{p} r^{hi}r^{lk}\left(r_{ik}r_{uv}+r_{iu}r_{kv}+r_{iv}r_{ku}\right),$$

where r^{ms} and r_{ms} denote the (m,s)th element of $\mathbf{R}^{-1}$ and $\mathbf{R}$, respectively. The second element of the information matrix $\mathbf{M}_{12}$ is zero. The third element $\mathbf{M}_{13}$ is formed by stacking the distinct elements of the symmetric matrix

$$\frac{n(p+2)}{(\nu-2)(\nu+p)(\nu+p+2)}\mathbf{R}^{-1}.$$

The remaining elements of the information matrix are given by

$$\mathbf{M}_{22} = \frac{n\nu(\nu+p)}{(\nu-2)(\nu+p+2)}\mathbf{R}^{-1},$$

$$\mathbf{M}_{23} = \mathbf{0},$$

and

$$M_{33} = n\left[\frac{1}{4}\psi'\left(\frac{\nu+p}{2}\right) - \frac{1}{4}\psi'\left(\frac{\nu}{2}\right) - \frac{\nu-4}{2(\nu-2)^2}\right] + \frac{1}{2}\frac{n\nu\left(\nu^2+\nu p-6p-2\nu-8\right)}{(\nu-2)^2(\nu+p)(\nu+p+2)}.$$

10.6 Multivariate t Model

Consider the following multivariate t model described in equation (9.2) of the preceding chapter

$$f(\mathbf{x}_1,\ldots,\mathbf{x}_n) = \frac{\Gamma((\nu+p)/2)}{(\pi^n\nu)^{p/2}\,\Gamma(\nu/2)\,|\mathbf{R}|^{n/2}} \times \left[1+\frac{1}{\nu}\sum_{i=1}^{n}(\mathbf{x}_i-\boldsymbol{\mu})^T\mathbf{R}^{-1}(\mathbf{x}_i-\boldsymbol{\mu})\right]^{-(\nu+np)/2}. \quad (10.12)$$

In this section, we consider estimation issues associated with the correlation matrix $\mathbf{R}$ and its trace $\mathrm{tr}(\mathbf{R})$.

10.6.1 Estimation of R

Joarder and Ali (1997) developed estimators of $\mathbf{R}$ (when the mean vector $\boldsymbol{\mu}$ is unknown) under the entropy loss function

$$L(u(\mathbf{A}),\mathbf{R}) = \mathrm{tr}\left(\mathbf{R}^{-1}u(\mathbf{A})\right) - \log\left|\mathbf{R}^{-1}u(\mathbf{A})\right| - p,$$

where $u(\mathbf{A})$ is any estimator of $\mathbf{R}$ based on the Wishart matrix $\mathbf{A}$ defined in (9.4). Based on the form of the likelihood function, the entropy loss function has been suggested in the literature by James and Stein (1961) and is sometimes known as the Stein loss function. Some important features of the entropy loss function are that it is zero if the estimator $u(\mathbf{A})$ equals the parameter $\mathbf{R}$, positive when $u(\mathbf{A}) \neq \mathbf{R}$, and invariant under translation as well as under a natural group of transformations of covariance matrices. Moreover, the loss function approaches infinity as the estimator approaches a singular matrix or when one or more elements (or one or more latent roots) of the estimator approaches infinity. This means that gross underestimation is penalized just as heavily as gross overestimation.

In estimating $\mathbf{R}$ by $u(\mathbf{A})$, Joarder and Ali (1997) considered the risk function $R(u(\mathbf{A}),\mathbf{R}) = E[L(u(\mathbf{A}),\mathbf{R})]$. An estimator $u_2(\mathbf{A})$ of $\mathbf{R}$ will be said to dominate another estimator $u_1(\mathbf{A})$ of $\mathbf{R}$ if, for all $\mathbf{R}$ belonging to the class of positive definite matrices, the inequality $R(u_2(\mathbf{A}),\mathbf{R}) \leq R(u_1(\mathbf{A}),\mathbf{R})$ holds and the inequality $R(u_2(\mathbf{A}),\mathbf{R}) < R(u_1(\mathbf{A}),\mathbf{R})$ holds for at least one $\mathbf{R}$.

Joarder and Ali (1997) obtained three estimators for $\mathbf{R}$, by minimizing the risk function of the entropy loss function among three classes of estimators.

- First, it is shown that the unbiased estimator $\widetilde{\mathbf{R}} = (\nu-2)\mathbf{A}/(\nu n)$ has the smallest risk among the class of estimators of the form $c\mathbf{A}$, where $c > 0$, and the corresponding minimum risk is given by

$$
\begin{aligned}
R\left(\widetilde{\mathbf{R}},\mathbf{R}\right) &= p\log n - \sum_{i=1}^{n} E\left[\log\left(\chi^2_{n+i-i}\right)\right] + p\log\left(\frac{\nu}{\nu-2}\right) \\
&\quad -2pE\left(\log\tau\right),
\end{aligned}
$$

  where τ has the inverted gamma distribution given by (9.3).

- Second, the estimator $\mathbf{R}^* = \mathbf{T}\mathbf{D}^*\mathbf{T}^T$, where $\mathbf{T}$ is a lower triangular matrix such that $\mathbf{A} = \mathbf{T}\mathbf{T}^T$ and $\mathbf{D}^* = \mathrm{diag}(d_1^*, \ldots, d_p^*)$ with d_i^* defined by

$$
d_i^* = \frac{\nu-2}{\nu}\frac{1}{n+p+1-2i},
$$

  has the smallest risk among the class of estimators $\mathbf{T}\boldsymbol{\Delta}\mathbf{T}^T$, where $\boldsymbol{\Delta}$ belongs to the class of all positive definite diagonal matrices and the corresponding minimum risk function of $\mathbf{R}^*$ is given by

$$
\begin{aligned}
R\left(\mathbf{R}^*,\mathbf{R}\right) &= \sum_{i=1}^{p}\log(n+1+p-2i) - \sum_{i=1}^{p} E\left[\log\left(\chi^2_{n+1-i}\right)\right] \\
&\quad +p\log\left(\frac{\nu}{\nu-2}\right) - 2pE\left(\log\tau\right),
\end{aligned}
$$

  where τ is as defined above. Furthermore, $\mathbf{R}^*$ dominates the unbiased estimator $\widetilde{\mathbf{R}} = (\nu-2)\mathbf{A}/(\nu n)$.

- Finally, consider the estimator $\widehat{\mathbf{R}} = \mathbf{S}\phi(\mathbf{M})\mathbf{S}$, where $\mathbf{A}$ has the spectral decomposition $\mathbf{A} = \mathbf{S}\mathbf{M}\mathbf{S}^T$, with $\phi(\mathbf{M}) = \mathbf{D}^*\mathbf{M}$. Then the estimator $\widehat{\mathbf{R}} = \mathbf{S}\mathbf{D}^*\mathbf{M}\mathbf{S}^T$ dominates the estimator $\mathbf{R}^* = \mathbf{T}\mathbf{D}^*\mathbf{T}^T$.

10.6.2 Estimation of $tr(\mathbf{R})$

Let $\delta = \mathrm{tr}(\mathbf{R})$ denote the trace of $\mathbf{R}$. Joarder (1995) considered the estimation of δ for the multivariate t model under a squared error loss function following Dey (1988). The usual estimator of δ is given by $\widetilde{\delta} = c_0\mathrm{tr}(\mathbf{A})$, where c_0 is a known positive constant and $\mathbf{A}$ is the Wishart matrix defined in (9.4). The estimator $\widetilde{\delta}$ defines an unbiased estimator of δ for $c_0 = (\nu - 2)/(\nu n)$ and a maximum likelihood estimator of $\widetilde{\delta}$ for $c_0 = 1/(n+1)$ (see, for example, Fang and Anderson, 1990, page 208). Joarder and Singh (1997) proposed an improved estimator of δ – based on a power transformation – given by

$$\widehat{\delta} = c_0 tr(\mathbf{A}) + c_0 c\left\{p|\mathbf{A}|^{1/p} - \mathrm{tr}(\mathbf{A})\right\},$$

where c_0 is a known positive constant and c is a constant chosen so that the mean square error (MSE) of $\widehat{\delta}$ is minimized. Calculations show that

$$MSE\left(\widehat{\delta}\right) = MSE\left(\widetilde{\delta}\right) + c\beta_1 + c^2\beta_2,$$

where

$$\beta_1 = 2c_0^2 E\left[(c_0\mathrm{tr}(\mathbf{A}) - \delta)\left(p|\mathbf{A}|^{1/p} - \mathrm{tr}(\mathbf{A})\right)\right] \tag{10.13}$$

and

$$\beta_2 = c_0^2 E\left[p|\mathbf{A}|^{1/p} - \mathrm{tr}(\mathbf{A})\right]. \tag{10.14}$$

Thus $MSE(\widehat{\delta})$ is minimized at $c = -\beta_1/(2\beta_2)$ and the minimum value is given by $MSE(\widetilde{\delta}) - \beta_1^2/(4\beta_2)$. This proves that $\widehat{\delta}$ is always better than the usual estimator in the sense of having a smaller MSE. The estimate of c is given by $\widehat{c} = -\widehat{\beta}_1/(2\widehat{\beta}_2)$, where $\widehat{\beta}_1$ and $\widehat{\beta}_2$ are obtained by calculating the expectations in (10.13) and (10.14) using the numerous properties given in Section 9.1 and then replacing $\mathbf{R}$ by the usual estimator $c_0\mathbf{A}$. It can be noted from Fang and Anderson (1990, page 208) that the estimators $\widehat{\beta}_1$ and $\widehat{\beta}_2$ are the maximum likelihood estimators of β_1 and β_2, respectively, provided $\mathbf{R} = c_0\mathbf{A}$ and $c_0 = 1/(n+1)$.

The following table taken from Joarder and Singh (1997) presents the percent relative efficiency of $\widehat{\delta}$ over $\widetilde{\delta}$.

ν	$\mathbf{R} = \text{diag}(1,1,1)$	$\mathbf{R} = \text{diag}(4,2,1)$	$\mathbf{R} = \text{diag}(25,1,1)$
5	105.32	130.31	153.90
10	102.13	117.56	148.76
15	101.53	112.07	127.15

The numbers are from a Monte Carlo study carried out by generating 100 Wishart matrices from the multivariate t-model with $n = 25$ and $p = 3$.

10.7 Generalized Multivariate t Model

Consider the generalized multivariate t model (9.7) discussed in the preceding chapter. The usual estimator of $\mathbf{R}$ is a multiple of the Wishart matrix of the form $\widetilde{\mathbf{R}} = c_0\mathbf{A}$, where $c_0 > 0$. Joarder and Ahmed (1998) proposed improved estimates for $\mathbf{R}$ as well as its trace and inverse under the quadratic loss function. The proposed estimators for $\mathbf{R}$ are

$$\widehat{\mathbf{R}} = c_0\mathbf{A} - c\left|\mathbf{A}\right|^{1/p}\mathbf{I}, \tag{10.15}$$

where $\mathbf{I}$ is an identity matrix and c is chosen such that $\widehat{\mathbf{R}}$ is positive definite. For an estimator $\mathbf{R}^*$ of $\mathbf{R}$, let $L(\mathbf{R}^*, \mathbf{R}) = \text{tr}[(\mathbf{R}^* - \mathbf{R})^2]$ denote the quadratic loss function and let $R(\mathbf{R}^*, \mathbf{R}) = EL(\mathbf{R}^*, \mathbf{R})$ denote the corresponding risk function. The relationship between $\widehat{\mathbf{R}}$ and $\widetilde{\mathbf{R}}$ is rather involved. Defining the dominance of one estimator over another in the same manner as in Section 10.6.1, Joarder and Ahmed (1998) established that $\widehat{\mathbf{R}}$ dominates $\widetilde{\mathbf{R}}$ for any c satisfying $d < c < 0$, where

$$d = \left(c_0\frac{np+2}{p} - \gamma\right)\frac{\Gamma_p\left((n-1)/2 + 1/p\right)}{\Gamma_p\left((n-1)/2 + 2/p\right)} \tag{10.16}$$

with $c_0 < p\gamma/((n-1)p+2)$ or $0 < c < d$, where d is given by (10.16) with $c_0 > p\gamma/(np+2)$ and γ by $\gamma = \gamma_2/\gamma_4$ and $\gamma_i = E(\tau^i)$, $i = 1, 2, 3, 4$. The risk functions of the two estimators are given by

$$\begin{aligned} R\left(\widehat{\mathbf{R}}, \mathbf{R}\right) &= 4p\gamma_4\frac{\Gamma_p\left(n/2 + 2/p\right)}{\Gamma_p\left(n/2\right)}\left|\mathbf{R}\right|^{2/p} c\left(c - \frac{d\text{tr}\left(\mathbf{R}/p\right)}{\left|\mathbf{R}\right|^{1/p}}\right) \\ &\quad + \left\{1 + (n-1)c_0\gamma_4\left(c_0 n - 2\gamma\right)\right\}\text{tr}\left(\mathbf{R}^2\right) \\ &\quad + (n-1)c_0^2\gamma_4\left(\text{tr}\mathbf{R}\right)^2 \end{aligned}$$

and

$$R\left(\widetilde{\mathbf{R}}, \mathbf{R}\right) = \left\{1 + (n-1)c_0\gamma_4\left(c_0 n - 2\gamma\right)\right\}\text{tr}\left(\mathbf{R}^2\right)$$

$$+(n-1)c_0^2\gamma_4\left(\text{tr}\mathbf{R}\right)^2,$$

respectively. Now consider estimating the trace $\delta = \text{tr}\mathbf{R}$. The usual and the proposed estimators are $\widetilde{\delta} = c_0\text{tr}\mathbf{A}$ and $\widehat{\delta} = c_0\text{tr}\mathbf{A} - cp \mid \mathbf{A} \mid^{1/p}$, respectively, where $c_0 > 0$ and c is such that the proposed estimator is positive. Joarder and Ahmed (1998) established that the corresponding risk functions are given by

$$\begin{aligned} R\left(\widetilde{\delta}, \delta\right) &= \left[(n-1)c_0\left\{(n-1)c_0\gamma_4 - 2\gamma_2\right\} + 1\right]\delta^2 \\ &\quad +2(n-1)c_0^2\gamma_4\text{tr}\left(\mathbf{R}^2\right) \end{aligned}$$

and

$$R\left(\widehat{\delta}, \delta\right) = R\left(\widetilde{\delta}, \delta\right) + 4p^2\gamma_4\frac{\Gamma_p(n/2 + 2/p)}{\Gamma_p(n/2)}|\mathbf{R}|^{2/p} c\left(c - \frac{\text{tr}(\mathbf{R}/p)}{|\mathbf{R}|^{1/p}}d\right),$$

respectively. It is evident that $\widehat{\delta}$ dominates $\widetilde{\delta}$. Finally, consider estimating the inverse $\mathbf{\Psi} = \mathbf{R}^{-1}$ with the usual and the proposed estimators given by $\widetilde{\mathbf{\Psi}} = c_0\mathbf{A}^{-1}$ and $\widehat{\mathbf{\Psi}} = c_0\mathbf{A}^{-1} - c_0 \mid \mathbf{A} \mid^{-1/p} \mathbf{I}$, respectively, where $c_0 > 0$ and c is such that the proposed estimator is positive definite. In this case, it turns out that $\widehat{\mathbf{\Psi}}$ dominates $\widetilde{\mathbf{\Psi}}$ for any c satisfying $d < c < 0$, where

$$d = 4\left(\frac{c_0}{n - 2/p - p - 2} - \frac{\gamma_{-2}}{\gamma_{-4}}\right)\frac{\Gamma_p((n-1)/2 - 1/p)}{\Gamma_p((n-1)/2 - 2/p)} \quad (10.17)$$

with $c_0 < (n - 2/p - p - 2)\gamma_{-2}/\gamma_{-4}$ or $0 < c < d$, where d is given by (10.17) with $c_0 > (n - 2/p - p - 2)\gamma_{-2}/\gamma_{-4}$ and $\gamma_i = E(\tau^i)$.

10.8 Simulation

Simulation is a key element in modern statistical theory and applications. In this section, we describe three known approaches for simulating from multivariate t distributions. Undoubtedly, many other methods will be proposed and elaborated in the near future.

10.8.1 Vale and Maureli's Method

Fleishman (1978) noted that the real-world distributions of (univariate) variables are typically characterized by their first four moments (that is, mean, variance, skewness, and kurtosis). He presented a procedure for generating nonnormal random numbers with these four moments specified. He accomplished this by taking a nonnormal variable X as a linear combination of the first three powers of a standard normal random variable Z

$$X = a + bZ + cZ^2 + dZ^3. \tag{10.18}$$

To determine the constants, Fleishman expanded (10.18) to express the first four moments of X in terms of the first 14 moments of Z. After considerable algebraic manipulation, Fleishman was able to represent the solution to the constants of (10.18) as a system of nonlinear equations. For a standard distribution (that is, with mean zero and variance one), the constants b, c, and d are found by simultaneously solving the following equations

$$b^2 + 6bd + 2c^2 + 15d^2 = 1, \tag{10.19}$$

$$2c\left(b^2 + 24bd + 105d^2 + 2\right) = \gamma_1, \tag{10.20}$$

and

$$24\left\{bd + c^2\left(1 + b^2 + 28bd\right) + d^2\left(12 + 48bd + 141c^2 + 225d^2\right)\right\} = \gamma_2, \tag{10.21}$$

where γ_1 is the desired skewness and γ_2 is the desired kurtosis. The constant a in (10.18) is determined by

$$a = -c. \tag{10.22}$$

Univariate nonnormal random numbers are then generated by drawing normal random numbers and transforming them using the constants a, b, c, and d in (10.18).

Vale and Maureli (1983) extended Fleishman's procedure for multivariate nonnormal distributions with specified intercorrelations as well as specified moments. The procedure begins by specifying the constants necessary for Fleishman's procedure. For each variable independently, these are given by the solution of (10.19)–(10.22). Define two variables Z_1 and Z_2 as from standard normal populations, and define the vector $\mathbf{z}$ as $\mathbf{z}^T = [1, Z, Z^2, Z^3]$. The weight vector $\mathbf{w}^T$ contains the power function weights a, b, c, and d: $\mathbf{w}^T = [a, b, c, d]$. The nonnormal variable X then becomes $X = \mathbf{w}^T\mathbf{z}$. If $r_{X_1X_2}$ denotes the correlation between two nonnormal variables X_1 and X_2 corresponding to the normal variables Z_1 and Z_2, it is then easily seen that $r_{X_1X_2} = \mathbf{w}_1^T\mathbf{R}\mathbf{w}_2$, where $X_1 = \mathbf{w}_1^T\mathbf{z}_1$, $X_2 = \mathbf{w}_2^T\mathbf{z}_2$ and $\mathbf{R}$ is the expected matrix product of $\mathbf{z}_1$

and $\mathbf{z}_2^T$ given by

$$\mathbf{R} = \begin{bmatrix} 1 & 0 & 1 & 0 \\ 0 & r_{Z_1Z_2} & 0 & 3r_{Z_1Z_2} \\ 1 & 0 & 2r_{Z_1Z_2}^2 + 1 & 0 \\ 0 & 3r_{Z_1Z_2} & 0 & 6r_{Z_1Z_2}^3 + 9r_{Z_1Z_2} \end{bmatrix}.$$

Collecting the terms and using (10.22), a third-degree polynomial in $r_{Z_1Z_2}$, the correlation between the normal variables Z_1 and Z_2, results

$$\begin{aligned} r_{X_1X_2} &= r_{Z_1Z_2}(b_1b_2 + 3b_1d_2 + 3d_1b_2 + 9d_1d_2) + 2c_1c_2r_{Z_1Z_2}^2 \\ &\quad +6d_1d_2r_{Z_1Z_2}^3. \end{aligned}$$

Solving this polynomial for $r_{Z_1Z_2}$ provides the correlation required to obtain the desired post-transformation correlation $r_{X_1X_2}$. These correlations can then be assembled into a matrix of intercorrelations, and this matrix can be decomposed to yield multivariate normal random numbers for input into Fleishman's transformation procedure.

10.8.2 Vaduva's Method

Vaduva (1985) provided a general algorithm for generating from multivariate distributions and illustrated its applicability for multivariate normal, Dirichlet, and multivariate t distributions. Here, we present a specialized version of the algorithm for generating the p-variate t distribution with the joint pdf

$$f(\mathbf{x}) = \frac{\Gamma((\nu+p)/2)}{(\pi\nu)^{p/2}\Gamma(\nu/2)\,|\mathbf{R}|^{1/2}}\left[1 + \frac{1}{\nu}\mathbf{x}^T\mathbf{R}^{-1}\mathbf{x}\right]^{-(\nu+p)/2}$$

over some domain D in $\Re^p$. It is as follows

(i) Initialize.
(ii) Determine an interval $I = [v_0^0, v_0^1] \times \cdots \times [v_p^0, v_p^1]$, where

$$v_0^0 = 0,$$

$$v_0^1 = 1,$$

$$v_i^0 = -\sqrt{\frac{\nu(p+1)}{\nu-1}}, \quad i = 1, \ldots, p,$$

and

$$v_i^1 = \sqrt{\frac{\nu(p+1)}{\nu-1}}, \quad i = 1, \ldots, p.$$

(iii) Generate the random vector $\mathbf{V}^*$ uniformly distributed over I. If RND is a uniform random number generator, then $\mathbf{V}^*$ may be generated as follows

  (a) Generate $U_0, U_1, \ldots, U_p$ uniformly distributed over $[0,1]$ and stochastically independent.
  (b) Calculate $V_i^* = v_i^0 + (v_i^1 - v_i^0)U_i$, $i = 0, 1, \ldots, p$.
  (c) Take $\mathbf{V}^* = (V_0^*, V_1^*, \ldots, V_p^*)$.

(iv) If $\mathbf{V}^* \notin D$, then go to step (iii).

(v) Otherwise, take $\mathbf{V} = \mathbf{V}^*$.

(vi) Calculate $Y_i = V_i/V_0$, $i = 1, \ldots, p$.

(vii) Take $\mathbf{X} = (Y_1, \ldots, Y_p)^T$. Stop.

Note that the steps from (iii) to (v) constitute a rejection algorithm. The performance of this algorithm is characterized by the probability to accept $\mathbf{V}^*$. This probability can be calculated in the form

$$p_a = \frac{\pi^{p/2}\Gamma(\nu/2)}{2^p(p+1)\Gamma((\nu+p)/2)\,|\mathbf{R}|^{1/2}}\left(\frac{\nu-1}{p+2}\right)^{p/2},$$

which yields

$$\lim_{\nu\to\infty} p_a = 0$$

and

$$\lim_{p\to\infty} p_a = 0,$$

indicating inadequate behavior of the algorithm for large values of p and/or ν.

10.8.3 Simulation Using BUGS

A relatively simple way to generate a multivariate t involves a sampling of z from gamma($\nu/2$, $\nu/2$) and then sampling a multivariate normal $\mathbf{y} \sim N_p(\boldsymbol{\mu}, \mathbf{R}/z)$. This mode of generation reflects the scale mixture form of the multivariate t pdf. In BUGS the multivariate normal is parameterized by the precision matrix $\mathbf{P}$; thus one programs a multivariate t pdf as follows to generate a sample of n cases (for Sigma[,], nu.2 and mu[] known)

```
for (i in 1:n)
{z[i] ~ dgamma(nu.2, nu.2)
y[i, 1:q] ~ dmnorm(mu, P.sc[,])}
```

```
for (i in 1:q) {for (j in 1:q)
{P[i, j] <- inverse(Sigma[,], i, j)
P.sc[i, j] <- z[i] * P[i, j]}}
```

If one has observed multivariate data and wishes to assume multivariate t sampling, then in BUGS the dmt() form is available

```
for (i in 1:n) {y[i, 1:q] ~ dmt(mu[], P[,], nu)}
```

where nu is assumed known.

11
Regression Models

There is a large number of contributions (scattered in the literature and many of them motivated by economic applications) dealing with regression models with the error term distributed according to the multivariate t distribution. In this chapter, we shall discuss several of them.

11.1 Classical Linear Model

Let the model for n observations $\mathbf{Y} = (y_1, \ldots, y_n)^T$ be

$$\mathbf{Y} = \mathbf{X}\boldsymbol{\beta} + \boldsymbol{\epsilon}, \tag{11.1}$$

where $\mathbf{X}$ is an $n \times p$ design matrix with rank p, $\boldsymbol{\beta}$ is a $p \times 1$ vector of regression parameters with unknown values, and $\boldsymbol{\epsilon}$ is an $n \times 1$ random error vector. For the usual t regression model it is assumed that the n elements of $\boldsymbol{\epsilon}$ have the multivariate t pdf

$$f(\boldsymbol{\epsilon}) = \frac{\nu^{\nu/2}\Gamma((n+\nu)/2)}{\pi^{n/2}\sigma^n\Gamma(\nu/2)}\left[\nu + \frac{\boldsymbol{\epsilon}^T\boldsymbol{\epsilon}}{\sigma^2}\right]^{-(n+\nu)/2}.$$

In practice, there are several situations in which the model (11.1) is useful. Under (11.1), the least squares estimate of $\boldsymbol{\beta}$ is

$$\widehat{\boldsymbol{\beta}} = \left(\mathbf{X}^T\mathbf{X}\right)^{-1}\mathbf{X}^T\mathbf{y}. \tag{11.2}$$

Zellner (1976) noted that this is also the maximum likelihood estimate of $\boldsymbol{\beta}$. From Singh (1991), $\widehat{\boldsymbol{\beta}}$ is a minimum variance linear unbiased estimator and also a minimum variance unbiased estimator. The variance-covariance matrix for $\widehat{\boldsymbol{\beta}}$ is

$$Var\left(\widehat{\boldsymbol{\beta}}\right) = E\left(\widehat{\boldsymbol{\beta}} - \boldsymbol{\beta}\right)\left(\widehat{\boldsymbol{\beta}} - \boldsymbol{\beta}\right)^T = \frac{\nu\sigma^2}{\nu - 2}\left(\mathbf{X}^T\mathbf{X}\right)^{-1}. \tag{11.3}$$

Note that as $\nu \to \infty$, the above variance approaches $(\mathbf{X}^T\mathbf{X})^{-1}\sigma^2$, which is the variance-covariance matrix in the normal case. Thus, for small and moderate values of ν, the variances of the elements of $\widehat{\beta}$ are inflated considerably, as compared to those for large values of ν.

Singh (1988) provided the following estimate of the degrees of freedom parameter

$$\widehat{\nu} = \frac{2\,(2\widehat{a}-3)}{\widehat{a}-3},$$

where

$$\widehat{a} = \frac{(1/n)\sum_{i=1}^{n}\left(y_i - \mathbf{x}_i^T\widehat{\beta}\right)^4}{\left\{(1/n)\sum_{i=1}^{n}\left(y_i - \mathbf{x}_i^T\widehat{\beta}\right)^2\right\}^2}$$

and $\widehat{\beta}$ is the least squares estimator given by (11.2).

The maximum likelihood estimate of σ^2 is

$$\widehat{\sigma}^2 = \frac{1}{n}\left(\mathbf{y} - \mathbf{X}\widehat{\beta}\right)^T\left(\mathbf{y} - \mathbf{X}\widehat{\beta}\right)$$

as in the normal case. For $\nu > 2$,

$$E\left(\widehat{\sigma}^2\right) = \frac{(n-p)\sigma_u^2}{n},$$

where $\sigma_u^2 = \nu\sigma^2/(\nu-2)$ is the common variance of the elements of ϵ. Thus, $\widehat{\epsilon}^T\widehat{\epsilon}/(n-p)$ is an unbiased estimator for σ_u^2 while

$$\widehat{\sigma}^2 = \frac{\nu-2}{\nu(n-p)}\widehat{\epsilon}^T\widehat{\epsilon} \qquad (11.4)$$

is an unbiased estimator for σ^2. In the class of estimators $q\widehat{\epsilon}^T\widehat{\epsilon}$, with q being a positive scalar, the minimal mean squared error estimator for σ^2 is (with $\nu > 4$)

$$\widetilde{\sigma}^2 = \frac{\nu-4}{\nu(n-p+2)}\widehat{\epsilon}^T\widehat{\epsilon}, \qquad (11.5)$$

while the minimal mean squared error estimator for σ_u^2 in this class is $(\nu-4)\widehat{\epsilon}^T\widehat{\epsilon}/\{(\nu-2)(n-p+2)\}$. The variances of the unbiased and the minimal mean squared error estimators of σ^2 are

$$Var\left(\widehat{\sigma}^2\right) = \frac{2\sigma^4}{n-p}\frac{n-p+\nu-2}{\nu-4} \qquad (11.6)$$

and

$$Var\left(\widetilde{\sigma}^2\right) = 2(n-p)\frac{(\nu-4)(n-p+\nu-2)}{(\nu-2)^2(n-p+2)^2}\sigma^4, \tag{11.7}$$

respectively. Since $\widehat{\sigma}^2$ is an unbiased estimator for σ^2, the variance (11.3) can be unbiasedly estimated by

$$\widetilde{Var}\left(\widehat{\beta}\right) = \left(\mathbf{X}^T\mathbf{X}\right)^{-1}\frac{\widehat{\epsilon}^T\widehat{\epsilon}}{n-p}\mathbf{I}.$$

Similarly, (11.4) and (11.5) can be estimated by

$$\widetilde{\widetilde{\sigma}^2} = \frac{\widehat{a}}{(2\widehat{a}-3)(n-p)}\widehat{\epsilon}^T\widehat{\epsilon} \tag{11.8}$$

and

$$\widetilde{\widetilde{\sigma}^2} = \frac{3}{(2\widehat{a}-3)(n-p+2)}\widehat{\epsilon}^T\widehat{\epsilon}, \tag{11.9}$$

respectively. The estimates for the variances given by (11.6)–(11.7) are

$$\widetilde{Var}\left(\widehat{\sigma}^2\right) = \frac{2\left(\sigma^{*2}\right)^2}{n-p}\frac{n-p+\widehat{\nu}-2}{\widehat{\nu}-4}$$

and

$$\widetilde{Var}\left(\widetilde{\sigma}^2\right) = 2(n-p)\frac{(\widehat{\nu}-4)(n-p+\widehat{\nu}-2)}{(\widehat{\nu}-2)^2(n-p+2)^2}\left(\sigma^{*2}\right)^2,$$

respectively, where σ^{*2} may be taken as $\widetilde{\widetilde{\sigma}^2}$ given in (11.8) or as $\widetilde{\widetilde{\sigma}^2}$ given in (11.9).

It is important to note that, even though the elements of ϵ have the nonnormal pdf and are not independent, tests and intervals based on the usual t and F statistics still remain valid. For example, $t = (\widehat{\beta}_i - \beta_i)/\{s\sqrt{m^{ii}}\}$, where m^{ii} is the (i,i)th element of $(\mathbf{X}^T\mathbf{X})^{-1}$ and $s^2 = \widehat{\epsilon}^T\widehat{\epsilon}/(n-p)$, has the usual Student's t distribution, and thus probability statements based on this statistic will be appropriate. Also, s^2/σ^2 has the usual F distribution with degrees of freedom ν and $n-k$. This fact can be used to construct confidence intervals for and test hypotheses about σ^2.

Singh et al. (1995) proposed the generalized estimator $\widehat{\beta}_g = g(t)\widehat{\beta}$ for β, where $t = \widehat{\epsilon}^T\widehat{\epsilon}/\widehat{\beta}^T\mathbf{X}^T\mathbf{X}\widehat{\beta}$ has at least the first $k \geq 6$ moments finite and $g(t)$, satisfying the validity conditions of Taylor's series expansion and having the first three derivatives with respect to t bounded, is a bounded function of t such that $g(0) = 1$ and $g(t) = O(1)$ as

$\theta = \beta^T \mathbf{X}^T \mathbf{X} \beta \to \infty$. It should be noted that the maximum likelihood estimator β and the estimators considered by Singh (1991) are all particular cases of $\widehat{\beta}_g$. Singh et al. (1995) investigated the bias $\rho(\widehat{\beta}_g) = E[(\widehat{\beta}_g - \beta)^T \mathbf{Q} (\widehat{\beta}_g - \beta)]$ of the generalized estimator when $\mathbf{Q}$ is a positive definite matrix. It was established that

$$E\left(\widehat{\beta}_g\right) = \beta + \frac{(n-p)\nu\sigma^2 g'(0)}{\theta(\nu-2)} \left[1 - \frac{\nu\sigma^2(p-2)}{\theta(\nu-4)} + \frac{\nu\sigma^2(n-p+2)g''(0)}{2\theta(\nu-4)g'(0)}\right]\beta + O\left(\frac{1}{\theta^3}\right)$$

and

$$\rho\left(\widehat{\beta}_g\right) = \frac{\nu\sigma^2}{\nu-2}\mathrm{tr}\left\{\left(\mathbf{X}^T\mathbf{X}\right)^{-1}\mathbf{Q}\right\} + \frac{(n-p)\gamma^2\sigma^4 g'(0)}{(\nu-2)(\nu-4)\theta} \times \left[2\mathrm{tr}\left\{\left(\mathbf{X}^T\mathbf{X}\right)^{-1}\mathbf{Q}\right\} + \frac{(n-p+2)g'(0)-4}{\theta}\beta^T\mathbf{Q}\beta\right] + O\left(\frac{1}{\theta^2}\right).$$

Since $\rho(\widehat{\beta}_g) < \rho(\beta)$, one observes that $\widehat{\beta}_g$ is dominant over the maximum likelihood estimator $\widehat{\beta}$. Also in the class $\widehat{\beta}_g$, there exists better estimators than those considered in Singh (1991).

Sutradhar (1988b) considered testing $H_0 : \mathbf{C}\beta = \mathbf{0}$ versus $H_1 : \mathbf{C}\beta \neq \mathbf{0}$, using the classical F statistic

$$W = \frac{W_2 - W_1}{W_1},$$

where

$$W_1 = \frac{\nu-2}{\nu\sigma^2}\left(\mathbf{I}_n - \mathbf{X}\left(\mathbf{X}^T\mathbf{X}\right)^{-1}\mathbf{X}^T\right)\mathbf{Y}$$

is the residual sum of squares of the full model (11.1) and

$$W_2 = \frac{\nu-2}{\nu\sigma^2}\left(\mathbf{I}_n - \mathbf{Z}\left(\mathbf{Z}^T\mathbf{Z}\right)^{-1}\mathbf{Z}^T\right)\mathbf{Y}$$

is the residual sum of squares of the reduced model

$$E(\mathbf{Y}) = \mathbf{Z}\alpha, \tag{11.10}$$

which is obtained from (11.1) by using the restriction under H_0. In H_0, $\mathbf{C}$ is an $r \times p$ matrix of known coefficients with rank($\mathbf{C}$) $= q$, and, in the reduced model (11.10), $\mathbf{Z}$ denotes the new design matrix of order

$n \times (p-q)$ and $\boldsymbol{\alpha}$ is a vector of $(p-q)$ parameters. Sutradhar (1988b) established that the pdf of W is given by

$$f(w) = \frac{2}{\nu-2}\sum_{k=0}^{\infty}\beta_{\delta/(\nu-2)}\left(k+1, \frac{\nu}{2}-1\right)\beta_w\left(k+\frac{q}{2}, \frac{n-p}{2}\right),$$

where

$$\beta_x(a,b) = \frac{\Gamma(a+b)x^{a-1}}{\Gamma(a)\Gamma(b)(1+x)^{a+b}}$$

and δ is the noncentrality parameter given by

$$\delta = \frac{\nu-2}{\nu\sigma^2}\boldsymbol{\beta}^T\left(\mathbf{X}^T\mathbf{X} - \mathbf{X}^T\mathbf{Z}\left(\mathbf{Z}^T\mathbf{Z}\right)^{-1}\mathbf{Z}^T\mathbf{X}\right)\boldsymbol{\beta}.$$

Sutradhar also computed the corresponding power of the test, yielding the expression

$$\frac{2}{\nu-2}\sum_{k=0}^{\infty}\beta_{\delta/(\nu-2)}\left(k+1, \frac{\nu}{2}-1\right)I_{u_0}\left(\frac{n-p}{2}, k+\frac{1}{2}\right),$$

where $u_0 = 1/[1+(q/(n-p))F_{q,n-p,\alpha}]$ and $I_u(a,b)$ denotes the incomplete beta function ratio. As $\nu \to \infty$, this expression reduces to the power of the F test under normality (Tiku, 1967).

The distribution of future responses given a set of data from an informative experiment is known as a predictive distribution. Haq and Khan (1990) derived the predictive distribution for (11.1). Rewrite (11.1) in the equivalent form $\mathbf{y} = \boldsymbol{\beta}\mathbf{X} + \sigma\mathbf{e}$ and let Y_f be a future response corresponding to the design vector $\mathbf{x}_f$, that is, $y_f = \boldsymbol{\beta}\mathbf{x}_f + \sigma e_f$. Haq and Khan (1990) showed that the predictive pdf of Y_f is given by

$$\begin{aligned} & f(y_f \mid \mathbf{y}) \\ \propto\ & \left[1 + s^{-2}(\mathbf{y})\left\{y_f - b(\mathbf{y})\mathbf{x}_f\right\}\left(1-\mathbf{x}_f^T\mathbf{A}^{-1}\mathbf{x}_f\right)\left\{y_f - b(\mathbf{y})\mathbf{x}_f\right\}^T\right]^{-(n-p+1)/2}, \end{aligned}$$

where $b(\mathbf{e}) = \mathbf{e}\mathbf{X}^T(\mathbf{X}\mathbf{X}^T)^{-1}$, $s^2(\mathbf{e}) = (\mathbf{e}-\widehat{\mathbf{e}})(\mathbf{e}-\widehat{\mathbf{e}})^T$, $\widehat{\mathbf{e}} = b(\mathbf{e})\mathbf{X}$, and $\mathbf{A} = \mathbf{X}\mathbf{X}^T + \mathbf{x}_f\mathbf{x}_f^T$. Thus, for the given informative data $\mathbf{y}$, the predictive distribution of Y_f is t with mean vector $b(\mathbf{y})\mathbf{x}_f$, variance-covariance matrix $(n-p)s^2(\mathbf{y})/\{(n-p-2)(1-\mathbf{x}_f^T\mathbf{A}^{-1}\mathbf{x}_f)\}$, and degrees of freedom $n-p$. A prediction interval of the desired coverage probability can easily be obtained by using the standard t-table. Note that the predictive distribution does not depend on the degrees of freedom parameter of the original t distribution. For a set of n' future responses given by

$\mathbf{Y}_f = \beta \mathbf{X}_f + \sigma \mathbf{e}_f$, Haq and Khan (1990) noted similarly that the predictive distribution of $\mathbf{y}_f$ is n'-variate t with mean vector $b(\mathbf{y})\mathbf{X}_f$, variance-covariance matrix $\mid \mathbf{I}_{n'} - \mathbf{X}_f \mathbf{Q}^{-1} \mathbf{X}_f \mid^{-1/2} s(\mathbf{y})$ (where $\mathbf{Q} = \mathbf{X}\mathbf{X}^T + \mathbf{X}_f \mathbf{X}_f^T$), and degrees of freedom $n-p$. It is to be noted that the distribution of $(n-p)s^{-2}(\mathbf{y})(\mathbf{Y}_f - b(\mathbf{y})\mathbf{X}_f)(\mathbf{I}_{n'} - \mathbf{X}_f \mathbf{Q}^{-1} \mathbf{X}_f)(\mathbf{Y}_f - b(\mathbf{y})\mathbf{X}_f)^T$ is F with degrees of freedom n', and $n-p$. This distribution can be utilized for determining the prediction region for a set of future responses with any desired coverage probability.

11.2 Bayesian Linear Models

In his classical paper, Zellner (1976) provided a Bayesian analysis of the linear model (11.1). Consider the diffusion prior for β and σ^2, that is,

$$p\left(\beta, \sigma^2\right) \quad \propto \quad \frac{1}{\sigma^2}, \tag{11.11}$$

where $0 < \sigma^2 < \infty$ and $\beta_i \in \Re$, $i = 1, \ldots, k$. Then, assuming that ν is known, the posterior pdf of the parameters is

$$p\left(\beta, \sigma^2 | \mathbf{y}, \nu\right) \quad \propto \quad \{A(\beta)\}^{-n/2} \left\{ \frac{\left[\nu\sigma^2 / A(\beta)\right]^{\nu/2-1}}{A(\beta)\left[1 + \nu\sigma^2/A(\beta)\right]^{(n+\nu)/2}} \right\},$$

where $A(\beta) = (\mathbf{y} - \mathbf{X}\beta)^T(\mathbf{y} - \mathbf{X}\beta)$. It follows that the conditional posterior pdf of β given σ^2 and ν is in the form of a multivariate t pdf with mean $\widehat{\beta}$ (the least squares estimate in (11.2)). The corresponding conditional posterior covariance matrix is given by

$$Var\left(\beta | \mathbf{y}, \sigma^2, \nu\right) \quad = \quad \frac{\nu\sigma^2 + (n-p)s^2}{\nu + n - p - 2} \left(\mathbf{X}^T\mathbf{X}\right)^{-1},$$

provided that $n - p + \nu > 2$, where $(n-p)s^2 = (\mathbf{y} - \mathbf{X}\widehat{\beta})^T(\mathbf{y} - \mathbf{X}\widehat{\beta})$. As $\nu \to \infty$, the conditional posterior pdf for β and σ^2 approaches a multivariate normal pdf with mean $\widehat{\beta}$ and covariance matrix $(\mathbf{X}^T\mathbf{X})^{-1}\sigma^2$, which is the usual result for the normal regression model with the diffuse prior pdf (11.11). The marginal posterior pdf for β is

$$p(\beta | \mathbf{y}, \nu) \quad \propto \quad \left\{ (n-p)s^2 + \left(\beta - \widehat{\beta}\right)^T \mathbf{X}^T\mathbf{X} \left(\beta - \widehat{\beta}\right) \right\}^{-n/2}, \tag{11.12}$$

which is in the form of a p-dimensional t pdf and does not depend on the value of ν. In fact, (11.12) is precisely the result that one obtains in the Bayesian analysis of the normal regression model with the diffuse prior for the parameters shown in (11.11). The marginal posterior pdf for σ^2 is

$$p\left(\sigma^2 \mid \mathbf{y}, \nu\right) \propto \left(\frac{\sigma^2}{s^2}\right)^{(\nu/2)-1} \left(1 + \frac{\nu\sigma^2}{(n-p)s^2}\right)^{-(\nu+n-p)/2},$$

from which it follows that σ^2/s^2 has the F pdf with degrees of freedom ν and $n-p$, a result paralleling the classical results mentioned in the preceding section. From properties of the F distribution, the modal value of σ^2/s^2 is $((n-p)/\nu)((\nu-2)/(n-p+2))$, when $\nu > 2$ and its mean is $(n-p)/(n-p-2)$ when $n-p > 2$. Also, as $\nu \to \infty$, the posterior distribution of $\nu s^2/\sigma^2$ approaches a chi-squared distribution with degrees of freedom $n-p$, a distributional result that holds for the Bayesian analysis of the usual normal regression model with diffuse prior assumptions. Finally, note that the posterior pdf for $n\sigma^2/(\mathbf{y}-\mathbf{X}\boldsymbol{\beta})^T(\mathbf{y}-\mathbf{X}\boldsymbol{\beta})$ is $F_{\nu,n}$.

The natural conjugate prior distribution for σ^2 and $\boldsymbol{\beta}$ is the product of the marginal F pdf for σ^2 times a conditional p-dimensional t pdf for $\boldsymbol{\beta}$ given σ^2, that is,

$$p\left(\boldsymbol{\beta}, \sigma^2 \mid \cdot\right) = p_F\left(\sigma^2 \mid \cdot\right) p_S\left(\boldsymbol{\beta} \mid \sigma^2, \cdot\right), \tag{11.13}$$

where

$$p_F\left(\sigma^2 \mid s_a^2, \nu_a, \nu\right) \propto \frac{\left(\nu\sigma^2/\nu_a s_a^2\right)^{(\nu-2)/2}}{\left(1 + \nu\sigma^2/\nu_a s_a^2\right)^{(\nu+\nu_a)/2}}$$

(where $\nu_a > 0$, $s_a > 0$, and $0 < \sigma < \infty$) and

$$p_S\left(\boldsymbol{\beta} \mid \sigma^2, \bar{\boldsymbol{\beta}}, \mathbf{A}, \bar{\nu}_a\right) \propto \bar{\sigma}_a^{-p} \left\{\bar{\nu}_a + \frac{1}{\bar{\sigma}_a^2}\left(\boldsymbol{\beta} - \bar{\boldsymbol{\beta}}\right)^T \mathbf{A} \left(\boldsymbol{\beta} - \bar{\boldsymbol{\beta}}\right)\right\}^{-(2\nu+p)/2},$$

where $\beta_i \in \Re$, $i = 1, \ldots, p$, $\mathbf{A}$ is symmetric and positive definite, $\bar{\boldsymbol{\beta}}$ is the prior mean vector, $\bar{\nu}_a = \nu + \nu_a$, and $\bar{\sigma}_a^2 = (\nu_a s_a^2 + \nu\sigma^2)/\bar{\nu}_a$. As with the natural conjugate for the usual normal regression model, it is seen that $\boldsymbol{\beta}$ and σ^2 are not independent in the natural conjugate prior distribution in (11.13). If the natural conjugate prior distribution is thought to represent the available prior information adequately, it can be used for obtaining the posterior distribution; see the appendix in Zellner (1976) for details.

11.3 Indexed Linear Models

Lange et al. (1989) and Fernandez and Steel (1999) provided a far-reaching extension of (11.1) to handle the situation when $\mathbf{y}_i$'s are assumed to have the t distribution with degrees of freedom ν_i and parameters $\boldsymbol{\mu} = g_i(\boldsymbol{\theta})$ and $\mathbf{R} = h_i(\boldsymbol{\phi})$ indexed by some unknown parameters $\boldsymbol{\theta}$ and $\boldsymbol{\phi}$. Lange et al. (1989) suggested an EM algorithm for estimation. They also considered methods for computing standard errors, developed graphical diagnostic checks, and provided applications to a variety of problems. The problems include linear and nonlinear regression, robust estimation of the mean and covariance matrix with missing data, unbalanced multivariate repeated-measures data, multivariate modeling of pedigree data, and multivariate nonlinear regression. They also derived the expected information matrix for $(\boldsymbol{\theta}, \boldsymbol{\phi}, \nu)$ for one observation in the form

$$E\left(\frac{\partial \log L}{\partial \theta_i \partial \theta_j}\right) = \frac{\nu + p}{\nu + p + 2} \frac{\partial \boldsymbol{\nu}^T}{\partial \theta_i} \mathbf{R}^{-1} \frac{\partial \boldsymbol{\nu}}{\partial \theta_j},$$

$$\begin{aligned} E\left(\frac{\partial \log L}{\partial \phi_i \partial \phi_j}\right) &= \frac{\nu + p}{2(\nu + p + 2)} \operatorname{tr}\left(\mathbf{R}^{-1} \frac{\partial \mathbf{R}}{\partial \phi_i} \mathbf{R}^{-1} \frac{\partial \mathbf{R}}{\partial \phi_j}\right) \\ &\quad - \frac{1}{2(\nu + p + 2)} \operatorname{tr}\left(\mathbf{R}^{-1} \frac{\partial \mathbf{R}}{\partial \phi_i}\right) \operatorname{tr}\left(\mathbf{R}^{-1} \frac{\partial \mathbf{R}}{\partial \phi_j}\right), \end{aligned}$$

$$E\left(\frac{\partial \log L}{\partial \phi_i \partial \nu}\right) = -\frac{1}{(\nu + p)(\nu + p + 2)} \operatorname{tr}\left(\mathbf{R}^{-1} \frac{\partial \mathbf{R}}{\partial \phi_i}\right),$$

and

$$\begin{aligned} E\left(\frac{\partial \log L}{\partial \nu \partial \nu}\right) &= -\frac{1}{2}\left\{\frac{1}{2}\psi''\left(\frac{\nu + p}{2}\right) - \frac{1}{2}\psi''\left(\frac{\nu}{2}\right) + \frac{p}{\nu(\nu + p)}\right. \\ &\quad \left. - \frac{1}{\nu + p} + \frac{\nu + 2}{\nu(\nu + p + 2)}\right\}, \end{aligned}$$

where $\psi''(x) = d^2 \log \Gamma(x)/d^2 x$ is the trigamma function. The remaining elements of the matrix are zero.

In an important paper, Fernandez and Steel (1999) revealed some pitfalls with a model of the above kind. Under a commonly used non-informative prior, they showed that Bayesian inference is precluded for certain samples, even though there exists a well-defined conditional distribution of the parameters given the observables. They also noted that global maximization of the likelihood function is a vacuous exercise since the latter becomes unbounded as one tends to the boundary of the parameter space. More specifically, let $l(\boldsymbol{\theta}, \sigma, \mathbf{R}, \nu)$ be the likelihood function for n independent observations $\mathbf{y}_i$ assumed to have the t distribution with mean vector $g_i(\boldsymbol{\theta})$, common covariance matrix $\sigma^2\mathbf{R}$, and common degrees of freedom ν. For given values of $\boldsymbol{\theta} = \boldsymbol{\theta}_0$, $\mathbf{R} = \mathbf{R}_0$, and $\nu = \nu_0$, let $0 \leq s(\boldsymbol{\theta}_0) \leq n$ denote the number of observations for which $y_i = g_i(\boldsymbol{\theta}_0)$. Then the following hold

(a) If

$$\nu_0 < \frac{ps(\boldsymbol{\theta}_0)}{n - s(\boldsymbol{\theta}_0)},$$

then

$$\lim_{\sigma \to 0} l(\boldsymbol{\theta}_0, \sigma, \mathbf{R}_0, \nu_0) = \infty.$$

(b) If

$$\nu_0 = \frac{ps(\boldsymbol{\theta}_0)}{n - s(\boldsymbol{\theta}_0)},$$

then

$$\lim_{\sigma \to 0} l(\boldsymbol{\theta}_0, \sigma, \mathbf{R}_0, \nu_0) \in (0, \infty).$$

(c) If

$$\nu_0 > \frac{ps(\boldsymbol{\theta}_0)}{n - s(\boldsymbol{\theta}_0)},$$

then

$$\lim_{\sigma \to 0} l(\boldsymbol{\theta}_0, \sigma, \mathbf{R}_0, \nu_0) = 0.$$

It is evident from this result that one can determine a value $\boldsymbol{\theta}_0$ such that $y_i = g_i(\boldsymbol{\theta}_0)$ holds for at least one observation and the likelihood function does not possess a global maximum. Indeed, for sufficiently small values of ν, one can make $l(\boldsymbol{\theta}_0, \sigma, \mathbf{R}_0, \nu_0)$ arbitrarily large by letting σ tend to zero. These pitfalls arise as a consequence of the (sometimes neglected) fact that the recorded data have zero probability under the assumed model. Fernandez and Steel (1999) proposed and illustrated a Bayesian analysis on the basis of set of observations that takes into account the precision with which the data were originally recorded.

11.4 General Linear Model

Rubin (1983) and Sutradhar and Ali (1986) considered the general linear model set up in the form

$$\mathbf{Y} = \boldsymbol{\beta}\mathbf{X} + \boldsymbol{\epsilon}, \tag{11.14}$$

where $\mathbf{X}$ is a $k \times n$ design matrix with rank k, $\boldsymbol{\beta}$ is a $p \times k$ matrix of regression parameters with unknown values, and $\boldsymbol{\epsilon}$ is a $p \times n$ random error matrix. It is assumed that the error variables ϵ_{ij} satisfy

$$E(\epsilon_{ij}) = 0, \quad \forall i, j = 1, \ldots, p,$$

$$E\left(\epsilon_{ij}^2\right) = \sigma^2 \Lambda_{ii}, \quad \forall i, j = 1, \ldots, p,$$

$$E(\epsilon_{ij}\epsilon_{lj}) = \sigma^2 \Lambda_{il}, \quad \forall i, j, l = 1, \ldots, p,$$

and

$$E(\epsilon_{ij}\epsilon_{lj'}) = 0, \quad \forall i, l, j \neq j',$$

where Λ_{ij} are unknown parameters. Furthermore, it is assumed that, for a given σ, the errors $\boldsymbol{\epsilon}_1, \ldots, \boldsymbol{\epsilon}_n$ are independently and normally distributed, with the distribution of $\boldsymbol{\epsilon}_j = (\epsilon_{1j}, \ldots, \epsilon_{pj})^T$ being $N(\mathbf{0}, \sigma^2 \boldsymbol{\Lambda})$ for $j = 1, \ldots, n$ while σ is assumed to be a random variable having an inverted gamma distribution with the pdf given by

$$f(\sigma) = \frac{2(\nu/2)^{(\nu+1)/2}}{\sqrt{\nu/2}\Gamma(\nu/2)} \sigma^{-(\nu+1)} \exp\left\{-\frac{\nu}{2\sigma^2}\right\},$$

where ν is an unknown parameter. Under these assumptions, one can show that the joint distribution of error variables is

$$f\left(\boldsymbol{\epsilon}_1^T, \ldots, \boldsymbol{\epsilon}_n^T\right) = \frac{(\nu-2)^{\nu/2}\Gamma((\nu+np)/2)}{\pi^{np/2}\Gamma(\nu/2)|\mathbf{R}|^{n/2}} \times \left[\nu - 2 + \sum_{j=1}^{n} \boldsymbol{\epsilon}_j^T \mathbf{R}^{-1} \boldsymbol{\epsilon}_j\right]^{-(\nu+np)/2},$$

where $\mathbf{R} = \nu\boldsymbol{\Lambda}/(\nu-2)$. It then follows that $E(\boldsymbol{\epsilon}_j) = \mathbf{0}$, $E(\boldsymbol{\epsilon}_j\boldsymbol{\epsilon}_j^T) = \mathbf{R}$ and $E(\boldsymbol{\epsilon}_j\boldsymbol{\epsilon}_s^T) = \mathbf{0}$ for $j \neq s$, $j, s = 1, \ldots, n$.

Sutradhar and Ali (1986) provided a least squares estimator for $\boldsymbol{\beta}$ as well as moment estimators for $\mathbf{R}$ and ν. The least squares estimator is

$$\widehat{\boldsymbol{\beta}} = \left(\mathbf{X}\mathbf{X}^T\right)^{-1} \mathbf{X}\mathbf{Y}^T$$

while the moment estimators are given by

$$\widehat{\mathbf{R}} = \frac{1}{n}\sum_{j=1}^{n}\left(\mathbf{y}_j - \widehat{\boldsymbol{\beta}}\mathbf{x}_j\right)\left(\mathbf{y}_j - \widehat{\boldsymbol{\beta}}\mathbf{x}_j\right)^T$$

and

$$\widehat{\nu} = 2\left(3\sum_{i=1}^{p}\widehat{\sigma}_{ii}^2 - \frac{2}{n}\sum_i\sum_j\widehat{\epsilon}_{ij}^4\right) \Big/ \left(3\sum_{i=1}^{p}\widehat{\sigma}_{ii}^2 - \frac{1}{n}\sum_i\sum_j\widehat{\epsilon}_{ij}^4\right),$$

where $\widehat{\epsilon}_{ij}$ are the so-called estimated residuals expressed as the difference

$$\widehat{\epsilon}_{ij} = y_{ij} - \sum_{r=1}^{k}\widehat{\theta}_{ir}x_{rj}.$$

All three estimators $\widehat{\boldsymbol{\beta}}$, $\widehat{\mathbf{R}}$, and $\widehat{\nu}$ are consistent as $n \to \infty$.

Let $\mathbf{Y} = (\mathbf{y}_1, \ldots, \mathbf{y}_n)^T$, where $\mathbf{y}_j = (y_{1j}, \ldots, y_{pj})^T$. Let $\mathbf{Y}^*$ denote the stacked random vector corresponding to $\mathbf{Y}$, so that $\mathbf{Y}^* = (y_{11}, \ldots, y_{p1}, y_{12}, \ldots, y_{p2}, \ldots, y_{1n}, \ldots, y_{pn})^T$. Let $\boldsymbol{\beta}^*$ and $\boldsymbol{\epsilon}^*$ be the corresponding stacked random vectors. Then the model (11.14) can be written in terms of Kroneckor products as

$$\mathbf{Y}^* = \left(\mathbf{I}_p \otimes \mathbf{X}^T\right)\boldsymbol{\beta}^* + \boldsymbol{\epsilon}^*. \tag{11.15}$$

Suppose one wishes to test the hypothesis that $H_0 : \boldsymbol{\theta}^* = \boldsymbol{\theta}_0^*$ versus $H_1 : \boldsymbol{\theta}^* \neq \boldsymbol{\theta}_0^*$. In the case where ν and $\mathbf{R}$ are known, Sutradhar and Ali (1986) showed that a suitable test statistic is

$$D = \frac{\nu}{\nu - 2}\left(\widehat{\boldsymbol{\theta}}^* - \boldsymbol{\theta}_0^*\right)^T\left\{\mathbf{R} \otimes \left(\mathbf{X}\mathbf{X}^T\right)^{-1}\right\}^{-1}\left(\widehat{\boldsymbol{\theta}}^* - \boldsymbol{\theta}_0^*\right).$$

Lower values of this statistic D will favor H_0 while higher values will direct the rejection of H_0. Actually, it can be shown that the pdf of D is

$$f(d) = \frac{\nu^{\nu/2} d^{kp/2-1}}{\Gamma(\nu/2)}\sum_{j=0}^{\infty}\frac{\Gamma\left((\nu + kp)/2 + 2j\right)}{\Gamma(kp/2 + j)\,\Gamma(j+1)} \times (\lambda d)^j(\lambda + \nu + d)^{-(\nu+kp)/2-2j},$$

where

$$\lambda = \frac{\nu}{\nu - 2}\left(\boldsymbol{\theta}^* - \boldsymbol{\theta}_0^*\right)^T\boldsymbol{\beta}^{-1}\left(\boldsymbol{\theta}^* - \boldsymbol{\theta}_0^*\right).$$

Note that, under $H_0 : \boldsymbol{\theta}^* = \boldsymbol{\theta}_0^*$, $D/(kp)$ has the usual F distribution with degrees of freedom kp and ν, whereas the analogous test for the classical

MANOVA model has the chi-squared distribution with degrees of freedom kp. Also note that the power of the test changes under H_1, whereas the similar statistic has the noncentral chi-squared distribution for the usual normal model. In the case where ν and $\mathbf{R}$ are not known, since $\widehat{\nu}$ and $\widehat{\mathbf{R}}$ are consistent estimators, an F test based on $D = \widehat{\nu}\widehat{\mathbf{U}}^T\widehat{\mathbf{U}}/(\widehat{\nu}-2)$, where $\widehat{\mathbf{U}} = \widehat{\boldsymbol{\beta}}^{-1/2}(\widehat{\boldsymbol{\theta}}^* - \boldsymbol{\theta}_0^*)$, may still be approximately valid.

Little (1988) extended the general linear model (11.14) to handle incomplete data. The methods for estimation employed are maximum likelihood (ML) for multivariate t and contaminated normal models. ML estimation was achieved by means of the EM algorithm and involves minor modifications to the EM algorithm for multivariate normal data.

11.5 Nonlinear Models

Nonlinear models involving multivariate t distributed errors have been studied relatively recently. Chib et al. (1991) considered nonlinear regression models with errors that follow the multivariate t distribution with degrees of freedom ν. For an $n \times 1$ vector of observations $\mathbf{y}$, the model is specified by

$$\mathbf{y} = h(\mathbf{X}, \boldsymbol{\beta}) + \boldsymbol{\epsilon}, \qquad (11.16)$$

where $\mathbf{X}$ is an $n \times r$ matrix of regressors, $\boldsymbol{\beta}$ is the regression coefficient vector, $h(\mathbf{X}, \boldsymbol{\beta})$ is a vector function of $(\mathbf{X}, \boldsymbol{\beta})$, and $\boldsymbol{\epsilon}$ is the error vector. It is assumed that $\boldsymbol{\epsilon} \mid \mathbf{X}, \boldsymbol{\beta}, \eta, \tau, \nu$ has an n-variate t distribution with zero mean vector, covariance matrix $(1/\tau)V(\mathbf{X}, \eta)$, and degrees of freedom ν. On can see that (11.16) reduces to (11.1) simply by setting $r = p$, $h(\mathbf{X}, \boldsymbol{\beta}) = \mathbf{X}\boldsymbol{\beta}$, and $V(\mathbf{X}, \eta) = \mathbf{I}_n$. The sampling density resulting from (11.16) is a t pdf, which can be represented as the following scale mixture of normal pdfs

$$f(\mathbf{y} \mid \mathbf{X}, w) = \int_0^\infty f(\mathbf{y} \mid \mathbf{X}, z, w)\, f(z \mid \mathbf{X}, w)\, dz,$$

where $f(\mathbf{y} \mid \mathbf{X}, z, w)$ is an n-variate normal pdf with mean vector $h(\mathbf{X}, \boldsymbol{\beta})$ and covariance matrix $1/(z\tau)V(\mathbf{X}, \eta)$ and $f(z \mid \mathbf{X}, w)$ is a gamma pdf with parameters $(\nu/2, \nu/2)$. Note that the proper pdf, $f(z \mid \mathbf{X}, w)$, is independent of $\mathbf{X}$ and does not involve parameters other than ν.

In the classical linear model due to Zellner (1976), the marginal posterior of the regression parameter, $\boldsymbol{\beta}$, is unaffected by the multivariate t assumption (see Section 11.2). This result was extended by Chib et al. (1998), Osiewalski (1991), and Osiewalski and Steel (1990) for elliptically distributed errors. For the nonlinear model above, Chib et al. (1991) provided the following sufficient conditions under which the posterior of ν, $p(\nu \mid \mathbf{y}, \mathbf{X})$, coincides with the prior, $p(\nu)$

- For proper priors $p(w)$, if ν is independent of $(\boldsymbol{\beta}, \eta, \tau z)$, then ν is independent of $(\mathbf{y}, \mathbf{X})$.
- For improper priors of the form $p(w) = p(\tau)p(\boldsymbol{\beta}, \eta)p(\nu)$, where $p(\tau) \propto 1/\tau$, $\tau > 0$ and $p(\nu)$ is proper and functionally independent of $(\tau, \boldsymbol{\beta}, \eta)$, if the posterior of ν exists, then $p(\nu \mid \mathbf{y}, \mathbf{X}) = p(\nu)$.

12
Applications

Due to limitations on the size of this book and since the aim is to collect and organize results on multivariate t distributions, in this short chapter we collect and present a small number of relatively recent applications of multivariate t distributions. The treatment is by no means exhaustive. Some other applications – in particular those related to Bayesian inference – are mentioned in the previous chapters (see Chapters 1, 3, 5, 10, and 11).

12.1 Projection Pursuit

Exploratory projection pursuit is a technique for finding "interesting" low p-dimensional projections of high P-dimensional multivariate data; see Jones and Sibson (1987) for an introduction. Typically, projection pursuit uses a projection index, a functional computed on a projected density (or data set), to measure the "interestingness" of the current projection and then uses a numerical optimizer to move the projection direction to a more interesting position. Loosely speaking, a robust projection index is one that prefers projections involving true clusters over those consisting of a cluster and an outlier. A good robust projection index should perform well even when specific assumptions required for "normal operation" fail to hold or hold only approximately. In a paper that was awarded the Royal Statistical Society Bronze Medal, Nason (2001) developed five new indices based on measuring divergence from the multivariate t distribution with the joint pdf

$$f(\mathbf{x}) = \frac{\Gamma\left((\nu+p)/2\right)}{\pi^{p/2}(\nu-2)^{p/2}\Gamma\left(\nu/2\right)}\left(1+\frac{\mathbf{x}^T\mathbf{x}}{\nu-2}\right)^{-(\nu+p)/2}$$

that are intended to be especially robust. The first three indices are all weighted versions of the L^2-divergences from f for $\nu \geq 3$. They are given by

$$I_{\nu,\alpha}^{\mathrm{TL2}} = \int \{g(\mathbf{x}) - f(\mathbf{x})\}^2 f^\alpha(\mathbf{x}) d\mathbf{x}$$

for $\alpha = 0, 1/2, 1$. Nason (2000) derived an explicit formula for the case $\alpha = 0$. The fourth index is the Student's t index defined by

$$I_\nu^{\mathrm{TI}} = -\int g^{1-2/(\nu+p)}(\mathbf{x}) d\mathbf{x}.$$

This index is minimized over all spherical densities by $f(\mathbf{x})$. Specifically, it satisfies the inequality

$$I_\nu^{\mathrm{TI}} \geq -\frac{(\nu+p-2)\Gamma\left((\nu+p)/2\right)}{\pi^{p/2}(\nu-2)^{1+p/2}\Gamma\left(\nu/2\right)}$$

for all spherical densities g with equality if and only if $g = f$ almost everywhere. The proof of this result uses the fact that the index can be represented as the sum of two F-divergences (Vajda, 1989). Through both numerical calculation and explicit analytical formulas, Nason (2001) found the the Student's t indices are generally more robust and that indices based on L^2-divergences are also the most robust in their class. A detailed analytical exploration of one of the indices ($I_{\nu,0}^{\mathrm{TL2}}$) showed that it acts robustly when outliers diverge from a main cluster but behaves like a standard projection index when two clusters diverge, that is, its behavior automatically changes depending on the degree of outlier contamination. The degree of sensitivity to outliers can be reduced by increasing the degrees of freedom ν of the $I_{\nu,0}^{\mathrm{TL2}}$-index to make it behave increasingly like Hall's index (Hall, 1989) as $\nu \to \infty$.

Using the transformation $x = \tan(\theta)$, Nason further developed the orthogonal expansion index given by

$$I_{3,1/2}^{\mathrm{TL2}} = \sqrt{\frac{2}{\pi}} \int_{-\pi/2}^{\pi/2} \left\{ g_\Theta(\theta) - \frac{2}{\pi}\cos^4\theta \right\}^2 d\theta,$$

where g_Θ is the pdf of the transformed projected data X. Using the Fourier series expansion of $g_\Theta(\theta)$ on $[-\pi/2, \pi/2]$,

$$g_\Theta(\theta) = \frac{a_0}{2} + \sum_{n=1}^{\infty} \left\{ a_n \cos(2n\theta) + a_n \sin(2n\theta) \right\},$$

where

$$a_n = \frac{2}{\pi}\int_{-\pi/2}^{\pi/2} g(\theta)\cos(2n\theta)d\theta$$

and

$$b_n = \frac{2}{\pi}\int_{-\pi/2}^{\pi/2} g(\theta)\sin(2n\theta)d\theta,$$

the index $I_{3,1/2}^{\text{TL2}}$ can be expanded as

$$I_{3,1/2}^{\text{TL2}} = \sqrt{\frac{\pi}{2}}\left\{\frac{1}{2}\left(a_0-\frac{3}{2\pi}\right)^2+\left(a_1-\frac{1}{\pi}\right)^2+\left(a_2-\frac{1}{4\pi}\right)^2 + \sum_{n=3}^{\infty}a_n^2+\sum_{n=3}^{\infty}b_n^2\right\}.$$

12.2 Portfolio Optimization

There are a number of places in finance where robust estimation has been used. For example, when a stock's returns are regressed on the market returns, the slope coefficient, called beta, is a measure of the relative riskiness of the stock in comparison to the market. Quite often, this regression will be performed using robust procedures. However, there appear to be fewer applications of robust estimation in the area of portfolio optimization. In the problem of finding a risk-minimizing portfolio subject to linear constraints, the classical approach assumes normality without exceptions. Lauprete et al. (2002) addressed the problem when the return data are generated by a multivariate distribution that is elliptically symmetric but not necessarily normal. They showed that when the returns have marginal heavy tails and multivariate tail-dependence, portfolios will also have heavy tails, and the classical procedures will be susceptible to outliers. They showed theoretically, and on simulated data, that robust alternatives have lower risks. In particular, they showed that when returns have a multivariate t distribution with degrees of freedom less than 6, the least absolute deviation (LAD) estimator has an asymptotically lower risk than the one based on the classical approach. The proposed methodology is applicable when heavy tails and tail-dependence in financial markets are documented especially at high sampling frequencies.

12.3 Discriminant and Cluster Analysis

In the past, there have been many attempts to modify existing methods of discriminant and cluster analyses to provide robust procedures. Some of these have been of a rather ad hoc nature. Recently the multivariate t distribution has been employed for robust estimation. Suppose, for simplicity, that one utilizes two samples in order to assign a new observation into one of two groups, and consider the joint distribution

$$\begin{aligned} & f\left(\mathbf{x}_1^*, \mathbf{x}_2^*\right) \\ = & \frac{\sqrt{\nu-2}\Gamma\left(\nu+np/2\right)}{\pi^{np/2}\left|\mathbf{R}\right|^{n/2}} \\ & \times\left[(\nu-2)+\sum_{i=1}^{2}\sum_{j=1}^{n_i}\left(\mathbf{x}_{ij}-\boldsymbol{\mu}_i\right)^T\mathbf{R}^{-1}\left(\mathbf{x}_{ij}-\boldsymbol{\mu}_i\right)\right]^{-(\nu+np)/2} \end{aligned} \tag{12.1}$$

of the two samples $\mathbf{X}_1^* = (\mathbf{X}_{11}, \ldots, \mathbf{X}_{1n_1})$ and $\mathbf{X}_2^* = (\mathbf{X}_{21}, \ldots, \mathbf{X}_{2n_2})$ of sizes n_1 and n_2, respectively. In (12.1), $n = n_1 + n_2$. The $(n_1 + n_2)p$-dimensional t distribution (12.1) was proposed by Sutradhar (1990). It is evident that the marginals are distributed according to

$$\begin{aligned} f\left(\mathbf{x}_{ij}\right) = & \frac{\sqrt{\nu-2}\Gamma\left(\nu+p/2\right)}{\pi^{p/2}\left|\mathbf{R}\right|^{n/2}} \\ & \times\left[(\nu-2)+\left(\mathbf{x}_{ij}-\boldsymbol{\mu}_i\right)^T\mathbf{R}^{-1}\left(\mathbf{x}_{ij}-\boldsymbol{\mu}_i\right)\right]^{-(\nu+p)/2} \end{aligned} \tag{12.2}$$

which is a slight reparameterization of the usual multivariate t pdf. Let π_1 and π_2 denote the two t-populations of the form (12.2) with parameters $(\boldsymbol{\mu}_1, \mathbf{R}, \nu)$ and $(\boldsymbol{\mu}_2, \mathbf{R}, \nu)$, respectively. Fisher's optimal discrimination criterion is robust against departure from normality (Sutradhar, 1990), and it assigns the new observation with measurement $\mathbf{X}$ to π_1 if

$$d(\mathbf{x}) = \left(\boldsymbol{\mu}_1-\boldsymbol{\mu}_2\right)^T\mathbf{R}^{-1}\mathbf{x}-\frac{1}{2}\left(\boldsymbol{\mu}_1-\boldsymbol{\mu}_2\right)^T\mathbf{R}^{-1}\left(\boldsymbol{\mu}_1+\boldsymbol{\mu}_2\right) \geq 0;$$

otherwise, it assigns the observation to π_2. But even though the classification is based on the robust criterion, the probability of misclassification depends on the degrees of freedom of the t distribution. If e_1 and e_2 are probabilities of misclassification of an individual observation from π_1 into π_2 and from π_2 into π_1, respectively, then

$$e_i = \frac{\sqrt{\nu-2}\Gamma\left(\nu+1/2\right)}{\sqrt{\pi}}\int_{-\infty}^{-\Delta/2}\left\{(\nu-2)+z^2\right\}^{-(\nu+1)/2}dz$$

for $i = 1, 2$, where $\Delta^2 = (\boldsymbol{\mu}_1 - \boldsymbol{\mu}_2)^T \mathbf{R}^{-1} (\boldsymbol{\mu}_1 - \boldsymbol{\mu}_2)$. Calculations of e_1 and e_2 for selected values of Δ and ν (Sutradhar, 1990) suggest that if a sample actually comes from a t-population (12.2) with degrees of freedom ν, then the evaluation of the classification error rates by normal-based probabilities would unnecessarily make an experimenter more suspicious. Sutradhar (1990) illustrated the use of the preceding discrimination approach by fitting the t distribution to some bivariate data on two species of flea beetles.

McLachlan and Peel (1998), McLachlan et al. (1999), and Peel and McLachlan (2000) used a mixture model of t distributions for a robust method of mixture estimation of clustering. They illustrated its usefulness by a cluster analysis of a simulated data set with added background noise and of an actual data set. For other recent methods for making cluster algorithms robust, see Smith et al. (1993), Davé and Krishnapuram (1995), Jolion et al. (1995), Frigui and Krishnapuram (1996), Kharin (1996), Rousseeuw et al. (1996), and Zhuang et al. (1996).

12.4 Multiple Decision Problems

The multivariate t distribution arises quite naturally in multiple decision problems. In fact, it is one of the earliest applications of this distribution in statistical inference. Suppose there are q dependent variates with means $\theta_1, \ldots, \theta_h, \ldots, \theta_q$, respectively, and that one has estimators $\hat{\theta}_h$ of θ_h, $h = 1, \ldots, q$ available, which are jointly distributed according to a q-variate normal distribution with mean θ_h, $h = 1, \ldots, q$, and covariance matrix $\sigma^2 \mathbf{R}$, where $\mathbf{R}$ is a $q \times q$ positive definite matrix and σ^2 is an unknown scale parameter. Let s^2 be an unbiased estimator of σ^2 such that s^2 is independent of the $\hat{\theta}_h$'s and $\nu s^2/\sigma^2$ has the chi-squared distribution with degrees of freedom ν. Consider $p \leq q$ linearly independent linear combinations of θ_h's,

$$m_i = \sum_{h=1}^{q} c_{ih} \theta_h = \mathbf{c}_i^T \boldsymbol{\theta},$$

for $i = 1, \ldots, p$, where $\mathbf{c}_i = (c_{i1}, \ldots, c_{ih}, \ldots, c_{iq})^T$ is a $q \times 1$ vector of known constants. The unbiased estimators of the m_i's are

$$\hat{m}_i = \sum_{h=1}^{q} c_{ih} \hat{\theta}_h = \mathbf{c}_i^T \hat{\boldsymbol{\theta}},$$

each of which is a normally distributed random variable with mean m_i and variance $\mathbf{c}_i^T \mathbf{R} \mathbf{c}_i$. Then

$$Y_i = \frac{\hat{m}_i - m_i}{s\sqrt{\mathbf{c}_i^T \mathbf{R} \mathbf{c}_i}}, \qquad i = 1, \ldots, p$$

is a Student's t-variate and $Y_1, \ldots, Y_p$ have the usual p-variate t distribution with degrees of freedom ν, zero means, and the correlation matrix $\{\delta_{iu}\}$ given by

$$\delta_{iu} = \frac{\mathbf{c}_i^T \mathbf{R} \mathbf{c}_u}{\sqrt{\mathbf{c}_i^T \mathbf{R} \mathbf{c}_i \mathbf{c}_u^T \mathbf{R} \mathbf{c}_u}}.$$

For multiple comparisons, one computes the one- and two-sided confidence interval estimates of m_i $(i = 1, \ldots, p)$ simultaneously with a joint confidence coefficient $1 - \alpha$, say. These estimates are given by (Dunnett, 1955)

$$\hat{m}_i \pm h_1 s \sqrt{\mathbf{c}_i^T \mathbf{R} \mathbf{c}_i}$$

and

$$\hat{m}_i \pm h_2 s \sqrt{\mathbf{c}_i^T \mathbf{R} \mathbf{c}_i},$$

respectively, where the constants h_1 and h_2 are determined so that the intervals in each case have a joint coverage probability of $1 - \alpha$. The constants h_1 and h_2 can be computed using the methods discussed in Chapter 8.

12.5 Other Applications

Bayesian prediction approaches using the multivariate t distribution have attracted wide-ranging applications in the last several decades, and many sources are available in periodic and monographic literature. Chien (2002) discusses applications in speech recognition and online environmental learning. In experiments of hands-free car speech recognition of connected Chinese digits, it was shown that the proposed approach is significantly better than conventional approaches. Blattberg and Gonedes (1974) were one of the first to discuss applications to security returns data. For other applications, we refer the reader to the numerous modern books on multivariate analysis and to the *Proceedings of the Valencia International Meetings*.

References

Abdel-Hameed, H. and Sampson, A. R. (1978). Positive dependence of the bivariate and trivariate absolute normal, t, χ^2, and F distributions, *Annals of Statistics* **6**, 1360–1368.

Abramowitz, M. and Stegun, I. A. (1965). *Handbook of Mathematical Functions* (Dover, New York).

Abusev, R. A. and Kolegova, N. V. (2001). On estimation of probabilities of linear inequalities for multivariate t distributions, *Journal of Mathematical Sciences* **103**, 542–546.

Aczel, J. (1966). *Lectures on Functional Equations and Their Applications* (Academic Press, New York).

Afonja, B. (1972). The moments of the maximum of correlated normal and t variates, *Journal of the Royal Statistical Society* B **34**, 251–262.

Ahmed, A. and Gokhale, D. (1989). Entropy expressions and their estimators for multivariate distributions, *IEEE Transactions on Information Theory* **35**, 688–692.

Ahner, C. and Passing, H. (1983). Berechnung der multivaiaten t-verteilung und simultane vergleiche gegen eine kontrolle bei ungleichen gruppenbesetzungen, *EDV in Medizin und Biologie* **14**, 113–120.

Amos, D. E. (1978). Evaluation of some cumulative distribution functions by numerical evaluation, *SIAM Review* **20**, 778–800.

Amos, D. E. and Bulgren, W. G. (1969). On the computation of a bivariate t distribution, *Mathematics and Computation* **23**, 319–333.

Anderson, D. N. and Arnold, B. C. (1991). Centered distributions with Cauchy conditionals, *Communications in Statistics—Theory and Methods* **20**, 2881–2889.

Anderson, T. W. (1984). *An Introduction to Multivariate Analysis*, second edition (John Wiley and Sons, New York).

Anderson, T. W. and Fang, K. T. (1987). Cochran's theorem for elliptically contoured distribution, *Sankhyā* A **49**, 305–315.

Ando, A. and Kaufman, G. W. (1965). Bayesian analysis of the independent multi-normal process – neither mean nor precision known, *Journal of the American Statistical Association* **60**, 347–358.

Arellano-Valle, R. and Bolfarine, H. (1995). On some characterization of the t-distribution, *Statistics and Probability Letters* **25**, 79–85.

Arellano-Valle, R., Bolfarine, H. and Iglesias, P. L. (1994). A predictivistic interpretation of the multivariate t distribution, *Test* **3**, 221–236.

Armitage, J. V. and Krishnaiah, R. R. (1965). Tables of percentage points of multivariate t distribution (abstract), *Annals of Mathematical Statistics* **36**, 726.

Arnold, B. C. and Beaver, R. J. (2000). Hidden truncation models, *Sankhyā* A **62**, 23–35.

Arnold, B. C. and Press, S. J. (1989). Compatible conditional distributions, *Journal of the American Statistical Association* **84**, 152–156.

Aroian, L., Taneja, V. and Cornwall, L. (1978). Mathematical forms of the distribution of the product of two two normal normal variates, *Communications in Statistics—Theory and Methods* **7**, 165–172.

Arsian, O., Constable, P. D. L. and Kent, J. T. (1995). Convergence behavior of the EM algorithm for the multivariate t distribution, *Communications in Statistics—Theory and Methods* **24**, 2981–3000.

Azzalini, A. and Capitanio, A. (1999). Statistical applications of the multivariate skew normal distribution, *Journal of the Royal Statistical Society* B **61**, 579–602.

Azzalini, A. and Capitanio, A. (2002). Distributions generated by perturbation of symmetry with emphasis on a multivariate skew t distribution. Submitted to *Journal of the Royal Statistical Society* B.

Azzalini, A. and Dalla Valle, A. (1996). The multivariate skew normal distribution, *Biometrika* **83**, 715-726.

Barlow, R. E., Bartholomew, D. J., Bremner, J. M. and Brunk, H. D. (1972). *Statistical Inference under Order Restrictions* (John Wiley and Sons, Chichester).

Bechhofer, R. E. and Dunnett, C. W. (1988). Tables of percentage points of multivariate t distributions, in *Selected Tables in Mathematical Statistics* **11**, ed. R. E. Odeh and J. M. Davenport (American Mathematical Society, Providence, Rhode Island).

Bechhofer, R. E., Dunnett, C. W. and Sobel, M. (1954). A two-sample multiple-decision procedure for ranking means of normal populations with a common unknown variance, *Biometrika* **41**, 170–176.

Bennett, B. M. (1961). On a certain multivariate normal distribution, *Proceedings of the Cambridge Philosophical Society* **57**, 434–436.

Bickel, P. J. and Lehmann, E. L. (1975). Descriptive statistics for nonparametric models I. Introduction, *Annals of Statistics* **3**, 1038–1044.

Birnbaum, Z. (1948). On random variables with comparable peakedness, *Annals of Mathematical Statistics* **19**, 76–81.

Blattberg, R. C. and Gonedes, N. J. (1974). A comparison of the stable and Student distributions as statistical models for stock prices, *Journal of Business* **47**, 224–280.

Bohrer, R. (1973). A multivariate t probability integral, *Biometrika* **60**, 647–654.

Bohrer, R. and Francis, G. K. (1972). Sharp one-sided confidence bounds over positive regions, *Annals of Mathematical Statistics* **43**, 1541–1548.

Bohrer, R., Schervish, M. and Sheft, J. (1982). Algorithm AS 184: Non-central studentized maximum and related multiple-t probabilities, *Applied Statistics* **31**, 309–317.

Bowden, D. C. and Graybill, F. A. (1966). Confidence bands of uniform and proportional width for linear models, *Journal of the American Statistical Association* **61**, 182–198.

Branco, M. D. and Dey, D. K. (2001). A general class of multivariate skew-elliptical distributions, *Journal of Multivariate Analysis* **79**, 99–113.

Bretz, F., Genz, A. and Hothorn, L. A. (2001). On the numerical availability of multiple comparison procedures, *Biometrical Journal* **43**, 645–656.

Bulgren, W. G. and Amos, D. E. (1968). A note on representation of the doubly non-central t distribution, *Journal of the American Statistical Association* **63**, 1013–1019.

Bulgren, W. G., Dykstra, R. L. and Hewett, J. E. (1974). A bivariate t distribution with applications, *Journal of the American Statistical Association* **69**, 525–532.

Cadwell, J. H. (1951). The bivariate normal integral, *Biometrika* **38**, 475–481.

Cain, M. (1996). Forecasting with the maximum of correlated components having bivariate t-distributed errors, *IMA Journal of Mathematics Applied in Business and Industry* **7**, 233–237.

Capitanio, A., Azzalini, A. and Stanghellini, E. (2002). Graphical models for skew-normal variates, *Scandinavian Journal of Statistics*, to appear.

Carlson, B. C. (1977). *Special Functions and Applied Mathematics* (Academic Press, New York).

Castillo, E. and Sarabia, J. M. (1990). Bivariate distributions with second kind Beta conditionals, *Communications in Statistics—Theory and Methods* **19**, 3433–3445.

Chapman, D. G. (1950). Some two-sample tests, *Annals of Mathematical Statistics* **21**, 601–606.

Chen, H. J. (1979). Percentage points of multivariate t distribution with zero correlations and their application, *Biometrical Journal* **21**, 347–360.

Chib, S., Osiewalski, J. and Steel, M. F. J. (1991). Posterior inference on the degrees of freedom parameter in multivariate-t regression models, *Economics Letters* **37**, 391–397.

Chib, S., Tiwari, R. C. and Jammalamadaka, S. R. (1988). Bayes prediction in regressions with elliptical errors, *Journal of Econometrics* **38**, 349–360.

Chien, J.-T. (2002). A Bayesian prediction approach to robust speech recognition and online environmental testing, *Speech Communication* **37**, 321–334.

Chow, Y. S. and Teicher, H. (1978). *Probability Theory* (Springer-Verlag, Berlin).

Constantine, A. G. (1963). Some noncentral distribution problems in multivariate analysis, *Annals of Mathematical Statistics* **34**, 1270–1285.

Corliss, G. F. and Rall, L. B. (1987). Adaptive, self-validating numerical quadrature, *SIAM Journal on Scientific and Statistical Computing* **8**, 831–847.

Cornish, E. A. (1954). The multivariate t distribution associated with a set of normal sample deviates, *Australian Journal of Physics* **7**, 531–542.

Cornish, E. A. (1955). The sampling distribution of statistics derived from the multivariate t distribution, *Australian Journal of Physics* **8**, 193–199.

Cornish, E. A. (1962). The multivariate t distribution associated with the general multivariate normal distribution, CSIRO Technical Paper No. 13, CSIRO Division in Mathematics and Statistics, Adelaide.

Cornish, E. A. and Fisher, R. A. (1950). Moments and cumulants in the specification of distributions, in *Contributions to Mathematical Statistics*

(John Wiley and Sons, New York).
Craig, C. (1936). On the frequency function of XY, *Annals of Mathematical Statistics* **7**, 1–15.
Cramér, H. (1951). *Mathematical Methods of Statistics* (Princeton University Press).
DasGupta, A., Ghosh, J. K. and Zen, M. M. (1995). A new general method for constructing confidence sets in arbitrary dimensions: with applications, *Annals of Statistics* **23**, 1408–1432.
Davé, R. N. and Krishnapuram, R. (1995). Robust clustering methods: A unified view, *IEEE Transactions on Fuzzy Systems* **5**, 270–293.
David, H. A. (1982). Concomitants of order statistics: theory and applications, in *Some Recent Advances in Statistics*, ed. Tiago de Oliveira, pp. 89–100 (Academic Press, New York).
Dawid, A. P. (1981). Some matrix-variate distribution theory: Notational considerations and a Bayesian application, *Biometrika* **68**, 265–274.
Deak, I. (1990). *Random Number Generators and Simulation* (Akademiai Kiado, Budapest).
Dempster, A. P., Laird, N. M. and Rubin, D. B. (1977). Maximum likelihood from incomplete data via the EM algorithm (with discussion), *Journal of the Royal Statistical Society* B **39**, 1–38.
Dempster, A. P., Laird, N. M. and Rubin, D. B. (1980). Iteratively weighted least squares for linear regression where errors are normal independent distributed, in *Multivariate Analysis* **5**, ed. P. R. Krishnaiah, pp. 35–37 (North-Holland, New York).
Dey, D. K. (1988). Simultaneous estimation of eigenvalues, *Annals of the Institute of Statistical Mathematics* **40**, 137–147.
Diaconis, P. and Ylvisaker, D. (1979). Conjugate priors for exponential families, *Annals of Statistics* **7**, 269–281.
Diaconis, P. and Ylvisaker, D. (1985). Quantifying prior opinion (with discussion), in *Bayesian Statistics* **2**, ed. J. M. Bernardo, M. H. DeGroot, D. V. Lindley and A. F. M. Smith, pp. 133–156 (North Holland, Amsterdam).
Dickey, J. M. (1965). Integrals of products of multivariate-t densities (abstract), *Annals of Mathematical Statistics* **36**, 1611.
Dickey, J. M. (1966a). Matric-variate generalizations of the multivariate t distribution and the inverted multivariate t distribution (abstract), *Annals of Mathematical Statistics* **37**, 1423.
Dickey, J. M. (1966b). On a multivariate generalization of the Behrens-Fisher distributions, *Annals of Mathematical Statistics* **37**, 763.
Dickey, J. M. (1967a). Expansions of t densities and related complete integrals, *Annals of Mathematical Statistics* **38**, 503–510.
Dickey, J. M. (1967b). Matric-variate generalizations of the multivariate t distribution and the inverted multivariate t distribution, *Annals of Mathematical Statistics* **38**, 511–518.
Dickey, J. M. (1968). Three multidimensional-integral identities with Bayesian applications, *Annals of Mathematical Statistics* **39**, 1615–1627.
Dickey, J. M., Dawid, A. and Kadane, J. B. (1986). Subjective-probability assessment methods for multivariate-t and matrix-t models, in *Bayesian Inference and Decision Techniques*, pp. 177–195 (North-Holland, Amsterdam).
Dreier, I. and Kotz, S. (2002). A note on the characteristic function of the

t-distribution, *Statistics and Probability Letters* **57**, 221–224.
Dunn, O. J. (1958). Estimation of the means of dependent variables, *Annals of Mathematical Statistics* **29**, 1095–1111.
Dunn, O. J. (1961). Multiple comparison among means, *Journal of the American Statistical Association* **56**, 52–64.
Dunn, O. J. (1965). A property of the multivariate t distribution, *Annals of Mathematical Statistics* **36**, 712–714.
Dunn, O. J. and Massey, F. J. (1965). Estimation of multiple contrasts using t-distributions, *Journal of the American Statistical Association* **60**, 573–583.
Dunnett, C. W. (1955). A multiple comparison procedure for comparing several treatments with a control, *Journal of the American Statistical Association* **50**, 1096–1121.
Dunnett, C. W. (1964). New tables for multiple comparisons with a control, *Biometrics* **20**, 482–491.
Dunnett, C. W. (1985). Multiple comparisons between several treatments and a specified treatment, in *Linear Statistical Inference*, Lecture Notes in Statistics No. **35**, ed. T. Caliński and W. Klonecki, pp. 39–46 (Springer-Verlag, New York).
Dunnett, C. W. (1989). Algorithm AS 251: Multivariate normal probability integrals with product correlation structure, *Applied Statistics* **38**, 564–579.
Dunnett, C. W. and Sobel, M. (1954). A bivariate generalization of Student's t-distribution with tables for certain special cases, *Biometrika* **41**, 153–169.
Dunnett, C. W. and Sobel, M. (1955). Approximations to the probability integral and certain percentage points of a multivariate analogue of Student's t-distribution, *Biometrika* **42**, 258–260.
Dunnett, C. W. and Tamhane, A. C. (1990). A step-up multiple test procedure, Technical Report 90-1, Department of Statistics, Northwestern University.
Dunnett, C. W. and Tamhane, A. C. (1991). Step-down multiple tests for comparing treatments with a control in unbalanced one-way layouts, *Statistics in Medicine* **10**, 939–947.
Dunnett, C. W. and Tamhane, A. C. (1992). A step-up multiple test procedure, *Journal of the American Statistical Association* **87**, 162–170.
Dunnett, C. W. and Tamhane, A. C. (1995). Step-up multiple testing of parameters with unequally correlated estimates, *Biometrics* **51**, 217–227.
Dutt, J. E. (1973). A representation of multivariate normal probability integrals by integral transforms, *Biometrika* **60**, 637–645.
Dutt, J. E. (1975). On computing the probability integral of a general multivariate t, *Biometrika* **62**, 201–205.
Dutt, J. E., Mattes, K. D., Soms, A. P. and Tao, L. C. (1976). An approximation to the maximum modulus of the trivariate T with a comparison to exact values, *Biometrics* **32**, 465–469.
Dutt, J. E., Mattes, K. D. and Tao, L. C. (1975). Tables of the trivariate t for comparing three treatments to a control with unequal sample sizes, G. D. Searle and Company, Math. and Statist. Services, TR-3.
Eaton, M. L. and Efron, B. (1970). Hotelling's T^2 test under symmetry conditions, *Journal of the American Statistical Association* **65**, 702–711.
Edwards, D. E. and Berry, J. J. (1987). The efficiency of simulation based

multiple comparisons, *Biometrics* **43**, 913–928.
Erdélyi, A., Magnus, W., Oberhettinger, F. and Tricomi, F. G. (1953). *Higher Transcendental Functions*, volumes 1 and 2 (McGraw-Hill, New York).
Esary, J. D., Proschan, F. and Walkup, D. W. (1967). Association of random variables with applications, *Annals of Mathematical Statistics* **38**, 1466–1474.
Fang, H.-B., Fang, K.-T. and Kotz, S. (2002). The meta-elliptical distributions with given marginals, *Journal of Multivariate Analysis* **82**, 1–16.
Fang, K.-T. and Anderson, T. W. (1990). *Statistical Inference in Elliptically Contoured and Related Distributions* (Alberton Press, New York).
Fang, K.-T., Kotz, S. and Ng, K. W. (1990). *Symmetric Multivariate and Related Distributions* (Chapman and Hall, London).
Fernandez, C. and Steel, M. F. J. (1999). Multivariate Student-t regression models: pitfalls and inference, *Biometrika* **86**, 153–167.
Fisher, R. A. (1925). Expansion of Student's integral in power of n^{-1}, *Metron* **5**, 109.
Fisher, R. A. (1935). The fiducial argument in statistical inference, *Annals of Eugenics* **6**, 391–398.
Fisher, R. A. (1941). The asymptotic approach to Behren's integral with further tables for the d-test of significance, *Ann. Eugen., Lond.* **11**, 141.
Fisher, R. A. and Healy, M. J. R. (1956). New tables of Behrens' test of significance, *Journal of the Royal Statistical Society* B **18**, 212–216.
Fisher, R. A. and Yates, F. (1943). *Statistical Tables for Biological, Agricultural and Medical Research*, second edition (Oliver and Boyd, London).
Fleishman, A. I. (1978). A method for simulating nonnormal distributions, *Psychometrika* **43**, 521–532.
Fraser, D. A. S. and Haq, M. S. (1969). Structural probability and prediction for the multivariate model, *Journal of the Royal Statistical Society* **31**, 317–331.
Freeman, H. and Kuzmack, A. (1972). Tables of multivariate t in six or more dimensions, *Biometrika* **59**, 217–219.
Freeman, H., Kuzmack, A. and Maurice, R. (1967). Multivariate t and the ranking problem, *Biometrika* **54**, 305–308.
Frigui, H. and Krishnapuram, R. (1996). A robust algorithm for automatic extraction of an unknown number of clusters from noisy data, *Pattern Recognition Letters* **17**, 1223–1232.
Fry, R. L. (ed). (2002). *Bayesian Inference and Maximum Entropy Methods in Science and Engineering: Proceedings of the 21st International Workshop on Bayesian Inference and Maximum Entropy Methods in Science and Engineering* (American Institute of Physics, New York).
Fujikoshi, Y. (1987). Error bounds for asymptotic expansions of scale mixtures of distributions, *Hiroshima Mathematics Journal* **17**, 309–324.
Fujikoshi, Y. (1988). Non-uniform error bounds for asymptotic expansions of scale mixtures of distributions, *Journal of Multivariate Analysis* **27**, 194–205.
Fujikoshi, Y. (1989). Error bounds for asymptotic expansions of the maximums of the multivariate t- and F-variables with common denominator, *Hiroshima Mathematics Journal* **19**, 319–327.

Fujikoshi, Y. (1993). Error bounds for asymptotic approximations of some distribution functions, in *Multivariate Analysis: Future Directions*, ed. C. R. Rao, pp. 181–208 (North-Holland, Amsterdam).

Fujikoshi, Y. (1997). An asymptotic expansion for the distribution of Hotelling's T^2-statistic under nonnormality, *Journal of Multivariate Analysis* **61**, 187–193.

Fujikoshi, Y. and Shimizu, R. (1989). Error bounds for asymptotic expansions of scale mixtures of univariate and multivariate distributions, *Journal of Multivariate Analysis* **30**, 279–291.

Fujikoshi, Y. and Shimizu, R. (1990). Asymptotic expansions of some distributions and their error bounds–the distributions of sums of independent random variables and scale mixtures, *Sugaku Expositions* **3**, 75–96.

Geisser, S. (1965). Bayesian estimation in multivariate analysis, *Annals of Mathematical Statistics* **36**, 150–159.

Geisser, S. and Cornfield, J. (1963). Posterior distributions for multivariate normal parameters, *Journal of the Royal Statistical Society* B **25**, 368–376.

Genz, A. (1992). Numerical computation of the multivariate normal probabilities, *Journal of Computational and Graphical Statistics* **1**, 141–150.

Genz, A. and Bretz, F. (1999). Numerical computation of multivariate t probabilities with application to power calculation of multiple contrasts, *Journal of Statistical Computation and Simulation* **63**, 361–378.

Genz, A. and Bretz, F. (2001). Methods for the computation of multivariate t-probabilities, *Journal of Computational and Graphical Statistics.*

Ghosh, B. K. (1973). Some monotonicity theorems for χ^2, F and t distributions with applications, *Journal of the Royal Statistical Society* B **35**, 480–492.

Ghosh, B. K. (1975). On the distribution of the difference of two t variables, *Journal of the American Statistical Association* **70**, 463–467.

Gill, M. L., Tiku, M. L. and Vaughan, D. C. (1990). Inference problems in life testing under multivariate normality, *Journal of Applied Statistics* **17**, 133–147.

Glaz, J. and Johnson, B. McK. (1984). Probability inequalities for multivariate distributions with dependence structures, *Journal of the American Statistical Association* **79**, 435–440.

Goldberg, H. and Levine, H. (1946). Approximate formulas for the percentage points and normalization of t and χ^2, *Annals of Mathematical Statistics* **17**, 216.

Goodman, M. R. (1963). Statistical analysis based on a certain multivariate complex Gaussian distribution (an introduction), *Annals of Mathematical Statistics* **34**.

Graybill, F. A. and Bowden, D. C. (1967). Linear segment confidence bands for simple linear models, *Journal of the American Statistical Association* **62**, 403–408.

Grosswald, E. (1976). The Student t-distribution of any degree of freedom is infinitely divisible, *Zeitschrift für Wahrscheinlichkeitstheorie und Verwandte Gebiete* **36**, 103–109.

Guerrero-Cusumano, J.-L. (1996a). A measure of total variability for the multivariate t distribution with applications to finance, *Information*

Sciences **92**, 47–63.

Guerrero-Cusumano, J.-L. (1996b). An asymptotic test of independence for multivariate t and Cauchy random variables with applications, *Information Sciences* **92**, 33–45.

Guerrero-Cusumano, J.-L. (1998). Measures of dependence for the multivariate t distribution with applications to the stock market, *Communications in Statistics—Theory and Methods* **27**, 2985–3006.

Gupta, A. K. (2000). Multivariate skew t distribution, Technical Report No. 00-04, Department of Mathematics and Statistics, Bowling Green State University, Bowling Green, Ohio.

Gupta, A. K. and Kollo. T. (2000). Multivariate skew normal distribution: some properties and density expansions, Technical Report No. 00-03, Department of Mathematics and Statistics, Bowling Green State University, Bowling Green, Ohio.

Gupta, R. P. (1964). Some extensions of the Wishart and multivariate t distributions in the complex case, *Journal of the Indian Statistical Association* **2**, 131–136.

Gupta, S. S. (1963). Probability integrals of multivariate normal and multivariate t, *Annals of Mathematical Statistics* **34**, 792–828.

Gupta, S. S., Nagel, K. and Panchapakesan, S. (1973). On the order statistics from equally correlated normal random variables, *Biometrika* **60**, 403–413.

Gupta, S. S., Panchapakesan, S. and Sohn, J. K. (1985). On the distribution of the studentized maximum of equally correlated normal random variables, *Communications in Statistics—Simulation and Computation* **14**, 103–135.

Gupta, S. S. and Sobel, M. (1957). On a statistic which arises in selection and ranking problems, *Annals of Mathematical Statistics* **28**, 957–967.

Hahn, G. J. and Hendrickson, R. W. (1971). A table of percentage points of the distribution of the largest absolute value of k Student t variates and its application, *Biometrika* **58**, 323–332.

Hahn, M. G. and Klass, M. J. (1980a). Matrix normalization of sums of random vectors in the domain of attraction of the multivariate normal, *Annals of Probability* **8**, 262–280.

Hahn, M. G. and Klass, M. J. (1980b). The generalized domain of attraction of spherically symmetric stable laws in $\Re^d$, in *Proceedings of the Conference on Probability in Vector Spaces II*, Lecture Notes in Mathematics **828**, pp. 52–81 (Springer-Verlag, New York).

Halgreen, C. (1979). Self-decomposability of the generalized inverse Gaussian and hyperbolic distributions, *Zeitschrift für Wahrscheinlichkeitstheorie und Verwandte Gebiete* **47**, 13–17.

Hall, P. (1989). On polynomial-based projection indices for exploratory projection pursuit, *Annals of Statistics* **17**, 589–605.

Halperin, M. (1967). An inequality on a bivariate t distribution, *Journal of the American Statistical Association* **62**, 603–606.

Halperin, M., Greenhouse, S. W., Cornfield, J. and Zalokar, J. (1955). Tables of percentage points for the studentized maximum absolute deviate in normal samples, *Journal of the American Statistical Association* **50**, 185–195.

Hammersley, J. M. and Handscomb, D. C. (1964). *Monte Carlo Methods* (Methuen & Co. Ltd, London).

Haq, M. S. and Khan, S. (1990). Prediction distribution for a linear regression model with multivariate Student-t error distribution, *Communications in Statistics—Theory and Methods* **19**, 4705–4712.

Harter, H. L. (1951). On the distribution of Wald's classification statistic, *Annals of Mathematical Statistics* **22**, 58–67.

Hayakawa, T. (1989). On the distributions of the functions of the F-matrix under an elliptical population, *Journal of Statistical Planning and Inference* **21**, 41–52.

Hochberg, Y. and Tambane, A. C. (1987). *Multiple Comparison Procedures* (John Wiley and Sons, New York).

Hsu, H. (1990). Noncentral distributions of quadratic forms for elliptically contoured distributions, in *Statistical Inference in Elliptically Contoured and Related Distributions*, pp. 97–102 (Allerton, New York).

Hsu, J. C. (1992). The factor analytic approach to simultaneous inference in the general linear model, *Journal of Computational and Graphical Statistics* **1**, 151–168.

Hsu, J. C. and Nelson, B. L. (1998). Multiple comparisons in the general linear model, *Journal of Computational and Graphical Statistics* **7**, 23–41.

Hutchinson, T. P. and Lai, C. D. (1990). *Continuous Bivariate Distributions, Emphasising Applications* (Rumsby, Adelaide).

Ifram, A. F. (1970). On the characteristic function of F and t distributions, *Sankhyā* A **32**, 350–352.

International Mathematical and Statistical Libraries (1987). *MATH/Library, Fortran Subroutines for Mathematical Applications* (International Mathematical and Statistical Libraries, Houston).

Iwashita, T. (1997). Asymptotic null and nonnull distribution of Hotelling's T^2-statistic under the elliptical distribution, *Journal of Statistical Planning and Inference* **61**, 85–104.

Iyengar, S. (1988). Evaluation of normal probabilities of symmetric regions, *SIAM Journal on Scientific and Statistical Computing* **9**, 418–423.

James, A. T. (1964). Distribution of matrix variates and latent roots derived from normal samples, *Annals of Mathematical Statistics* **35**, 475–501.

James, W. and Stein, C. (1961). Estimation with quadratic loss, in *Proceedings of the Fourth Berkeley Symposium on Mathematical Statistics and Probability* **1**, pp. 361–379.

Javier, W. R. and Gupta, A. K. (1985). On matric variate-t distribution, *Communications in Statistics—Theory and Methods* **14**, 1413–1425.

Javier, W. R. and Srivastava, T. N. (1988). On the multivariate t distribution, *Pakistan Journal of Statistics* **4**, 101–109.

Jaynes, E. T. (1957). Information theory and statistical mechanics, *Physics Review* **106**, 620–630.

Jensen, D. R. (1994). Closure of multivariate t and related distributions, *Statistics and Probability Letters* **20**, 307–312.

Joarder, A. H. (1995). Estimation of the trace of the scale matrix of a multivariate t-model, in *Proceedings of the Econometrics Conference*, pp. 467–474 (Monash University, Australia).

Joarder, A. H. (1998). Some useful Wishart expectations based on the multivariate t-model, *Statistical Papers* **39**, 223–229.

Joarder, A. H. and Ahmed, S. E. (1996). Estimation of the characteristic roots of the scale matrix, *Metrika* **44**, 259–267.

Joarder, A. H. and Ahmed, S. E. (1998). Estimation of the scale matrix of a class of elliptical distributions, *Metrika* **48**, 149–160.

Joarder, A. H. and Ali, M. M. (1992). On some generalized Wishart expectations, *Communications in Statistics—Theory and Methods* **21**, 283–294.

Joarder, A. H. and Ali, M. M. (1996). On the characteristic function of the multivariate t distribution, *Pakistan Journal of Statistics* **12**, 55–62.

Joarder, A. H. and Ali, M. M. (1997). Estimation of the scale matrix of a multivariate t-model under entropy loss, *Metrika* **46**, 21–32.

Joarder, A. H. and Singh, S. (1997). Estimation of the trace of the scale matrix of a multivariate t-model using regression type estimator, *Statistics* **29**, 161–168.

Joe, H. (1989). Relative entropy measures of multivariate dependence, *Journal of the American Statistical Association* **84**, 157–164.

Joe, S. (1990). Randomization of lattice rules for numerical multiple integration, *Journal of Computational and Applied Mathematics* **31**, 299–304.

Jogdeo, K. (1977). Association of probability inequalities, *Annals of Statistics* **5**, 495–504.

John, S. (1961). On the evaluation of the probability integral of the multivariate t distribution, *Biometrika* **48**, 409–417.

John, S. (1964). Methods for the evaluation of probabilities of polygonal and angular regions when the distribution is bivariate t, *Sankhyā* A **26**, 47–54.

John, S. (1966). On the evaluation of probabilities of convex polyhedra under multivariate normal and t distributions, *Journal of the Royal Statistical Society* B **28**, 366–369.

Johnson, M. (1987). *Multivariate Statistical Simulation* (John Wiley and Sons, New York).

Johnson, N. L. and Kotz, S. (1972). *Distributions in Statistics: Continuous Multivariate Distributions* (John Wiley and Sons, New York).

Johnson, N. L., Kotz, S. and Balakrishnan, N. (1995). *Continuous Univariate Distributions*, volume 2, second edition (John Wiley and Sons, New York).

Johnson, R. A. and Weerahandi, S. (1988). A Bayesian solution to the multivariate Behrens-Fisher problem, *Journal of the American Statistical Association* **83**, 145–149.

Jolion, J.-M., Meer, P. and Bataouche, S. (1995). Robust clustering with applications in computer vision, *IEEE Transactions on Pattern Analysis and Machine Intelligence* **13**, 791–802.

Jones, M. C. (2001a). A skew t distribution, in *Probability and Statistical Models with Applications*, ed. C. A. Charalambides, M. V. Koutras and N. Balakrishnan, pp. 269–278 (Chapman and Hall, London).

Jones, M. C. (2001b). Multivariate t and beta distributions associated with multivariate F distribution, *Metrika* **54**, 215–231.

Jones, M. C. (2002a). A bivariate distribution with support above the diagonal and skew t marginals. Submitted.

Jones, M. C. (2002b). A dependent bivariate t distribution with marginal on different degrees of freedom, *Statistics and Probability Letters* **56**, 163–170.

Jones, M. C. (2002c). Marginal replacement in multivariate densities, with application to skewing spherically symmetric distributions, *Journal of Multivariate Analysis* **81**, 85–99.

Jones, M. C. and Faddy, M. J. (2002). A skew extension of the t distribution with applications. Submitted.

Jones, M. C. and Sibson, R. (1987). What is projection pursuit (with discussion)? *Journal of the Royal Statistical Society* A **150**, 1–36.

Kabe, D. G. and Gupta, A. K. (1990). Hotelling's T^2-distribution for a mixture of two normal populations, *South African Statistical Journal* **24**, 87–92.

Kano, Y. (1994). Consistency property of elliptical probability density functions, *Journal of Multivariate Analysis* **51**, 139–147.

Kano, Y. (1995). An asymptotic expansion of the distribution of Hotelling's T^2-statistic under general distributions, *American Journal of Mathematical and Management Sciences* **15**, 317–341.

Kappenman, R. F. (1971). A note on the multivariate t ratio distribution, *Annals of Mathematical Statistics* **42**, 349–351.

Kass, R. E. and Steffey, D. (1989). Approximate Bayesian in conditionally independent hierarchical models, *Journal of the American Statistical Association* **84**, 717–726.

Kelejian, H. H. and Prucha, I. R. (1985). Independent or uncorrelated disturbances in linear regression: An illustration of the difference, *Economic Letters* **19**, 35–38.

Kelker, D. (1970). Distribution theory of spherical distributions and location scale parameters, *Sankhyā* A **32**, 419–430.

Kelker, D. (1971). Infinite divisibility and variance mixtures of the normal distribution, *Annals of Mathematical Statistics* **42**, 802–808.

Kendall, M. G. and Stuart, A. (1958). *The Advanced Theory of Statistics* (Hafner, New York).

Kent, J. T., Tyler, D. E. and Vardi, Y. (1994). A curious likelihood identity for the multivariate t-distribution, *Communications in Statistics—Simulation and Computation* **23**, 441–453.

Kharin, Y. (1996). *Robustness in Statistical Pattern Recognition* (Kluwer, Dordrecht).

Khatri, C. G. (1967). On certain inequalities for normal distributions and their applications to simultaneous confidence bands, *Annals of Mathematical Statistics* **38**, 1853–1867.

Kiefer, J. and Schwarz, R. (1965). Admissible Bayes character of T^2-, R^2-, and other fully invariant tests for classical multivariate normal problems, *Annals of Mathematical Statistics* **36**, 747–770.

Kopal, Z. (1955). *Numerical Analysis* (Chapman and Hall, London).

Kottas, A., Adamidis, K. and Loukas, S. (1999). Bivariate distributions with Pearson type VII conditionals, *Annals of the Institute of Statistical Mathematics* **51**, 331–344.

Kotz, S., Balakrishnan, N. and Johnson, N. L. (2000). *Continuous Multivariate Distributions, Volume 1: Models and Applications*, second edition (John Wiley and Sons, New York).

Kotz, S., Lumelski, Y. and Pensky, M. (2003). *Strength Stress Models with Applications* (World Scientific Press, Singapore).

Kozumi, H. (1994). Testing equality of the means in two independent multivariate t distributions, *Communications in Statistics—Theory and Methods* **23**, 215–227.

Krishnaiah, P. R. and Armitage, J. V. (1966). Tables for multivariate t distribution, *Sankhyā* B **28**, 31–56.

Krishnan, M. (1959). Studies in statistical inference, Ph.D. Thesis, Madras University, India.

Krishnan, M. (1967a). The moments of a doubly noncentral t distribution, *Journal of the American Statistical Association* **62**, 278–287.

Krishnan, M. (1967b). The noncentral bivariate chi distribution, *SIAM Review* **9**, 708–714.

Krishnan, M. (1968). Series representations of the doubly noncentral t distribution, *Journal of the American Statistical Association* **63**, 1004–1012.

Krishnan, M. (1970). The bivariate doubly noncentral t distribution (abstract), *Annals of Mathematical Statistics* **41**, 1135.

Krishnan, M. (1972). Series representations of a bivariate singly noncentral t distribution, *Journal of the American Statistical Association* **67**, 228–231.

Kshirsagar, A. M. (1961). Some extensions of the multivariate generalization t distribution and the multivariate generalization of the distribution of the regression coefficient, in *Proceedings of the Cambridge Philosophical Society* **57**, pp. 80–85.

Kudô, A. (1963). A multivariate analogue of the one-sided test, *Biometrika* **50**, 403–418.

Kullback, S. (1968). *Information Theory and Statistics* (John Wiley and Sons, New York).

Kunte, S. and Rattihalli, R. N. (1984). Rectangular regions of maximum probability content, *Annals of Statistics* **12**, 1106–1108.

Kurths, J., Voss, A. and Saparin, P., Witt, A., Kleiner, H. J. and Wessel, N. (1995). Quantitative analysis of heart rate variability, *Chaos* **1**, 88–94.

Kwong, K.-S. (2001a). A modified Dunnett and Tamhane step-up approach for establishing superiority/equivalence of a new treatment compared with k standard treatments, *Journal of Statistical Planning and Inference* **97**, 359–366.

Kwong, K.-S. (2001b). An algorithm for construction of multiple hypothesis testing, *Computational Statistics* **16**, 165–171.

Kwong, K.-S. and Iglewicz, B. (1996). On singular multivariate normal distribution and its applications, *Computational Statistics and Data Analysis* **22**, 271–285.

Kwong, K.-S. and Liu, W. (2000). Calculation of critical values for Dunnett and Tamhane's step-up multiple test procedure, *Statistics and Probability Letters* **49**, 411–416.

Landenna, G. and Ferrari, P. (1988). The k-variate student distribution and a test with the control of type I error in multiple decision problems, Technical Report, Istituto di Scienze Statistiche e Matematiche, Università di Milano, Italy.

Lange, K. and Sinsheimer, J. S. (1993). Normal/independent distributions and their applications in robust regression, *Journal of Computational and Graphical Statistics* **2**, 175–198.

Lange, K. L., Little, R. J. A. and Taylor, J. M. G. (1989). Robust statistical modeling using the t distribution, *Journal of the American Statistical Association* **84**, 881–896.

Lauprete, G. J., Samarov, A. M. and Welsch, R. E. (2002). Robust portfolio optimization, *Metrika* **55**, 139–149.

Lazo, A. and Rathie, P. (1978). On the entropy of continuous probability distributions, *IEEE Transactions on Information Theory* **24**, 120–122.

Lebedev, N. N. (1965). *Special Functions and Their Applications* (Prentice-Hall Inc., New Jersey).

Lee, R. E. and Spurrier, J. D. (1995). Successive comparisons between ordered treatments, *Journal of Statistical Planning and Inference* **43**, 323–330.

Lehmann, E. L. (1966). Some concepts of dependence, *Annals of Mathematical Statistics* **37**, 1137–1153.

Leonard, T. (1982). Comment on "A simple predictive density function" by M. Lejeune and G. D. Faukkenberry," *Journal of the American Statistical Association* **77**, 657–658.

Leonard, T., Hsu, J. S. J. and Ritter, C. (1994). The Laplacian T-approximation in Bayesian inference, *Statistica Sinica* **4**, 127–142.

Leonard, T., Hsu, J. S. J. and Tsui, K. W. (1989). Bayesian marginal inference, *Journal of the American Statistical Association* **84**, 1051–1058.

Lin, P. (1972). Some characterizations of the multivariate t distribution, *Journal of Multivariate Analysis* **2**, 339–344.

Linfoot, E. (1957). An informational measure of correlation, *Information and Control* **1**, 85–89.

Little, R. J. A. (1988). Robust estimation of the mean and covariance matrix from data with missing values, *Applied Statistics* **37**, 23–39.

Little, R. J. A. and Rubin, D. B. (1987). *Statistical Analysis with Missing Data* (John Wiley and Sons, New York).

Liu, C. (1993). Bartlett's decomposition of the posterior distribution of the covariance for normal monotone ignorable missing data, *Journal of Multivariate Analysis* **46**, 198–206.

Liu, C. (1995). Missing data imputation using the multivariate t distribution, *Journal of Multivariate Analysis* **53**, 139–158.

Liu, C. (1996). Bayesian robust multivariate linear regression with incomplete data, *Journal of the American Statistical Association* **91**, 1219–1227.

Liu, C. (1997). ML estimation of the multivariate t distribution and the EM algorithm, *Journal of Multivariate Analysis* **63**, 296–312.

Liu, C. and Rubin, D. B. (1995). ML estimation of the multivariate t distribution with unknown degrees of freedom, *Statistica Sinica* **5**, 19–39.

Liu, C., Rubin, D. B. and Wu, Y. N. (1998). Parameter expansion to accelerate EM: the PX-EM algorithm, *Biometrika* **85**, 755–770.

Liu, W., Miwa, T. and Hayter, A. J. (2000). Simultaneous confidence interval estimation for successive comparisons of ordered treatment effects, *Journal of Statistical Planning and Inference* **88**, 75–86.

Magnus, J. R. and Neudecker, H. (1979). The commutation matrix: Some properties and applications, *Annals of Statistics* **7**, 381–394.

Magnus, W., Oberhettinger, F. and Soni, R. P. (1966). *Formulas and Theorems for the Special Functions of Mathematical Physics* (Springer-Verlag, New York).

Mann, N. R. (1982). Optimal outlier tests for a Weibull model – To identify process changes or to predict failure times, *TIMS/Studies in the Management Sciences* **19**, 261–279.

Mardia, K. V. (1970a). *Families of Bivariate Distributions* (Griffin, London).
Mardia, K. V. (1970b). Measures of multivariate skewness and kurtosis with applications, *Biometrika* **57**, 519–530.
Maronna, R. A. (1976). Robust M-estimators of multivariate location and scatter, *Annals of Statistics* **4**, 51–67.
Marsaglia, G. (1965). Ratios of normal variables and ratios of sums of uniform variables, *Journal of the American Statistical Association* **60**, 193–204.
Marshall, A. W. and Olkin, I. (1974). Majorization in multivariate distributions, *Annals of Statistics* **2**, 1189–1200.
McLachlan, G. J. and Peel, D. (1998). Robust cluster analysis via mixtures of multivariate t-distributions, in Lecture Notes in Computer Science **1451**, ed. A. Amin, D. Dori, P. Pudil amd H. Freeman, pp. 658–666 (Springer-Verlag, Berlin).
McLachlan, G. J., Peel, D., Basford, K. E. and Adams, P. (1999). Fitting of mixtures of normal and t components, *Journal of Statistical Software* **4**.
Meng, X. L. and van Dyk, D. (1997). The EM algorithm—An old folk song sung to fast new tune (with discussion), *Journal of the Royal Statistical Society* B **59**, 511–567.
McCann, M. and Edwards, D. (1996). A path inequality for the multivariate t distribution, with applications to multiple comparisons, *Journal of the American Statistical Association* **91**, 211–216.
Miller, K. S. (1968). Some multivariate t distributions, *Annals of Mathematical Statistics* **39**, 1605–1609.
Milton, R. C. (1963). Tables of the equally correlated multivariate normal probability integral, Technical Report No. 27, University of Minnesota, Minneapolis.
Morales, D., Pardo, L. and Vajda, I. (1997). Some new statistics for testing hypotheses in parametric models, *Journal of Multivariate Analysis 62*, 137–168.
Nagarsenker, B. N. (1975). Some distribution problems connected with multivariate t distribution, *Metron* **33**, 66–74.
Nason, G. P. (2000). Analytic formulae for projection indices in a robustness experiment, Technical Report 00:06, Department of Mathematics, University of Bristol.
Nason, G. P. (2001). Robust projection indices, *Journal of the Royal Statistical Society* B **63**, 551–567.
Neyman, J. (1959). Optimal asymptotic tests for composite hypotheses, in *Probability and Statistics*, ed. U. Grenander, pp. 213–234 (John Wiley and Sons, New York).
Nicholson, C. (1943). The probability integral for two variables, *Biometrika* **33**, 59–72.
Osiewalski, J. (1991). A note on Bayesian inference in a regression model with elliptical errors, *Journal of Econometrics* **48**, 183–193.
Osiewalski, J. and Steel, M. F. J. (1990). Robust Bayesian inference in elliptical regression models, Center Discussion Paper 9032, Tilburg University.
Owen, D. B. (1956). Tables for computing bivariate normal probabilities, *Annals of Mathematical Statistics* **27**, 1075–1090.
Owen, D. B. (1965). A special case of a bivariate non-central t distribution, *Biometrika* **52**, 437–446.

Patil, S. A. and Kovner, J. L. (1968). On the probability of trivariate Student's t distribution (abstract), *Annals of Mathematical Statistics* **39**, 1784.

Patil, S. A. and Kovner, J. L. (1969). On the bivariate doubly noncentral t distributions (abstract), *Annals of Mathematical Statistics* **40**, 1868.

Patil, S. A. and Liao, S. H. (1970). The distribution of the ratios of means to the square root of the sum of variances of a bivariate normal sample, *Annals of Mathematical Statistics* **41**, 723–728.

Patil, V. H. (1965). Approximation to the Behrens-Fisher distributions, *Biometrika* **52**, 267–271.

Patnaik, P. B. (1955). Hypotheses concerning the means of observations in normal samples, *Sankhyā* **15**, 343–372.

Paulson, E. (1952). On the comparison of several experimental categories with a control, *Annals of Mathematical Statistics* **23**, 239–246.

Pearson, K. (1923). On non-skew frequency surfaces, *Biometrika* **15**, 231.

Pearson, K. (1931). *Tables for Statisticians and Biometricians*, Part II (Cambridge University Press for the Biometrika Trust, London).

Peel, D. and McLachlan, G. J. (2000). Robust mixture modelling using the t distribution, *Statistics and Computing* **10**, 339–348.

Pestana, D. (1977). Note on a paper of Ifram, *Sankhyā* A **39**, 396–397.

Pillai, K. C. S. and Ramachandran, K. V. (1954). Distribution of a Studentized order statistic, *Annals of Mathematical Statistics* **25**, 565–571.

Press, S. J. (1969). The t ratio distribution, *Journal of the American Statistical Association* **64**, 242–252.

Press, S. J. (1972). *Applied Multivariate Analysis* (Holt, Rinehart and Winston, Inc, New York).

Press, W. H. (1986). *Numerical Recipes: The Art of Scientific Computing* (Cambridge University Press, Cambridge).

Raiffa, H. and Schlaifer, R. (1961). *Applied Statistical Decision Theory* (Harvard University Press, Cambridge, MA).

Rattihalli, R. N. (1981). Regions of maximum probability content and their applications, Ph.D. Thesis, University of Poona, India.

Rausch, W. and Horn, M. (1988). Applications and tabulations of the multivariate t distribution with $\rho = 0$, *Biometrical Journal* **30**, 595–605.

Rényi, A. (1959). On the dimension and entropy of probability distributions, *Acta Mathematica Academiae Scientiarum Hungaricae* **10**, 193–215.

Rényi, A. (1960). A few fundamental problems of information theory (in Hungarian), *A Magyar Tudományos Akadémia Matematikai és Fizikai Tudományok Osztályának Közleményei* **10**, 251–282.

Rényi, A. (1961). On measures of entropy and information, in *Proceedings of the Fourth Berkeley Symposium on Mathematical Statistics and Probability* **I**, pp. 547–561 (University of California Press, Berkeley).

Robbins, H. (1948). The distribution of Student's t when the population means are unequal, *Annals of Mathematical Statistics* **19**, 406–410.

Rousseeuw, P. J., Kaufman, L. and Trauwaert, E. (1996). Fuzzy clustering using scatter matrices, *Computational Statistics and Data Analysis* **23**, 135–151.

Ruben, H. (1960). On the distribution of weighted difference of two independent Student variates, *Journal of the Royal Statistical Society* B **22**, 188–194.

Rubin, D. B. (1983). Iteratively reweighted least squares, in *Encyclopedia of Statistical Sciences* **4**, ed. S. Kotz and N. L. Johnson, pp. 272–275 (John Wiley and Sons, New York).

Rubin, D. B. (1987). *Multiple Imputation for Nonresponse in Surveys* (John Wiley and Sons, New York).

Rubin, D. B. and Schafer, J. L. (1990). Efficiently creating multiple imputations for incomplete multivariate normal data, in *Proceedings of the Statistical Computing Section of the American Statistical Association*, pp. 83–88 (American Statistical Association, Washington, DC).

Sahu, S. K., Dey, D. K. and Branco, M. D. (2000). A new class of multivariate skew distributions with applications to Bayesian regression models, Research Report RT-MAE 2000-16, Department of Statistics, University of Sao Paulo, Sao Paulo, Brasil.

Sarabia, J. M. (1995). The centered normal conditionals distribution, *Communications in Statistics—Theory and Methods* **24**, 2889–2900.

Schafer, J. L. (1997). *Analysis of Incomplete Multivariate Data* (Chapman and Hall, London).

Scott, A. (1967). A note on conservative confidence regions for the mean of a multivariate normal, *Annals of Mathematical Statistics* **38**, 278–280. Correction: *Annals of Mathematical Statistics* **39**, 1968, 2161.

Seal, K. C. (1954). On a class of decision procedures for ranking means, Institute of Statistics Mimeograph Series No. 109, University of North Carolina at Chapel Hill.

Seneta, E. (1993). Probability inequalities and Dunnett's test, in *Multiple Comparisons, Selection, and Applications in Biometry*, pp. 29–45 (Marcel Dekker, New York).

Sepanski, S. J. (1994). Asymptotics for multivariate t-statistic and Hotelling's T^2-statistic under infinite second moments via bootstrapping, *Journal of Multivariate Analysis* **49**, 41–54.

Sepanski, S. J. (1996). Asymptotics for multivariate t-statistic for random vectors in the generalized domain of attraction of the multivariate normal law, *Statistics and Probability Letters* **30**, 179–188.

Shampine, L. F. and Allen, R. C. (1973). *Numerical Computing: An Introduction* (Saunders, Philadelphia).

Shimizu, R. and Fujikoshi, Y. (1997). Sharp error bounds for asymptotic expansions of the distribution functions for scale mixtures, *Annals of the Institute of Statistical Mathematics* **49**, 285–297.

Šidák, Z. (1965). Rectangular confidence regions for means of multivariate normal distributions, *Bulletin of the Institute of International Statistics* **41**, 380–381.

Šidák, Z. (1967). Rectangular confidence regions for the means of multivariate normal distributions, *Journal of the American Statistical Association* **62**, 626–633.

Šidák, Z. (1971). On probabilities of rectangles in multivariate Student distributions: their dependence and correlations, *Annals of Mathematical Statistics* **42**, 169–175.

Šidák, Z. (1973). A chain of inequalities for some types of multivariate distributions, with nine special cases, *Applications of Mathematics* **18**, 110–118.

Siddiqui, M. M. (1967). A bivariate t distribution, *Annals of Mathematical*

Statistics **38**, 162–166.

Singh, R. K. (1988). Estimation of error variance in linear regression models with errors having multivariate student-t distribution with unknown degrees of freedom, *Economics Letters* **27**, 47–53.

Singh, R. K. (1991). James-Stein rule estimators in linear regression models with multivariate-t distributed error, *Australian Journal of Statistics* **33**, 145–158.

Singh, R. K., Mistra, S. and Pandey, S. K. (1995). A generalized class of estimators in linear regression models with multivariate-t distributed error, *Statistics and Probability Letters* **23**, 171–178.

Siotani, M. (1959). The extreme value of the generalized distances of the individual points in the multivariate normal sample, *Annals of the Institute of Statistical Mathematics* **10**, 183–208.

Siotani, M. (1964). Interval estimation for linear combinations of means, *Journal of the American Statistical Association* **59**, 1141–1164.

Siotani, M. (1976). Conditional and stepwise multivariate t distributions, in *Essays in Probability and Statistics*, pp. 287–303 (Tokyo).

Singh, R. S. (1991). James-Stein rule estimators in linear regression models with multivariate-t distributed error, *Australian Journal of Statistics* **33**, 145–158.

Sloan, I. H. and Joe, S. (1994). *Lattice Methods for Multiple Integration* (Clarendon Press, Oxford).

Smith, D. J., Bailey, T. C. and Munford, G. (1993). Robust classification of high-dimensional data using artificial neural networks, *Statistics and Computing* **3**, 71–81.

Somerville, P. N. (1993a). Simultaneous confidence intervals (General linear model), *Bulletin of the International Statistical Institute* **2**, 427–428.

Somerville, P. N. (1993b). Exact all-pairwise multiple comparisons for the general linear model, in *Proceedings of the 25th Symposium on the Interface, Computing Science and Statistics*, pp. 352–356 (Interface Foundation, Virginia)

Somerville, P. N. (1993c). Simultaneous multiple orderings, Technical Report TR-93-1, Department of Statistics, University of Central Florida, Orlando.

Somerville, P. N. (1994). Multiple comparisons, Technical Report TR-94-1, Department of Statistics, University of Central Florida, Orlando.

Somerville, P. N. (1997). Multiple testing and simultaneous confidence intervals: Calculation of constants, *Computational Statistics and Data Analysis* **25**, 217–233.

Somerville, P. N. (1998a). A Fortran 90 program for evaluation of multivariate normal and multivariate-t integrals over convex regions, *Journal of Statistical Software*, http://www.stat.ucla.edu/journals/jss/v03/i04.

Somerville, P. N. (1998b). Numerical computation of multivariate normal and multivariate t probabilities over convex regions, *Journal of Computational and Graphical Statistics* **7**, 529–544.

Somerville, P. N. (1999a). Numerical evaluation of multivariate integrals over ellipsoidal regions, *Bulletin of the International Statistical Institute.*

Somerville, P. N. (1999b). Critical values for multiple testing and comparisons: one step and step down procedures, *Journal of Statistical Planning and Inference* **82**, 129–138.

Somerville, P. N. (2001). Numerical computation of multivariate normal and multivariate t probabilities over ellipsoidal regions, *Journal of Statistical Software*, http://www.stat.ucla.edu/www.jstatsoft.org/v06/i08.

Somerville, P. N. and Bretz, F. (2001). Fortran 90 and SAS-IML programs for computation of critical values for multiple testing and simultaneous confidence intervals, *Journal of Statistical Software*, http://www.stat.ucla.edu/www.jstatsoft.org/v06/i05.

Somerville, P. N., Miwa, T., Liu, W. and Hayter, A. (2001). Combining one-sided and two-sided confidence interval procedures for successive comparisons of ordered treatment effects, *Biometrical Journal* **43**, 533–542.

Song, K.-S. (2001). Rényi information, loglikelihood and an intrinsic distribution measure, *Journal of Statistical Planning and Inference* **93**, 51–69.

Spainer, J. and Oldham, K. B. (1987). *An Atlas of Functions* (Hemisphere Publishing Company, Washington, DC).

Spurrier, J. D. and Isham, S. P. (1985). Exact simultaneous confidence intervals for pairwise comparisons of three normal means, *Journal of the American Statistical Association* **80**, 438–442.

Srivastava, M. S. and Awan, H. M. (1982). On the robustness of Hotelling's T^2-test and distribution of linear and quadratic forms in sampling from a mixture of two multivariate normal populations, *Communications in Statistics—Theory and Methods* **11**, 81–107.

Steffens, F. E. (1969a). A stepwise multivariate t distribution, *South African Statistical Journal* **3**, 17–26.

Steffens, F. E. (1969b). Critical values for bivariate Student t-tests, *Journal of the American Statistical Association* **64**, 637–646.

Steffens, F. E. (1970). Power of bivariate studentized maximum and minimum modulus tests, *Journal of the American Statistical Association* **65**, 1639–1644.

Steffens, F. E. (1974). A bivariate t distribution which occurs in stepwise regression (abstract), *Biometrics* **30**, 385.

Steyn, H. S. (1993). One the problem of more than one kurtosis parameter in multivariate analysis, *Journal of Multivariate Analysis* **44**, 1–22.

Stone, M. (1964). Comments on a posterior distribution of Geisser and Cornfield, *Journal of the Royal Statistical Society* B **26**, 274–276.

Sukhatme, P. V. (1938). On Fisher and Behrens' test of significance for the difference in means of two normal samples, *Sankhyā* **4**, 39–48.

Sultan, S. A. and Tracy, D. S. (1996). Moments of the complex multivariate normal distribution. Special issue honoring Calyampudi Radhakrishna Rao, *Linear Algebra and Its Applications* **237/238**, 191–204.

Sun, L., Hsu, J. S. J., Guttman, I. and Leonard, T. (1996). Bayesian methods for variance component models, *Journal of the American Statistical Association* **91**, 743–752.

Sutradhar, B. C. (1986). On the characteristic function of the multivariate Student t-distribution, *Canadian Journal of Statistics* **14**, 329–337.

Sutradhar, B. C. (1988a). Author's revision, *Canadian Journal of Statistics* **16**, 323.

Sutradhar, B. C. (1988b). Testing linear hypothesis with t error variable, *Sankhyā* B 175–180.

Sutradhar, B. C. (1990). Discrimination of observations into one of two t populations, *Biometrics* **46**, 827–835.

Sutradhar, B. C. (1993). Score test for the covariance matrix of the elliptical t-distribution, *Journal of Multivariate Analysis* **46**, 1–12.

Sutradhar, B. C. and Ali, M. M. (1986). Estimation of the parameters of a regression model with a multivariate t error variable, *Communications in Statistics—Theory and Methods* **15**, 429–450.

Sutradhar, B. C. and Ali, M. M. (1989). A generalization of the Wishart distribution for the elliptical model and its moments for the multivariate t model, *Journal of Multivariate Analysis* **29**, 155–162.

Sweeting, T. J. (1984). Approximate inference in location-scale regression models, *Journal of the American Statistical Association* **79**, 847–852.

Sweeting, T. J. (1987). Approximate Bayesian analysis of censored survival data, *Biometrika* **74**, 809–816.

Takano, K. (1994). On Bessel equations and the Lévy representation of the multivariate t distribution, Technical Report, Department of Mathematics, Ibaraki University, Japan.

Tan, W. Y. (1969a). Note on the multivariate and the generalized multivariate beta distributions, *Journal of the American Statistical Association* **64**, 230–241.

Tan, W. Y. (1969b). Some distribution theory associated with complex Gaussian distribution, *Tamkang Journal* **7**, 263–302.

Tan, W. Y. (1973). On the complex analogue of Bayesian estimation of a multivariate regression model, *Annals of the Institute of Statistical Mathematics* **25**, 135–152.

Tiao, G. C. and Zellner, A. (1964). On the Bayesian estimation of multivariate regression, *Journal of the Royal Statistical Society* B **26**, 277–285.

Tierney, L. and Kadane, J. (1986). Accurate approximations for posterior moments and marginal densities, *Journal of the American Statistical Association* **81**, 82–86.

Tiku, M. L. (1967). Tables of the power of the F-test, *Journal of the American Statistical Association* **62**, 525–539.

Tiku, M. L. and Gill, P. S. (1989). Modified maximum likelihood estimators for the bivariate normal based on Type II censored samples, *Communications in Statistics—Theory and Methods* **18**, 3505–3518.

Tiku, M. L. and Kambo, N. S. (1992). Estimation and hypothesis testing for a new family of bivariate nonnormal distributions, *Communications in Statistics—Theory and Methods* **21**, 1683–1705.

Tiku, M. L. and Suresh, R. P. (1992). A new method of estimation for location and scale parameters, *Journal of Statistical Planning and Inference* **30**, 281–292.

Tong, Y. L. (1970). Some probability inequalities of multivariate normal and multivariate t, *Journal of the American Statistical Association* **65**, 1243–1247.

Tong, Y. L. (1982). Rectangular and elliptical probability inequalities for Schur-concave random variables, *Annals of Statistics* **10**, 637–642.

Tranter, C. J. (1968). *Bessel Functions with Some Physical Applications* (English Universities Press Ltd., London).

Trout, J. R. and Chow, B. (1972). Table of the percentage points of the trivariate t distribution with an application to uniform confidence

bands, *Technometrics* **14**, 855–879.
Vaduva, I. (1985). Computer generation of random vectors based on transformation of uniformly distributed vectors, in *Proceedings of the Seventh Conference on Probability Theory*, ed. M. Iosifescu, pp. 589–598 (NU Science Press, Utrecht).
Vajda, I. (1989). *Theory of Statistical Inference and Information* (Kluwer Academic Publishers, Dordrecht).
Vale, C. D. and Maurelli, V. A. (1983). Simulating multivariate nonnormal distributions, *Psychometrika* **48**, 465–471.
van Dijk, H. K. (1985). Existence conditions for posterior moments of simultaneous equation model parameters, Report 8551 of the Econometric Institute, Erasmus University, Rotterdam.
van Dijk, H. K. (1986). A product of multivariate T densities as upper bound for the posterior kernel of simultaneous equation model parameters.
Vijverberg, W. P. M. (1995). Monte Carlo evaluation of multivariate normal probabilities, *Journal of Econometrics.*
Vijverberg, W. P. M. (1996). Monte Carlo evaluation of multivariate Student's t probabilities, *Economics Letters* **52**, 1–6.
Vijverberg, W. P. M. (1997). Monte Carlo evaluation of multivariate normal probabilities, *Journal of Econometrics* **76**, 281–307.
Vijverberg, W. P. M. (2000). Rectangular and wedge-shaped multivariate normal probabilities, *Economics Letters* **68**, 13–20.
Wald, A. (1944). On a statistical problem arising in the classification of an individual into one of two groups, *Annals of Mathematical Statistics* **15**, 145–162.
Wallgren, C. M. (1980). The distribution of the product of two correlated t variates, *Journal of the American Statistical Association* **75**, 996–1000.
Walker, G. A. and Saw, J. G. (1978). The distribution of linear combinations of t variables, *Journal of the American Statistical Association* **73**, 876–878.
Wang, O. and Kennedy, W. J. (1990). Comparison of algorithms for bivariate normal probability over a rectangle based on self-validating results from interval analysis, *Journal of Statistical Computation and Simulation* **37**, 13–25.
Wang, O. and Kennedy, W. J. (1997). Application of numerical interval analysis to obtain self-validating results for multivariate probabilities in a massively parallel environment, *Statistics and Computing* **7**, 163–171.
Watson, G. N. (1958). *A Treatise on the Theory of Bessel Functions* (Cambridge University Press, Cambridge).
Whittaker, E. T. and Watson, G. N. (1952). *Modern Analysis* (Cambridge University Press, Cambridge).
Weir, J. B. de V. (1966). Table of $0 \cdot 1$ percentage points of Behrens's d, *Biometrika* **53**, 267–268.
Wooding, R. A. (1956). The multivariate distribution of complex normal variables, *Biometrika* **43**, 212–215.
Wu, C. F. J. (1983). On the convergence properties of the EM algorithm, *Annals of Statistics* **11**, 95–103.
Wynn, H. P. and Bloomfield, P. (1971). Simultaneous confidence bands for regression analysis (with discussion), *Journal of the Royal Statistical Society* B **33**, 202–217.
Yang, Z. Q. and Zhang, C. M. (1997). Dimension reduction and

L_1-approximation for evaluations of multivariate normal integrals, *Chinese Journal of Numerical Mathematics and Applications* **19**, 82–95.

Zellner, A. (1971). *An Introduction to Bayesian Inference in Econometrics* (John Wiley and Sons, New York).

Zellner, A. (1976). Bayesian and non-Bayesian analysis of the regression model with multivariate Student-t error terms, *Journal of the American Statistical Association* **71**, 400–405.

Zhuang, X., Huang, Y., Palaniappan, K. and Zhao, Y. (1996). Gaussian density mixture modeling, decomposition and applications, *IEEE Transactions on Image Processing* **5**, 1293–1302.

Zografos, K. (1999). On maximum entropy characterization of Pearson's type II and type VII multivariate distributions, *Journal of Multivariate Analysis* **71**, 67–75.

Index

Printed in the United States
by Baker & Taylor Publisher Services